MATERIALS FOR BIOFUELS

World Scientific Series in Materials and Energy* **ISSN: 2335-6596**

Series Editor: Leonard C. Feldman *(Rutgers University)*

Published

Vol. 1 Handbook of Instrumentation and Techniques for Semiconductor Nanostructure Characterization
edited by Richard Haight (IBM TJ Watson Research Center, USA), Frances M. Ross (IBM TJ Watson Research Center, USA), James B. Hannon (IBM TJ Watson Research Center, USA) and Leonard C. Feldman (Rutgers University, USA)

Vol. 2 Handbook of Instrumentation and Techniques for Semiconductor Nanostructure Characterization
edited by Richard Haight (IBM TJ Watson Research Center, USA), Frances M. Ross (IBM TJ Watson Research Center, USA), James B. Hannon (IBM TJ Watson Research Center, USA) and Leonard C. Feldman (Rutgers University, USA)

Vol. 3 The World Scientific Handbook of Energy
edited by Gerard M. Crawley (University of South Carolina, USA)

Vol. 4 Materials for Biofuels
edited by Arthur J. Ragauskas (Georgia Institute of Technology, USA)

Forthcoming

Batteries
edited by Jack Vaughey (Argonne National Lab., USA) and David Schroeder (Northern Illinois, USA)

Encyclopedia of Practical Semiconductors (In 4 Volumes)
edited by Eugene A. Fitzgerald (Massachusetts Institute of Technology, USA)

Graphene
edited by Michael G. Spencer (Cornell University, USA)

Handbook of Green Materials: Properties, Technologies and Applications (In 4–5 Volumes)
edited by Kristiina Oksman (Lulea University of Technology, Sweden)

Handbook of Silicon Surfaces and Formation of Interfaces: Basic Science in the Industrial World (2nd Edition)
Jarek Dabrowski and Hans-Joachim Mussig (Institute for Semiconductor Research, Germany)

*The complete list of titles in the series can be found at
http://www.worldscientific.com/series/mae

MATERIALS AND ENERGY – Vol. 4

MATERIALS FOR BIOFUELS

Editor

Arthur J Ragauskas

Georgia Institute of Technology, USA

NEW JERSEY • LONDON • SINGAPORE • BEIJING • SHANGHAI • HONG KONG • TAIPEI • CHENNAI

Published by

World Scientific Publishing Co. Pte. Ltd.
5 Toh Tuck Link, Singapore 596224
USA office: 27 Warren Street, Suite 401-402, Hackensack, NJ 07601
UK office: 57 Shelton Street, Covent Garden, London WC2H 9HE

Library of Congress Cataloging-in-Publication Data
Ragauskas, Arthur J. (Arthur Jonas), 1957–
Materials for biofuels / Arthur J. Ragauskas, Georgia Institute of Technology, USA.
pages cm -- (World Scientific Series in Materials and Energy ; v. 4)
ISBN 978-9814513272 (hardcover : alk. paper)
1. Biomass energy--Materials. I. Title.
TP339.R34 2013
333.95'3--dc23

2013018608

British Library Cataloguing-in-Publication Data
A catalogue record for this book is available from the British Library.

In-house Editor: Song Yu

Typeset by Stallion Press
Email: enquiries@stallionpress.com

Printed in Singapore

PREFACE

The synthesis of this is text addresses humanities' call for sustainable technologies and products, especially as it applies to biofuels. It provides a broad and detailed introduction to the compelling subject of transforming biomass to biofuels. The evolution of this field of study has grown so large that no one text can fully cover all aspects of biofuel production and as a result, this study covers the broad field of biological and thermal conversion technologies for lignocellulosics. The chemical structure of terrestrial plants is explored and its relationship to biomass recalcitrance is reviewed. This book provides an enlightening view of the frontiers in leading pretreatments, downstream enzymatic hydrolysis, fermentation technology, pyrolysis, upgrading of bio-oils, corrosion issues related to biofuels and merging biofuels technology into a pulp mill to pave the way for future large-scale biofuel production. These and related subjects are examined in twelve chapters contributed by researchers actively involved in biorefining biomass to biofuels. It gives extensive coverage for different types of biomass (e.g., hardwood, softwood, herbaceous species, and agricultural residues) and the contributing cell wall factors that impact biomass recalcitrance. Different types of pretreatment approaches (e.g., water/acid, organosolv and ionic liquid) to enhance enzymatic deconstruction of lignocellulose to fermentable sugars and their fermentation processes to bioethanol and related biofuels are introduced and discussed. Furthermore, the pyrolysis of biomass to bio-oils and the upgrading of bio-oils to biofuel is presented in great detail. The corrosion issue associated with biofuels and the incorporation of biofuels technology into a pulp mill are also extensively discussed.

This book offers a review of state-of-the-art research and provides guidance for future paths for bio-conversion of biomass to biofuels in the fields of biotechnology, microbiology, chemistry, materials science and engineering. It is an accessible reference book for students, researchers, academicians and industrialists in biorefineries. In summary, I would like to thank all the contributing authors for their many efforts to insure the breath and accuracy of the information given in their chapters. Their enthusiasm and support made this project possible. Apart from the efforts of authors, I would also like to acknowledge the referees for carefully reading the book chapters and giving constructive comments that significantly improved the quality of the book. Finally, I want to thank my family for their unwavering support through all the years.

Arthur J. Ragauskas
Institute of Paper Science and Technology
School of Chemistry and Biochemistry
Georgia Institute of Technology
Atlanta, GA, USA

CONTENTS

CHAPTER 1

WHAT IS BIOMASS

FANG HUANG

School of Chemistry and Biochemistry/Institute of Paper Science and Technology, Georgia Institute of Technology
10th St. NW, Atlanta, GA, 30318, USA
fang.huang@ipst.gatech.edu

Lignocellulosic materials, such as wood, grass, agricultural and forest residues, are potential resources for the production of bioethanol. This chapter introduces the chemical compositions of common lignocellulosic biomasses including softwood, hardwood, herbaceous species and agricultural residues. The key points are focused on the physical and chemical properties of cellulose, hemicellulose and lignin in these common biomasses.

1. Introduction

Biomass is the plant material derived from the reaction with CO_2 in the air, water and sunlight, via photosynthesis, to produce carbohydrates and lignins that form the building blocks of biomass. Typically photosynthesis converts less than 1% of the available sunlight to stored, chemical energy.[1] The solar energy driving photosynthesis is stored in the chemical bonds of the structural components of biomass. If biomass is processed efficiently, either chemically or biologically, by extracting the energy stored in the chemical bonds and the subsequent 'energy' product combined with oxygen, the carbon is oxidized to produce CO_2 and water. The process is cyclical, as the CO_2 is then available to produce new biomass.

The value of a particular type of biomass depends on the chemical and physical properties of the large molecules from which it is made. Man for millennia has exploited the energy stored in these chemical bonds, by burning biomass as a fuel and by eating plants for the nutritional content such as sugar and starch. More recently, fossilized biomass has been exploited as coal and oil. However, since it takes millions of years to convert biomass into fossil fuels, these fuels are not renewable within a timescale mankind can use. Burning fossil fuels uses "old" biomass and converts it into "new" CO_2, which contributes to the "greenhouse" effect and depletes a non-renewable resource. Burning new biomass does not contribute new CO_2 to the atmosphere, because replanting harvested biomass ensures CO_2 to be absorbed and returned for a cycle of new growth.

One important factor that is often overlooked when considering the use of biomass to contribute alleviate global warming, is the time delay between the instantaneous release of CO_2 from burning fossil fuels and its eventual uptake as biomass, which can take many years. One of the dilemmas facing the developed world is the need to recognize this time delay and take appropriate action to mitigate against the lag period.

Numerous crops have been proposed or testing for commercial energy farming. Potential energy crops include all perennial crops (e.g., woody crops and grasses/herbaceous plants), starch and sugar crops and oilseeds.

In general, the characteristics of the ideal energy crop are[2,3]:

- high yield (maximum production of dry matter per hectare),
- low energy input to produce,
- low cost,
- composition with the least contaminants,
- use of secondary land,
- low nutrient requirements.

Desired characteristics will also depend on local climate and soil conditions. Water consumption can be a major constraint in many areas of the world, which makes the drought resistance of the crop an important factor.

2. Drivers for Biomass

In the past 10 years, there has been renewed interest, world-wide, in biomass as an energy source. There are several reasons for this situation:

- Firstly, technological developments relating to the conversion, crop production, etc. promise the application of biomass at lower cost but with higher conversion efficiency than that was possible previously. More advanced options to produce electricity are looking promising and allow a cost-effective use of energy crops like production of methanol and hydrogen by means of gasification processes.
- Secondly, the amount of fossil deposits on Earth is finite and the source to refill fossil energy is remote. The time for the complete depletion of fossils lies in the foreseeable future despite new discoveries of deposits. Therefore, it is imperative to find alternatives to fossil sources since the time scale for fossils to recycle or naturally replenish is, at best, on the order of 280 Ma,[4] as shown in Table 1. Fossils are replenished if at all when a carbonaceous age occurs on the Earth, preserving a large amount of organic matter. Given the advanced forms of life existing now, the chances of such an event occurring is unlikely. Nevertheless, the 280 million years-rotation is far too long and the amount of possible reserves is negligible compared to the recharge duration from a human use (or need) standpoint, even if mankind could survive a fossil replenishment era. Thus, fossil resources in general are deemed non-renewable. Societal awareness of environmental impact as well as problems in the stability and sustainability of the energy supply have made the development and implementation of bio-based chemicals and energy urgent and in the same time, more questions were raised. Regional energy security and rural economies both benefit from a plant biomass-derived chemical/energy economic base. Reverting to reliance on biomass as chemical and energy source is thus a preferred destiny.[5]
- Thirdly, the potential threat posed by climate change, due to high emission levels of greenhouse gases (CO_2 being the most important one), has become a major stimulus for renewable energy sources in general. When produced by sustainable means, biomass emits roughly

Table 1. Recycle times and productivities of chemical/energy feedstock (Reproduced from Ref. 6 with permission of Elsevier)

Feed stock	Recycle time	Standing biomass tons/ha	Biomass production tons/(ha.year)
Algae	1 month	0.9	11.25
Agricultural grasses	3 months-1 year	4.5	2.93
Savanna	1 year	18.0	4.05
Temperate grasses	1 year	7.2	2.70
Shrubs	1–5 year	27.0	3.15
Tropical season forest	5–25 year	157.5	7.20
Tropical forest	5–25 year	202.5	9.90
Temperate deciduous	10–50 year	135.0	5.40
Boreal forest	25–80 year	90.0	3.60
Temperate evergreen	10–80 year	157.5	5.85
Oil, gas and coal	280 Ma	(38.4×10^{21})	(0)

the same amount of carbon during conversion as is taken up during plant growth. The use of biomass therefore does not contribute to a build-up of CO_2 in the atmosphere.

However, these three main issues are not the only stimuli: biomass is also an indigenous energy source, available in most countries and its application may diversify the fuel-supply in many situations, which in turn may lead to a more secure energy supply. Biomass production can generate employment and if intensive agriculture is replaced by less intensively managed energy crops, there are likely to be environmental benefits, like reduced leaching of fertilizers and reduced use of pesticides. Moreover, if appropriate crops are selected, restoration of degraded lands may be possible. Depending on the crops used and the way the biomass is cultivated, increased biodiversity can be obtained, compared to current agricultural practice.

Biomass is available on a renewable basis, either through natural processes, or it can be made available as a by-product of human activities i.e., organic wastes. The potential of biomass energy derived from forest and agricultural residues worldwide is estimated to be about 30 EJ/yr, compared to an annual world-wide energy demand of over 400 EJ. If biomass is to contribute to a larger extent to the world's energy supply, then energy farming, the cultivation of dedicated crops for energy purposes, will be required, using fallow land and marginal lands where the latter are largely unsuited for food crops. When energy crops are considered to be sources of biomass, the total energy potential of biomass for energy production may be considerably larger than the energy potential of biomass residues. In 1992 at the Rio United Nations Conference on environment and development, the renewable intensive global energy scenario (RIGES) suggested that, by 2050, approximately half the world's current primary energy consumption of about 400 EJ/yr, could be met by biomass so that 60% of the world's electricity market could be supplied by renewables, of which biomass is a significant component.[7]

Biomass can be converted into three main types of products:

- electrical/heat energy
- transport fuel
- chemical feedstock

Of particular interest in this chapter is the generation of transport biofuel or bio-ethanol, the other two end-products are to be discussed in the future studies.

3. Biomass Types

Lignocellulosic biomasses were generally grouped into six categories, namely;[8]

- energy crops
- agricultural residues
- logging residues
- mill residues
- forest resources
- urban wastes

Other feedstocks reported by some studies, such as aquatic plants and manures, are not the focus of this chapter. The energy crop category includes lignocellulosic crops grown to be converted to energy, such as herbaceous crops (e.g., switchgrass) and short rotation woody crops (e.g., hybrid poplar, willow). The agricultural residue category consists of the lignocellulosic portions of the plant remaining after primary harvest. Lignocellulosic biomass originating from the forest is disaggregated into three categories by different sources and operations from which the biomass originates. The first two categories are comprised of residues produced by industry. Logging residues are generated from forest management and harvesting activities. Other removals such as downed stock or those from cultural and land-clearing operations are also included in this category, because these materials are commonly grouped together in the studies, or, because they could not be disaggregated from logging residues. Mill residues are produced by wood processing operations and include primary and secondary mill residues, as well as pulping residues. In the forest resources category, we grouped fuel treatment, thinning activities, and the standing forest biomass harvested for "fuelwood." Lastly, the urban waste category consists of the lignocellulosic portion of residential, commercial, and industrial waste typically disposed by landfilling and also includes urban wood wastes such as woody yard trimmings and construction and demolition debris.

This chapter introduces the chemical compositions of common lignocellulosic biomasses including softwood, hardwood, herbaceous species and agricultural residues. The key points are focused on the physical and chemical properties of cellulose, hemicellulose and lignin in these common biomasses.

4. Understanding Lignocellulosic Biomass

4.1. *Composition of lignocellulosic biomass*

The term "lignocellulosic biomass" is used to refer to higher plants, such as grasses, softwoods (SW) and hardwoods (HW). Understanding lignocellulosic biomass, particularly its chemical composition, is a prerequisite for developing effective pretreatment technologies to deconstruct its rigid structure, designing enzymes to liberate sugars, particularly cellulase to release glucose from recalcitrant cellulose, as well as engineering microorganisms to convert sugars into ethanol and other bio-based chemicals. The main components of the lignocellulosic materials are cellulose, hemicellulose, lignin and a remaining smaller part (extractives and ash). The composition of lignocellulose highly depends on its source. There is a significant variation of the lignin and (hemi) cellulose content of lignocellulose depending on whether it is derived from hardwood, softwood, or grasses. Table 2 summarizes the composition of lignocellulose encountered in the most common sources of biomass.

4.2. *Physical and chemical characteristics of lignocellulosic biomass*

Lignocellulosic biomass has a complex internal structure. The major components of lignocellulosic, i.e., cellulose, hemicellulose and lignin, also have intricate structures. To obtain a clear picture of the material, an analysis of the structure of each main component is made in this section, concluding with the description of the structure of lignocellulose itself. The physical properties of each component and how these components contribute to the behavior of the complex structure are also addressed.

Table 2. Typical lignocellulosic biomass compositions (% dry basis).[9–12]

	Cellulose	Hemicellulose	Lignin
Pine	43.3	20.5	28.3
Spruce	45.0	22.9	27.9
Douglas fir	45.0	19.2	30.0
Poplar	44.7	18.5	26.4
Eucalyptus	49.5	13.1	27.7
Corn stover	36.8	30.6	23.1
Miscanthus	39.3	24.8	22.7
Wheat straw	42.1	23.8	20.5
Switchgrass	42.0	25.2	18.1

4.2.1. *Cellulose*

Cellulose is the β-1,4-polyacetal of cellobiose (4-O- β-D-glucopyranosyl-D-glucose). Cellulose is more commonly considered a polymer of glucose because it consists of two molecules of glucose. The chemical formula of cellulose is $(C_6H_{10}O_5)_n$ and the structure of one chain of the polymer is presented in Fig. 1. Many properties of cellulose depend on its degree of polymerization (DP), i.e., the number of glucose units that make up of one polymer molecule. The DP of cellulose varies from 5000 in native wood to 1000 in bleached wood pulp,[13] and 500–1000 in the herbaceous cellulose, as shown in Table 3. Each D-anhydroglucopyranose unit possesses hydroxyl groups at C2, C3, and C6 positions, capable of undergoing the typical reactions known for primary and secondary alcohols. The molecular structure imparts cellulose with its characteristic properties: hydrophylicity, chirality, degradability, and broad chemical variability initiated by the high donor reactivity of hydroxyl groups.

The nature of the bonding between the glucose molecules (β-1,4 glycosidic) allows the polymer to be arranged in linear chains. Cellulose has a strong tendency to form intra- and inter-molecular hydrogen bonds by hydroxyl groups between the molecules of cellulose. The hydrogen bonds in turn result in the formation of an ultrastructure that is comprised of several parallel chains hydrogen-bonded to each other.[14] The coalescence of several polymer chains leads to the formation of microfibrils, which in turn are united to form fibers.

The hydrogen bonds in the linear cellulose chains promote aggregation into a crystalline structure and give cellulose a multitude of partially crystalline fiber structures and morphologies.[15] The average degree of crystallinity of native cellulose ranges 50–70%.[16,17] The ultrastructure of

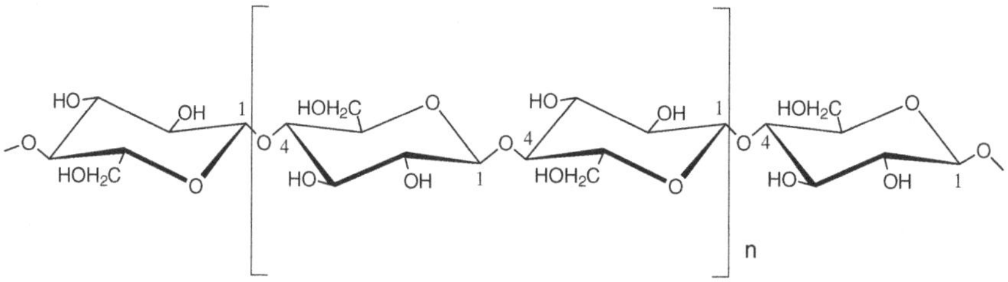

Fig. 1. The structure of cellulose.

Table 3. DP of native wood and non-woody cellulose after nitration using viscosmetric method. (Reproduced from Ref. 13 with permission of John Wiley & Sons Inc.)

Biomass species	DP
Trembling aspen	5000
Red maple	4450
Beech	4050
Eastern white cedar	4250
Jack pine	5000
Eastern hemlock	3900
Tamarack	4350
Balsam fir	4400
White spruce	4000
White birch	5500
Eucalyptus regnans	1510
Pinusradiata	3063
Wheat straw	1045
Sugarcane bagasse	925

native cellulose (cellulose I) has been discovered to possess unexpected complexity in the form of two crystal phases: I_α and I_β.[18] Electron diffraction and nuclear magnetic resonance (NMR) studies have shown that cellulose I_α is an allomorph with triclinic unit cells, whereas cellulose I_β is an allomorph with two-chain monoclinic units.[19] The relative amounts of I_α and I_β have been found to vary between samples from different origins. The I_α-rich specimens have been found in the cell wall of some algae and in bacterial cellulose, whereas I_β-rich specimens have been found in cotton, wood, and ramie fibers.[20,21] Native cellulose also contains para-crystalline and amorphous portion. Para-crystalline cellulose is loosely described as chain segments having more-order and less mobility than amorphous chains segments but less-ordered and more mobile than the crystalline domain.[22,23] The presence of crystalline cellulose, with regions of less order, and the size of the elementary fibrils work together to produce interesting combination of contrary properties such as stiffness and rigidity on one hand and flexibility on the other hand.[24]

There are different techniques to measure cellulose crystallinity. The frequently used method to determine the crystallinity of cellulose is based on X-ray diffraction (XRD). This method accounted for 70 to 85% in these studies.[25] The crystallinity index (CrI) is usually employed to describe the crystalline degree of biomass and pulps, which is defined as[26,27]:

$$CrI = [(I_{002} - I_{amorphous})/I_{002}] \times 100,$$

where I_{002} is the diffraction intensity of 002 peak (crystallinity portion) at $2\theta = 22.5°$ and $I_{amorphous}$ is the scattering intensity of amorphous region at $2\theta = 18.7°$.

In addition to the XRD method, the Cross Polarization/Magic Angle Spinning ^{13}C Nuclear Magnetic Resonance (CP/MAS ^{13}C-NMR)[28,29] and Fourier-Transform Infrared Spectroscopy (FTIR)[30] have been employed to determine CrI values. In the CP/MAS ^{13}C-NMR method, the CrI can be determined from the area of the crystalline and amorphous C4 signals by using the following equation, based on the solid NMR analysis:[31]

$$CrI = [A_{86-92ppm}/(A_{79-86ppm} + A_{86-92ppm})] \times 100,$$

where $A_{86-92ppm}$ is the area of the crystalline C4 signal, $A_{79-86ppm}$ is the area of the amorphous C4 signal. Results from cellulose crystallinity and structure from CP/MAS NMR of cellulose from hybrid poplar, Loblolly pine and switchgrass are given in Table 4. Cellulose from poplar is 63% crystalline with cellulose I_{β} as the predominant crystalline form. The less ordered amorphous region of poplar comprises 18% solvent inaccessible fibril surfaces. The intermediate *para*-crystalline form of cellulose also accounts for a significant proportion of poplar cellulose structure (Table 4). Crystallinity of poplar cellulose is comparable to that from Loblolly pine, but about 20% higher than switchgrass (Table 4). The amorphous region of switchgrass cellulose is mostly in the form of inaccessible surfaces, however, it may hider enzymatic hydrolysis.

In the FTIR methods, the ratio of amorphous to crystalline cellulose associates with the ratio of intensities of the bands at 900 cm^{-1} and

Table 4. Cellulose crystallinity and structure determined from CP/MAS ^{13}C NMR.

	Crystallinity	I_α	$I_{\alpha+\beta}$	Para-crystalline	I_β	Accessible fibril surface	Inaccessible fibril surface
Poplar[32]	63	5.0	14.2	31.1	19.8	10.2	18.3
Loblolly pine[33]	63	0.1	30.7	24.8	6.9	33.1	15.6
Switchgrass[34]	44	2.3	8.8	27.3	4.5	5.7	51.3

1098 cm^{-1}.[35] and the crystalline cellulose polymorphs (I_α/I_β) ratio can be measured by comparing the intensity of the bands at 750 cm^{-1} and 710 cm^{-1}.[36–38]

In the cellulose crystallinity measurement, although different methods use different techniques,[25–29] the basic principles are the same: CrI measures the relative fraction of crystalline cellulose in the combination of crystalline and amorphous fractions. However, it has been shown that different measurement techniques give different CrI values, but the order of crystallinity is relatively constant within each measurement.[25] Furthermore, it should be noted that the presence of lignin and hemicelluloses in the lignocellulosic biomass may have interference with the intensities at FTIR bands or introduce overlaps in the peaks during the XRD and NMR measurements.[39] Thus, it is advisable to remove lignin and hemicelluloses in the biomass prior to the cellulose crystallinity measurement.[40,41] Moreover, it was also reported that drying of the sample before analysis always cause the change of cellulose crystallinity.[42,43] Therefore, care must be taken in the cellulose sample preparation for the CrI analysis.

Crystalline cellulose has a very limited accessibility to water and chemicals. Chemical attack can therefore be expected to occur primarily on amorphous cellulose and crystalline surface. Cellulose is a relatively hygroscopic material absorbing 8–14% water under normal atmospheric conditions (20°C, 60% relative humidity).[44] Nevertheless, it is insoluble in water, where it swells. Cellulose is also insoluble in dilute acid solutions at low temperatures. The solubility of the polymer is strongly related to the degree of hydrolysis achieved. As a result, factors that affect the hydrolysis rate of cellulose also affect its solubility that takes place. In alkaline solutions extensive swelling of cellulose takes place as well as dissolution of the low molecular weight fractions of the polymer (DP < 200).[45]

4.2.2. *Hemicellulose*

The term hemicellulose is a collective term. It is used to represent a family of polysaccharides that are found in the plant cell wall and have different composition and structure depending on their source and the extraction method. Unlike cellulose, hemicellulose is composed of combinations of pentose (xylose (Xyl) and arabinose (Ara)) and/or hexoses (mannose (Man), galactose (Gal) and glucose (Glu)); and it is frequently acetylated and has side chain groups such as uronic acid and its 4-O-methyl ester. The chemical nature of hemicellulose varies from species to species. In general, the main hemicelluloses of softwood are galactoglucomannans and arabinoglucuronoxylan, while in hardwood is glucuronoxylan (Fig. 2).[46] Table 5 summarizes the main structural features of hemicelluloses appearing in both softwood and hardwood.[47,48]

Important aspects of the structure and composition of hemicellulose are the lack of crystalline structure mainly due to the highly branched structure, and the presence of acetyl groups on the polymer chain. Hemicellulose extracted from plants possesses a high degree of polydispersity, polydiversity and polymolecularity (a broad range of size, shape and mass characteristics). However, the degree of polymerization does not usually exceed 300 units whereas the minimum limit can be around 50 monomers, which are much lower than cellulose.

Table 6 lists the monosaccharide of several lignocellulosic biomasses. It can be seen that except the glucose, softwood contains more mannose than the hardwood, while the latter has more xylose.

In addition, most sugar components in the hemicellulose can take part in the formation of lignin-carbohydrate complexes (LCC) by covalent linkages between lignin and carbohydrates.[57,58] The most frequently suggested LCC-linkages in native wood are benzyl ester, benzyl ether, and glycosidic linkages.[59] The benzyl ester linkage is alkali-labile and may therefore be hydrolyzed during the alkaline pretreatment. The latter two

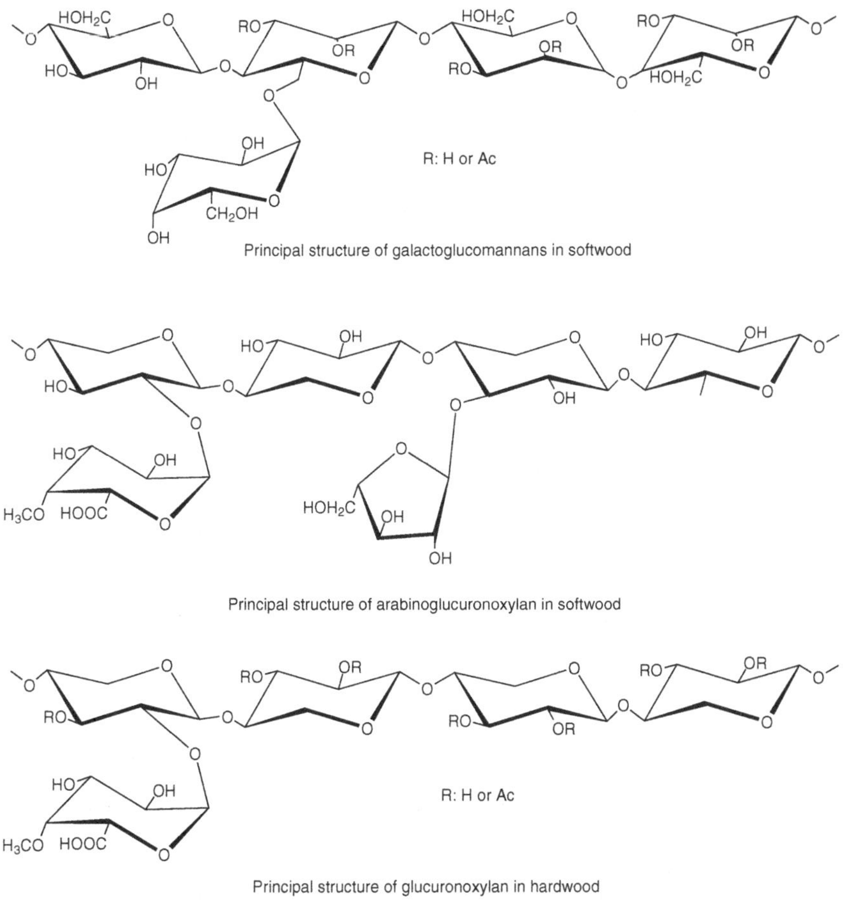

Principal structure of galactoglucomannans in softwood

Principal structure of arabinoglucuronoxylan in softwood

Principal structure of glucuronoxylan in hardwood

Fig. 2. Principal polysaccharides in woody hemicellulose (Reproduced from Ref. 46 with permission of John Wiley & Sons Inc.).

linkages are alkali-stable and may survive from the hydrolysis during alkaline pretreatment.

4.2.3. *Lignin*

Among the three major biopolymers that constitute wood, lignin is distinctly different from the other macromolecular polymers.[60] Lignin is an amorphous, cross-linked, and three-dimensional polyphenolic polymer that is synthesized by enzymatic dehydrogenative polymerization of

Table 5. Major hemicellulose component in softwood and hardwood (Reproduced from Ref. 46 with permission of John Wiley & Sons Inc.)

			Composition			
Wood	Hemicellulose type	Amount (% on wood)	Units	Molar ratio	Linkage	DP
Hardwood	glucuronoxylan	15–30	β-D-Xyl*p*	10	1→4	200
			4-O-Me-α-D-Glc*p*A	1	1→2	
			Acetyl	7		
	glucomannan	2–5	β-D-Man*p*	1–2	1→4	200
			β-D-Glc*p*	1	1→4	
Softwood	galactoglucomannans	10–15	β-D-Man*p*	4	1→4	100
			β-D-Glc*p*	1	1→4	
			β-D-Gal*p*	0.1	1→6	
			Acetyl	1		
	arabinoglucuronoxylan	7–10	β-D-Xyl*p*	10	1→4	100
			4-O-Me-α-D-Glc*p*A	2	1→2	
			β-L-Ara*f*	1.3	1→3	

Table 6. Content of monosaccharides for several lignocellulosics biomass.

	Composition (%, dry basis)				
	Arabinose	Galactose	Glucose	Mannose	Xylose
Loblolly pine[49]	1.4	2.1	48.0	7.3	9.7
Poplar[50]	0.4	0.7	45.5	2.7	17.0
Switchgrass[51]	2.1	0.6	43.7	trace	18.5
Miscanthus[52]	1.8	0.4	39.5	trace	19.0
Corn stover[53]	3.4	1.8	38.9	0.4	23.0
Wheat straw[54]	6.7	2.0	44.5	23.8	0.2
Rice straw[54]	4.2	1.4	52.5	12.0	0.1
Sugarcane bagasse[55]	3.4	2.0	n.a	28.6	0.2
Cotton stalk[56]	1.3	1.1	31.1	8.3	trace

n.a.: data not available.

4-hydroxyphenyl propanoid units.[61,62] The biosynthesis of lignin stems from the polymerization of three types of phenylpropane units as monolignols: coniferyl, sinapyl, and *p*-coumaryl alcohol.[63,64] Figure 3 depicts these three structures.

The polymerization process is initiated by an enzyme-catalyzed oxidation of the monolignol phenolic hydroxyl groups to yield free radicals. A monolignol free radical can then couple with another monolignol free radical to generate a dilignol. Subsequent nucleophilic attack by water, alcohols, or phenolic hydroxyl groups on the benzyl carbon of the quinone methide intermediate restores the aromaticity of the benzene ring. The generated dilignols then undergo further polymerization to form protolignin.

The relative abundance of these units depends on the contribution of a particular monomer to the polymerization and biosynthesis has been hotly debated in the literature and details of this complex pathway.[65–67] Hardwood and herbaceous lignins are composed mainly of syringyl (S) and guaiacyl (G) units with minor amounts of p-hydroxyphenyl (H), whereas softwood lignin is composed mainly of guaiacyl units and trace amounts of H.[65–70]

Figure 4 shows some common lignin inter-unit linkages. β-O-4 (β aryl ether) linkages are the most frequently occurring inter-unit linkage and are also the ones most easily cleaved by chemical processes such as pulping and biomass pre-treatments. The other linkages β-5, β-β, 5-5, 4-O-5 and

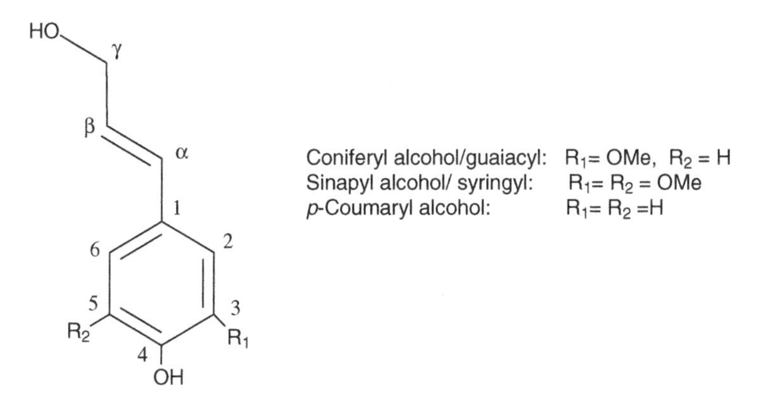

Fig. 3. Three building blocks of lignin (Reproduced from Ref. 46 with permission of John Wiley & Sons Inc.).

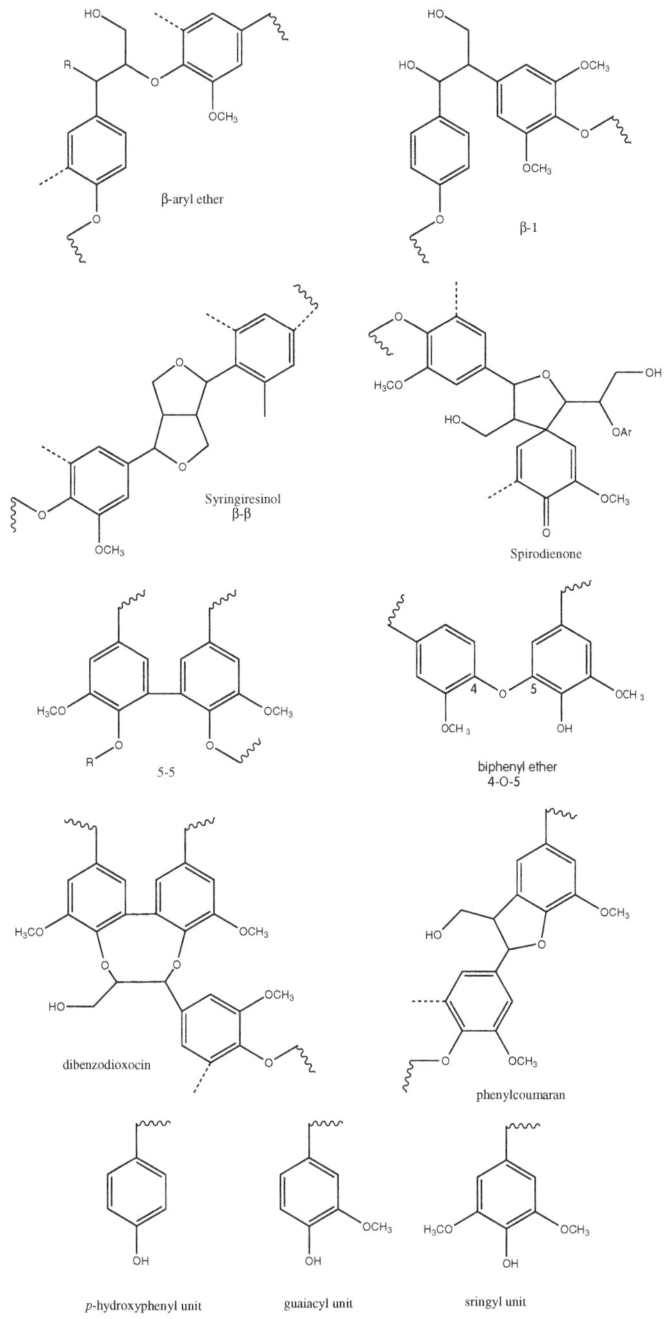

Fig. 4. Lignin sub-unit structures (Reproduced from Ref. 65 with permission of Springer).

β-1 are all more resistant to chemical degradation.[67] Hardwood lignins with a higher proportion of S units have fewer β-5, 5-5 and 4-O linkages than softwood lignin with more G units.[65]

Although the exact structure of protolignin is unknown, improvements in methods for identifying lignin-degradation products and advancements in spectroscopic methods have enabled scientists to elucidate the predominant structural features of lignin. Table 7 showed the typical abundance of common linkages and functional groups found in softwood lignin.[49,71]

Wet chemistry techniques coupled with gas chromatography, such as thioacidolysis and nitrobenzene oxidation, have been traditionally used to study lignin structure. While these methods can be very precise for specific functional groups and structure moieties, each technique can only provide limited information and does not give a general picture of the entire lignin structure. The thioacidolysis procedure cleaves β-O-4 linkages in lignin, giving rise to monomers and dimers, which are then used to calculate the S and G content. Similar formation can be obtained by using nitrobenzene oxidation, but it can lead to overestimated S/G ratios.[72] In general, the S/G ratio of lignin is a good indicator of its overall composition and response to pulping and biomass pretreatment. S/G ratios of lignin from different biomass species are summarized in Table 8. It can be seen that the S/G ratio of the hardwood species (i.e., poplar) is higher than the herbaceous biomass, such as switchgrass and *Miscanthus*. This is to be expected given the higher H contents in grass. In softwood species (i.e., Loblolly pine), the lignin is made up of 99% G unit and trace of H unit.

The advantage of spectroscopic methods over degradation techniques is their ability to analyze the whole lignin structure and directly detect lignin moieties. The development of quantitative ^{13}C NMR for lignin analysis[79] is an important advance in lignin chemistry. Multidimentional NMR spectroscopy has also been successfully utilized to elucidate details of lignin structure.[80] While two-dimensional NMR can help provide unambiguous structural assignments, performing quantitative two-dimensional NMR experiments requires special precautions and is typically only semi-quantitative at best.[81] A combination of quantitative ^{13}C and two-dimensional HMQC NMR has been shown to provide comprehensive structural information on lignin from a variety of source.[81,82] For example, 80% of the side chains of eucalyptus lignin were estimated at the structural level by

Table 7. Proportions of different types of linkages connecting the phenylpropane units in softwood lignin.[49,70,72]

Linkage type	Dimer structure	~Percentage
β-O-4	Phenylpropane β-aryl ether	50
β-5	Phenylcoumaran	9–12
5–5	Biphenyl	15–25
5–5/α-O-4	Dibenzodioxicin	10–15
4-O-5	Diaryl ether	4
β-1	1,2-Diaryl propane	7
β-β	β-β-linked structures	2

Table 8. The G/S/H lignin ratio of common biomass feedstock.

	G lignin	S lignin	H lignin	S/G ratio
Switchgrass[73]	51	41	8	0.80
Poplar[74]	32	68	trace	2.12
Miscanthus[75]	52	44	4	0.85
Corn stover[76]	51	3.6	46	0.07
Wheat straw[77]	49	46	5	0.94
Rice straw[77]	45	40	15	0.89
Sugarcane bagasse[78]	30	37	33	1.23
Loblolly pine[49]	99	—	1	—

using these methods. Other NMR techniques, such as phosphitylation of lignin hydroxyl groups, followed by ^{31}P NMR, ^{1}H and HSQC NMR can also yield valuable structural information. The most frequently used procedure to isolate lignin for NMR analysis is ball milling wood to a fine meal, followed by lignin extraction with aqueous dioxane. This milled wood lignin (MWL) is regarded as being fairly similar to nature lignin in wood.

The property of polydispersity, just as with hemicellulose, characterizes lignin as well. The degree of polymerization (DP) for softwood lignin is approximately 60–100 and the molecular weight is in excess of 10000.[83,84]

Lignin in wood behaves as an insoluble three-dimensional network. It plays an important role in the cell's endurance and development, as it affects the transport of water, nutrients and metabolites in the plant cell. It acts as binder between cells creating a composite material that has a remarkable resistance to impact, compression and bending.

Lignin is much less hydrophilic than either cellulose or hemicelluloses and it has a general effect of inhibiting water adsorption and fiber swelling. Solvents that have been identified to significantly dissolve lignin include low molecular alcohols, dioxane, acetone, pyridine, and dimethyl sulfoxide. Furthermore, it has been observed that at elevated temperatures, thermal softening of lignin takes place, which allows depolymerisation reactions of acidic or alkaline nature to accelerate.[85]

4.2.4. *Ash content and inorganic element profiles*

The inorganic elements present in biomass collectively constitute its ash content and act as a waste stream during its conversion to biofuels. They are the source of biochar and slagging during thermomechanical conversion. Knowledge of the ash content and composition is essential regardless of the conversion pathway or end product. In addition, several studies have highlighted that soil productivity requirements may necessitate this valuable inorganic resource to be returned to the soils.[86,87] Also, some inorganic elements such as P, K, Ca, and Mg, act as macronutrients. Knowledge of their contents in the biomass can provide information on nutrient depletion of the soil, which can be used to maintain soil fertility in subsequent rotations.[88] A compilation of available data on ash content and selected inorganic element distributions of some softwood, hardwood and herbaceous biomass are given in Table 9. The data presented in Table 9 shows that with a significant variation in ash content ranging from 0.3% to 4.3%, the distribution of inorganic elements shows very little variation among the different species. The data shows that the ash content of hybrid poplar is slightly higher than softwood biomass (i.e., Loblolly pine), but substantially (4×) lower than other biofuels feedstocks such as switchgrass.

4.2.5. *Extractive content*

Non-structural material is often removed from biomass prior to chemical analysis due to its potential interference with analytical techniques. This includes solvent-soluble, non-volatile compounds such as fatty acids,

Table 9. Ash content and inorganic elements (wt% dry weight) of common biomass.

		Inorganic elements (% dry wt.)				
Biomass species	Ash (% dry wt.)	P	K	Na	Ca	Mg
Hybrid poplar[89]	1.80	0.06	0.21	0.01	0.56	0.04
Willow[89]	2.29	0.49	1.83	0.15	6.76	0.48
Oak[90]	n.a.	n.a.	0.09	0.01	0.08	0.02
Switchgrass[91]	4.30	0.05	0.07	0.02	0.62	0.05
Loblolly pine[49,92]	0.30	0.01	0.04	n.a.	0.08	n.a.

n.a.: data not available.

resins, sterols, terpenes, waxes, etc. and usually comprises a minor proportion of biomass. For large-scale lignocellulosic biorefinery operations, however, extractives can be a potential source of value-added co-products. The compounds present in the extractives fraction are a function of the solvent, which is usually ethanol, acetone, dichloromethane, or a mixture of ethanol/benzene or ethanol/toluene. The organic solvent extractives content of some biomass species are presented in Table 10. It can be seen that switchgrass contain much higher extractive contents than the woody biomass since it includes a large amount of low-molecular weight of carbohydrates besides the fatty acids, sterols and alkanols, etc. In addition, the extractives content of softwood species (i.e., Scots pine) contains much more resin acids than those of the hardwood species (i.e., aspen).

4.2.6. *Elemental composition*

The major elemental composition of biomass on a gravimetric basis, which is commonly referred to as ultimate analysis is very important in performing mass balances on biomass conversion processes. These results can also be used to calculate empirical molecular formulae. Elemental compositions for some common biomasses compiled with the literature are given in Table 11. The sulfur content of poplar wood is lower than that of wheat straw and switchgrass (Table 11), which is advantageous in terms of strict environmental regulations limiting the sulfur content of

Table 10. Extractive contents of selected lignocellulosic biomass (wt% dry weight).

	Switchgrass[93]	Wheat straw[94]	Aspen[95,96]	Scots pine[97]
Extraction method	95% ethanol	Toluene/ethanol (2:1, v/v)	Acetone	Acetone
Total extractive yield	13.00	2.38	3.80	4.58
Carboxylic acids	1.56	0.76	0.24	0.21
Sugars and their derivatives	7.8	—	—	—
Alkanol	1.95	—	—	—
Glycerol	0.13	—	—	1.71
Alkane	0.26	—	—	—
Sterol	0.78	0.10	0.22	0.05
Wax	—	0.03	—	—
Phenolics	—	—	0.07	—
Flavonoids	—	—	0.07	—
Steryl Triterpene esters	—	0.33	0.46	—
Lipids	—	—	0.08	—
Resin acids	—	0.56	—	0.33

Table 11. Elemental contents of some biomass species.

	Ultimate analysis (% dry wt.)					
Biomass species	C	H	O	N	S	Si
Hybrid poplar[98]	50.20	6.06	40.40	0.60	0.02	—
Eastern cottonwood[99]	50.29	6.45	—	—	—	—
P. deltoides[50]	49.65	5.85	41.88	0.08	0.05	—
Corn stover[101]	43.65	5.56	43.31	0.61	0.01	—
Switchgrass[101]	47.75	5.75	42.37	0.74	0.08	—
Wheat straw[101]	43.20	5.00	39.40	0.61	0.11	—
Ponderosa pine[101]	49.25	5.99	44.36	0.06	0.03	—
Rice straw[100]	41.00	4.00	36.00	—	—	12.00

transportation fuels. As expected, there is not much variation in the elemental composition of different biomass species. It should be noted that the rice straw contains higher amount of silicon than other biomass species.

Table 12. Heating values (HHV) of common biomass feedstock (Reproduced from Ref. 102 with permission of John Wiley & Sons Inc).

Biomass species	Heating value (dry) (MJ/kg)
Hybrid poplar	19.38
Black cottonwood	15.00
Ponderosa pine	20.02
Douglas fir	20.37
Corn stover	17.65
Wheat straw	17.51
Switchgrass	18.64

4.2.7. *Heating values*

Heating values is the net enthalpy released upon reacting a material with oxygen under isothermal conditions. If water vapor formed during the reaction condenses at the end of the process, the latent enthalpy of condensation contributes to what is termed the higher heating value (HHV). These measurements are typically performed in the bomb calorimeter. Table 12 shows the HHVs of some common biomass species. It can be found that the softwood (Ponderosa pine and Douglas fir) yield slightly higher HHV than the hardwood (hybrid poplar and cottonwood), herbaceous (switchgrass) biomass and agricultural residues (corn stover, wheat straw), as shown in Table 12.

Acknowledgments

The author thanks the US Department of Energy (DOE) for providing financial support (Project: DE-EE0003144) for these studies.

References

1. P. McKendry, *Bioresour. Technol.*, **83**, 37 (2002).
2. H. J. Huang and S. Ramaswamy, *Biomass Bioenergy*, **33**, 234 (2009).
3. R. Kumar, S. Singh and O. V. Singh, *J. Ind. Microbiol. Biotechnol.*, **35**, 377 (2008).
4. T. E. Amidon and S. J. Liu, *Biotechnol. Adv.*, **27**, 542 (2009).

5. S. Liu, H. Lu, R. Hu, A, Shupe, L. Lu and B. Liang, *Biotechnol. Adv.*, **30**, 785 (2012).
6. S. J. Liu, *Biotechnol. Adv.*, **28**, 563 (2010).
7. B. Price. In *Electricity from Biomass* (Financial Times Energy, London, 1998).
8. M. Gronowska, S. Joshi and H. L. MacLean, *Bioresources*, **4(1)**, 341 (2009).
9. M. Galbe and G. Zacchi, Advan. *Biochem Eng/Biotechnol.*, **108**, 41 (2007).
10. C. N. Hamelinck, G. V. Hooijdonk and A. P. C. Faaij, *Biomass Bioenergy*, **28**, 384 (2005).
11. S. J. Kim, M. Y. Kim, S. J. M. S. Jeong and I. M. Jang, *Ind Crops Products*, **38**, 46 (2012).
12. T. Ingram, K. Wormeyer, J. C. Lima, V. Bockemuhl, G. Antranikian, G. Brunner and I. Smirnova, *Bioresource Technol.* **102**, 5221 (2011).
13. B. B. Hallac and A. J. Ragauskas, *Biofuels Bioprod. Biorefin.*, **5(2)**, 215 (2011).
14. J. Faulon, G. A. Carlson and P. G. Hatcher, *Org. Geochem.*, **21**, 1169 (1994).
15. D. Klemm, B. Heublein, H. P. Fink and A. Bohn, *Angew. Chem.*, **44**, 3358 (2005).
16. A. Thygesen, J. Oddershede, H. Lilholt, A. B. Thomsen and K. Stahl, *Cellulose*, **12(6)**, 563 (2005).
17. R. H. Newman, *Holzforschung*, **58(1)**, 91 (2004).
18. R. H. Atalla and D. L. VanderHart, *Science*, **223(4633)**, 283 (1984).
19. J. Sugiyama, R. Vuong and H. Chanzy, *Macromolecules*, **24**, 4168 (1991).
20. F. Horii, A. Hirai and R. Kitamaru, *Macromolecules*, **20(6)**, 1440 (1987).
21. J. Sugiyama, J. Persson and H. Chanzy, *Macromolecules*, **24(6)**, 2461 (1991).
22. P. T. Larsson, E. L. Hult, K. Wickholm, E. Pettersson and T. Iversen, *Solid State Nucl. Magn. Reson.*, **15(1)**, 31 (1999).
23. C. H. Stephens, P. M. Whitmore, H. R. Morris and M. E. Bier, *Biomacromolecules*, **9**, 1093 (2008).
24. L. Wagberg and G. O. Annergren, in *Physicochemical Characterization of Paper-making Fibers, in: The Fundamentals of Papermaking Materials: Transaction of the 11th Fundamental Re-search Symposium in Held at Cambridge, Ed. C. F. Baker*, (Pira International, Surrey, UK, 1997), p1.
25. S. Park, J. O. Baker, M. E. Himmel, P. A. Parilla and D. K. Johnson, *Biotechnol. Biofuels*, **3**, 10 (2010).
26. L. Segal, J. J. Creely, A. E. Martin Jr and C. M. Conrad, *Text. Res. J.*, **29**,786 (1959).
27. X. Zhao, L. Wang and D. Liu, *J. Chem. Technol. Biot.*, **83**, 950 (2008).
28. R. Newman, *Solid State Nucl. Magn. Reson.*, **15(1)**, 21 (1999).
29. S. Park, D. Johnson, C. Ishizawa, P. Parilla and M. Davis, *Cellulose*, **16(4)**, 641 (2009).
30. S. Oh, D. Yoo, Y. Shin and G. Seo, *Carbohydr. Res.*, **340(3)**, 417 (2005).
31. P. T. Larsson, U. Westermark and T. Iversen, *Carbohydr. Res.*, **278**, 339 (1995).
32. M. Foston, C. A. Hubbell, M. Davis and A. J. Ragauskas, *Bioenerg. Res.*, **2**, 193 (2009).
33. P. Sannigrahi, A. J. Ragauskas and S. J. Miller, *Bioenerg. Res.*, **1**, 205 (2008).
34. R. Samuel, Y. Pu, M. Foston and A. J. Ragauskas, *Cellulose*, **1**, 85 (2010).
35. L. Laureano-Perez, F. Teymouri, H. Alizadeh, B. E. Dale, *Appl. Biochem. Biotechnol.*, **121–124**, 1081 (2005).

36. C. Boisset, H. Chanzy, B. Henrissat, R. Lamed, Y. Shoham and E. A. Bayer, *Biochem. J.*, **340**, 829 (1999).
37. M. Wada, T. Okano, *Cellulose*, **83**, 183 (2001).
38. C. Tokoh, K. J. Takabe and M. Fujita, *Cellulose*, **91**, 65 (2002).
39. X. Zhao, L. Zhang, D. Liu, *Biofuels, bioproducts & biorefinin.*, **6(4)**, 465 (2012).
40. Z. Yu, H. Jameel, H. M. Chang and S. Park, *Bioresource. Technol.*, **102**, 9083, (2011).
41. X. Zhao, E. van der Heide, T. Zhang and D. Liu, *BioResources*, **5**, 1565 (2010).
42. L. T. Fan, Y. H. Lee, D. H. Beardmore, *Biotechnol. Bioeng.*, **23**, 419 (1981).
43. L. T. Fan, Y. H. Lee, D. H. Beardmore, *Biotechnol. Bioeng.*, **22**, 177 (1980).
44. P. F. H. Harmsen, W. J. J. Huijgen, L. M. Bermúdez López, R. R. C. Bakker, in *Literature Review of Physical and Chemical Pretreatment Processes for Lignocellulosic Biomass* (Energy center of the Netherlands, 2010).
45. H. Krassig and J. Schurz, in *Ullmann's Encyclopedia of Industrial Chemistry, Sixth Edition*, (Wiley-VCH, Weinheim, Germany, 2002).
46. Y. Pu, D. Zhang, P. M. Singh and A. J. Ragauskas, *Biofuels Bioprod Biorefin.*, **2**, 58 (2008).
47. S. Willför, A. Sundberg, A. Pranovich and B. Holmbom, *Wood Sci. Technol.*, **39**, 601 (2005).
48. S. Willför, A. Sundberg, J. Hemming and B. Holmbom, *Wood Sci. Technol.* **39**, 245 (2005).
49. F. Huang, P. M. Singh and A. J. Ragauskas, *J. Agr. Food Chem.*, **59**, 12910 (2011).
50. USDOE-Office of Energy efficiency and Renewable energy. Biomass Feedstock and Composition Database (2006). Available at http://www1.eere.energy.gov/biomass/printable_versions/feedstock_databases.html (Consulted on September 1, 2012).
51. Z. Hu, R. Sykes, M. F. Davis, E. C. Brummer and A. J. Ragauskas, *Bioresource Technol.*, **101**, 3253 (2010).
52. T. de Vrije, G. G. de Haas, G. B. Tan, E. R. P. Keijsers and P. A. M. Claassen, *Intl. J. Hydrogen Energy*, **27**, 1381 (2002).
53. D. W. Templeton, A. D. Sluiter, T. K. Hayward, B. R. Hames and S. R. Thomas, *Cellulose*, **16**, 621 (2009).
54. J. X. Sun, F. C. Mao, X. F. Sun and R. C. Sun, *J. Wood Chem. Technol.*, **24(3)**, 239 (2004).
55. X. F. Sun, R. C. Sun, J. Tomkinson and M. S. Baird, *Carbohydrate Poly.*, **53**, 483 (2003).
56. R. A. Silverstein, Y. Chen, R. R. Sharma-Shivappa, M. D. Boyette and J. Osborne, *Bioresource Technol.*, **98**, 3000 (2007).
57. A. Barakat, H. Winter, C. Rondeau-Mouro, B. Saake, B. Chabbert and B. Cathala, Planta, **226**, 267 (2007).
58. M. Bunzel, J. Ralph, F. Lu, R. D. Hatfield and H. Steinhart, *J Agric. Food Chem.*, **52**, 6496 (2004).
59. M. Lawoko, G. Henriksson and G. Gellerstedt, *Holzforschung*, **60**, 156 (2006).

60. G. Brunow, K. Lundquist and G. Gellerstedt, in *Lignin: Analytical Methods in Wood Chemistry. In: Pulping and Papermaking. Ed., E Sjostrom,* (Springer, Berlin, Germany, 1999), p. 77.
61. L. B. Davin and N. G. Lewis, *Curr Opin Biotechnol.*, **16**, 407 (2005).
62. C. Halpin, *Biotechnol Genet Eng Rev.*, **21**, 229 (2004).
63. W. Boerjan, J. Ralph and M. Baucher, *Annu Rev Plant Biol.*, **54**, 519 (2003).
64. Y. Pu, N. Jiang and A. J. Ragauskas, *J. Wood Chem. Technol.*, **27**, 23 (2007).
65. J. Ralph, K. Lundquist, G. Brunow, F. Lu, H. Kim and P. F. Schatz, *Phytochem. Rev.*, **3**, 29 (2004).
66. N. G. Lewis and E. Yamamoto, *Annu. Rev. Plant Physiol. Plant Mol. Biol.*, **41**, 455 (1990).
67. W. Boerjan, J. Ralph and M. Baucher, *Annu. Rev. Plant Biol.*, **54**, 519 (2003).
68. Y. Pu, D. Zhang, P. M. Singh and A. J. Ragauskas, *Biofuels Bioprod. Bioref.*, **2**, 58, (2008).
69. A. J. Ragauskas, M. Nagy, D. H. Kim, C. A. Eckert, J. P. Hallett and C. L. Liotta, *Indust. Biotechnol.*, **2**, 55 (2006).
70. E. A. Capanema, M. Y. Balakshin and J. F. Kadla, *J. Agric. Food Chem.*, **53**, 9639 (2005).
71. F. S. Chakar and A. J. Ragauskas, *Ind. Crops Products*, **20**, 131 (2004).
72. J. J. Stewart, J. F. Kadla and S. D. Mansfield, *Holzforschung*, **60**, 111 (2006).
73. R. Samuel, Y. Pu, B. Raman and A. J. Ragauskas, *Appl. Biochem. Biotechnol.*, **162**, 62 (2010).
74. A. R. Robinson and S. D. Mansfield, *Plant J.*, **58**, 706 (2009).
75. R. El Hage, N. Brosse, L. Chrusciel, C. Sanchez, P. Sannigrahi and A. J. Ragauskas, *Poly. Degrada. Stab.*, **94**, 1632 (2009).
76. S. C. Fox and A. G. McDonald, *Bioresources*, **5**, 990 (2010).
77. A. Buranov and G. Mazza, *Indust. Crops and Products*, **28**, 237 (2008).
78. E. C. Ramires, J. D. Megiatto, C. Gardrat, A. Castellan and E. Frollini, *Biotechnol. Bioeng.*, **107**, 612 (2010).
79. D. Robert and D. Gagnaire, in *Quantitative Analysis of Lignins by 13C NMR. Proceedings of the International Symposium on Wood and Pulping Chemistry*, (Stockholm, Sweden, 1981), p. 86.
80. E. Ammalahti, G. Brunow, A. Bardet, D. Robert and I. Kilpelainen, *J. Agric. Food Chem.*, **46**, 5113 (1998).
81. E. A. Capanema, M. Y. Balakshin and J. F. Kadla, *J. Agric. Food Chem.*, **52**, 1850 (2004).
82. E. A. Capanema, M. Y. Balakshin and J. F. Kadla, *J. Agric. Food Chem.*, **53**, 9639 (2005).
83. J. J. Meister, *J. Macromol Sci. Poly. Rev.*, **C42(2)**, 235 (2002).
84. A. Guerra, A. Xavier, A. Hai, I. Filpponen, L. Lucia and D. Argyropoulos, *J. Agrc. Food Chem.*, **54**, 9696 (2006).
85. M. P. Pandey and C. S. Kim, *Chem. Eng. Technol.*, **34**, 29 (2011).

86. R. Lal, *Science*, **304**, 1623 (2004).
87. L. K. Paine, T. L. Peterson, D. J. Undersander, K. C. Rineer, G. A. Bartlelt and S. A. Temple, *Biomass Bioenerg.*, **10**, 231 (1996).
88. T. W. Bowersox, *Wood Science*, **11**, 257 (1979).
89. P. J. Tharakan, T. A. Volk, L. P. Abrahamson and E. H. White, *Biomass Bioenerg.*, **25**, 571 (2003).
90. L. Allison, A. J. Ragauskas and J. S. Hsieh, *Tappi J.*, **83(8)**, 1 (2000).
91. R. Fahmi, A. V. Bridgewater, L. I. Darvell, J. M. Jones, N. Yates and S. Thain, *Fuel*, **86**, 1560 (2006).
92. D. H. van Lear, J. B. Waide and M. J. Tueke, *Forest Sci.*, **30(2)**, 395 (1984).
93. J. Yan, Z. Hu, Y. Pu, E. C. Brummer and A. J. Ragauskas, *Biomass Bioenergy*, **34**, 48 (2010).
94. R. C. Sun and J. Tomkinson, *J. Wood. Sci.*, **49**, 47 (2003).
95. M. P. Fernandez, P. A. Watson and C. Breuil, *J. Chromato. A.*, **922**, 225 (2001).
96. A. D. Yanchunk, I. Spilda and M. M. Micko, *Wood Sci. Technol.*, **22**, 67 (1988).
97. J. Dorado, F. W. Classsen, T. A. van Beek, G. Lenon, J. B. P. A. Wijnberg, R. Sierra-Alvarez, *J. Biotechnol.*, **80**, 231 (2000).
98. T. R. Miles, R. W. Bryers, B. M. Jenkins and L. L. Oden, in Alkali deposits found in biomass power plants. A preliminary investigation of their extent and nature: National Renewable Energy Laboratory Report NREL/TP-433-8142, (Oakridge, TN, 1995).
99. P. Chow and G. L. Rolfe, *Wood. Fiber Sci.*, **21**, 30 (1980).
100. M. D. Summers, in *Fundamental Properties of Rice Straw in Comparison with Softwoods*, (University of California Davis, Davis, CA, 2000).
101. R. C. Brown, in *Biorenewable Resources: Engineering New Products from Agriculture*, (Iowa State press, Ames, IA, 2003).
102. P. Sannigrahi, A. J. Ragauskas and G. A. Tuskan, Biofules Bioprod. Bioref., 4(2), 209 (2010).

CHAPTER 2

BIOMASS RECALCITRANCE AND THE CONTRIBUTING CELL WALL FACTORS

MARCUS FOSTON

Department of Energy, Environmental and Chemical Engineering,
Washington University in St. Louis
One Brookings Drive, St. Louis, MO 63130, USA
mfoston@wustl.edu

The conversion of lignocellulosic biomass offers the potential to produce fuels and chemicals in a renewable and economical fashion. However, to accomplish this economical production of renewable material from lignocellulosic feedstock on a large-scale at costs competitive with petroleum production streams, it is vital to improve the fundamental strategies of overcoming biomass recalcitrance, and the innate ability of plant cell walls to resist chemical and biological deconstruction. This chapter begins with defining biomass recalcitrance and illustrating why this complex property is the major barrier to entry of biomass derived biofuels into the market. We further summarize the effect various chemical, molecular, and structural features have on biomass recalcitrance.

1. Introduction

Recently, widespread concerns have been raised regarding the need to develop more sustainable and economic technologies to meet rising global manufacturing and energy demands. Renewed emphasis regarding the reduction of the world's dependency on conventional natural resources and the associated environmental impact has led to considerable research focused on producing fuels and chemicals from feedstocks other than petroleum. Lignocellulosic biomass represents an abundant and renewable resource, which not only facilitates the efficient conversion of sunlight into chemical potential energy but also fixes atmospheric CO_2. It has been estimated that there is a global annual production of dry lignocellulosic biomass of ~10–50 billion tons. Considering the massive volumes of fuel and chemical needed annually, lignocellulosic biomass is therefore an ideal renewable source to displace or even replace select petroleum feedstocks.

A major part of biomass utilization has included efforts to develop "bio-refineries", which would facilitate an integrated conversion process producing foods, fuels, chemicals, feeds, materials, heat, and power. A bio-refinery would efficiently utilize the materials and energy locked within biomass in a fashion analogous to modern petroleum refineries. Currently, the conversion of biomass can be separated into two general classes, technologies for production of power and heat (e.g., combustion, gasification, and anaerobic digestion) or fuel and chemicals (e.g., fermentation, pyrolysis, liquefaction, and gasification). Within that second class of technologies, two processes: (1) the production of bio-oil from biomass pyrolysis and (2) the generation of bioalcohol (e.g., bioethanol) by enzymatic hydrolysis and fermentation are some of the most heavily studied routes being considered.

Although supplementing petroleum use with biomass utilization has been a subject of intense investigation for over half a century, the production of a sufficient supply of biofuels, in particular bioethanol, in a cost-effective manner continues to be a challenge to this day. This has been attributed to cost associated with an innate property of biomass, known as biomass recalcitrance. Biomass recalcitrance is the ability of the plant cell wall to resist chemical and biological deconstruction, and originates from the composite of several plant substrate and cell wall characteristics. Bioethanol production currently relies on enzymatic hydrolysis to convert the major cell wall polysaccharides, mainly found in cellulose and hemicellulose, into fermentable sugars. These fermentable sugars are then used to produce ethanol, typically by microorganisms such as yeast. The high cost of dedicated enzyme production and the modest kinetic and yield profiles, typically associated with cell wall polysaccharide enzymatic hydrolysis, greatly affect the economics of bioethanol production, with hydrolysis kinetic and yield profiles correlated directly to biomass recalcitrance.

The plant cell wall of lignocellulosic biomass contains three major biopolymers: cellulose, lignin, and hemicelluloses. Cellulose, a linear polysaccharide composed of β-(1$\rightarrow$4) linked D-glucopyranosyl units, is found in the cell wall in both crystalline and amorphous morphologies, and is the most abundant terrestrial source of carbon. Hemicellulose, another cell wall polysaccharide, is a broad class of branched heteropolymer which have a variety of 5- and 6-carbon monomeric sugars. Lignin is constructed of hydroxycinnamyl monomers with various degrees of methoxylation, i.e., *p*-hydroxyphenyl (H), guaiacyl (G), and syringyl (S), on its monomer units. Lignin is also a racemic, cross-linked, and highly heterogeneous aromatic macromolecule, whose main function is defensive and structural. The plant cell wall contains other compounds, e.g., pectin, cell wall proteins, and ash.

The unique chemical and physical properties of the plant cell wall are due in part to its highly heterogeneous and multi-component microstructure. This cell wall microstructure is believed to exist as a lignin and hemicellulose matrix encapsulating and supporting cellulose microfibrils packed into bundles. Reinforcing this encapsulating matrix potential sites for covalent and/or non-covalent cross-linking between cell wall polysaccharides and lignin, and are also commonly cited as a major contributing factor in biomass recalcitrance. These cross-linking sites are known as lignin carbohydrate complexes (LCCs). Millions of years of evolution have developed cell wall features like this complex microstructure to serve as key plant structural elements and to protect cell wall polysaccharides, a common source of energy for a variety of fungi and microorganism. As a consequence, the very property that makes the plant cell wall so useful in nature as structural and protective elements, makes it difficult and expensive to industrially deconstruct.

Biomass recalcitrance is a complex multi-variant and multi-scale problem. Substrate characteristics relevant to determining recalcitrance can extend over several decades of length-scale. Cell wall features ranging from chemical features, characterized by nanometer and sub-nanometer dimensions, to plant structural features on the meter and millimeter scale can all affect biomass recalcitrance. For this reason, the relationship between many substrate characteristics and biomass recalcitrance has been sometimes over-simplified. Further complicating the problem, there are few single principle investigator laboratories that have the capability to comprehensively analyze the numerous and multi-scale recalcitrant related substrate characteristics. Biomass recalcitrance has been described in terms of seven naturally occurring factors: (1) plant epidermal tissue, i.e., cuticle and epicuticular waxes, (2) vascular bundle arrangement and density, (3) the relative high proportion of sclerenchymatous tissues that give support to the plant, (4) the degree of lignification, (5) plant cell wall structural/chemical heterogeneity and complexity, (6) the inefficiency associated with the catalysis of an insoluble substrate, and (7) the propensity of biomass to generate inhibitory compounds or recalcitrance structures during thermochemical processing.[1]

In addition to the practical problems related to analyzing complex and multi-scale substrates such as biomass, deconvoluting effects of multiple and inter-related substrate characteristics that contribute to biomass recalcitrance is also difficult. The relationship between biomass recalcitrance and various substrate characteristics is not a simple first-order linear correlation, but more than likely the change in an individual characteristic changes the relationship between biomass recalcitrance and a collection of other characteristics. Many studies acknowledge the substrate complexity of biomass. In an effort to simplify the problem and analyze the effect of various characteristics, isolated cell wall components or mixtures of those components are used. This approach however ignores the complex inter-play between substrate characteristics and the importance of the intact cell wall microstructure. Collectively, the above issues have led to confusion in the literature about biomass recalcitrance, and made developing an improved understanding more challenging.

The substrate characteristics that affect biomass recalcitrance can be divided into three categories, distinguished by length scale: structural, molecular, and chemical. Generally, the characteristics in these categories inhibit enzymatic hydrolysis via two mechanisms, features which impede access or limit enzymatic-cellulose interaction due to the biomass structure or though the mechanisms cellulolytic enzymes. This chapter will highlight the major plant cell wall structural, molecular, and chemical characteristics that affect biomass recalcitrance.

2. Structural Features

The plant cell wall is typically deposited in three layers. The first cell wall layer to form is the middle lamella, generally rich in pectins, is considered the interface that bonds adjacent plant cells together. The next layer to deposit during cell growth is the primary cell wall, largely a thin and flexible layer consisting of the major cell wall polysaccharides (cellulose, hemicellulose, and pectin). There are two types of primary cell walls, which are defined by the types of cross-linked polysaccharides. Dicotyledonous plants have Type I walls, consisting of equal amounts of glucan and xyloglucan embedded in a matrix of pectin. Type II primary walls, present in grasses, are characterized by glucuronoarabinoxylans and glucans, and have low levels of pectin and structural proteins. On the other hand, the secondary cell wall includes carbohydrates, lignin, and a variety of other biomolecules. The exact composition depends highly on cellular function, and is used to modify mechanical properties and permeability. This secondary cell wall is deposited inside the primary cell wall after the cell is fully grown and is not found in all cell types. The secondary cell wall is the predominate structure in woody biomass and is sometimes further composed of three distinct sub-layers: S_1 (outer), S_2 (middle), and S_3 (inner) layers. These sub-layers differ in the directional alignment of the cellulose microfibrils. Cellulose, hemicellulose, and lignin also have different distributions within these sub-layers, and can even display different distributions of lignin and hemicellulose monomeric units. Cell corners typically have the highest lignin concentration. This cell wall organization changes in different plant tissues and organs, being mainly characterized by structural heterogeneity, which evolved as part of a plant's defensive mechanism and clearly affects biomass recalcitrance.

Other larger scale anatomical and structural characteristics of lignocellulosic biomass such as particle size, fiber modulus, fiber length, cell wall thickness, and cell wall roughness have been identified as potential contributing factors in biomass recalcitrance; however, researchers often ignore these differences, focusing on molecular and chemical length-scale features. In fact, most laboratory research utilizes small particle sizes to accommodate smaller pretreatment and bio-reactors and to differentiate kinetics from transfer effects. There are a number of studies that have suggested average particle size of the lignocellulosic substrates may influence enzymatic hydrolysis. Since cellulase adsorption is a prerequisite step in the hydrolytic process, it might seem intuitive that particle size is a significant factor in biomass recalcitrance. Smaller particle sizes equate to higher surface area-to-volume ratios, and means more available adsorption sites per mass of substrate. In addition, as lignocellulosic biomass particle size decreases, the inherent chemical and structural heterogeneity is reduced, and as a result may also reduce biomass recalcitrance. Though, no direct causation between particle size and biomass recalcitrance has been shown, one might infer a possible connection. Increased pretreatment severity was shown to reduce particle size, while increasing enzymatic hydrolysis rates. In the initial stages of the hydrolysis, smaller sized fractions within wood pulps are hydrolyzed preferentially.[2,3] However, analysis of cellulosic substrates, e.g., cotton linters and microcrystalline cellulose, has shown little evidence to support any clear correlation between average particle size and hydrolysis rates, specifying that materials sized below 0.420 mm exhibited similar enzymatic hydrolysis profiles.[4]

In highly porous systems such as biomass, particle size and specific surface area may not have a linear relationship. Many of the studies focusing on the effect of particle size measured enzymatic hydrolysis activity as a function of average particle size rather than specific surface area. Moreover, many of these same studies utilize milling or pretreatment to alter and control particle size. These size reduction methods have significant effects on a variety of recalcitrant related molecular and chemical "level" features, making deconvoluting the effect of particle size challenging.

3. Molecular Features

3.1. *Cellulose crystallinity*

Cellulose within the plant cell wall can be found in both crystalline and amorphous forms. It has been suggested that the elementary cellulose fibril has a core-shell configuration consisting of 36 glucan chains. The most inner core has six true crystalline chains, which are surrounded by 12 *para*-crystalline chains, followed by another shell consisting of 18 more non-crystalline (amorphous or disordered) chains at the cellulose fibril surface. The mean cross-sectional size of the cellulose fibrils or crystallites is 2–5 nm, while cellulose aggregates or fibrils bundle have a dimension on the order of 15–25 nm.

Crystalline cellulose is considered to be highly ordered and has significant intra- and inter-molecular hydrogen bonding. Adding to the structural complexity of cellulose, various diffraction studies have provided information suggesting multiple crystalline allomorphs, i.e., cellulose I_α and I_β, exist in native cellulose. These allomorphs represent different chain packing (parallel) and unit-cell dimensions. There are several other non-native crystalline forms of cellulose, which in some instances also contain further sub-classification, including: (1) cellulose II — obtained by regeneration or mercerization processes, (2) cellulose III — prepared with anhydrous liquid ammonia, and (3) cellulose IV — prepared by and thermal annealing in glycerol. The substantial hydrogen bonding in crystalline cellulose makes these regions particularly recalcitrant, resisting deconstruction by both chemical and biological means. Although, recent computational efforts have shown that cellulose III has a lower number of intra-chain hydrogen bonds, which was correlated to then increases in enzymatic hydrolysis rates.[5]

The concept of *para*-crystalline cellulose has been used to explain the "poor" three-dimensional order in a crystalline cellulose lattice. The exact structure and nature of *para*-crystalline cellulose remains unknown. Some research suggest that *para*-crystalline structure in cellulose is a phase defined by a gradual transition between fully crystalline and amorphous phases, whereas other studies suggest it to be distortions or imperfection in crystal structure. Nevertheless, a number of analytical techniques have proven the existence of *para*-crystalline cellulose in native systems, suggesting that it comprises ~30–40% of the cellulose ultrastructure.

Commonly, amorphous cellulose degradation caused by either acid- or auto-catalyzed hydrolysis has been shown to occur at a much faster rate than *para*-crystalline and crystalline cellulose. This effect is obvious during acid hydrolysis, alcholysis, or acetolysis with degree of polymerization level-off. As a result, many researchers postulated a similar trend would be observed for enzymatic hydrolysis of cellulose. This idea was so pervasive that an often cited goal of thermochemical pretreatment prior to enzymatic hydrolysis, a process intended to reduce biomass recalcitrance, was the reduction of cellulose crystallinity. And indeed some studies seem to suggest during enzymatic hydrolysis of cellulose, amorphous regions are hydrolyzed first, followed by the more recalcitrant crystalline cellulose. This observation was even extended to different types of crystalline cellulose allomorphs (cellulose I_α and I_β) and *para*-crystalline cellulose. It was found that in an initial rapid phase of hydrolysis, *para*-crystalline and non-crystalline cellulose regions were more readily enzymatically deconstructed than cellulose I_α, which in turn displayed a faster rate of hydrolysis than cellulose I_β.[6] However, other studies directly contradict these findings, showing either no significant change in crystallinity occurred with enzymatic hydrolysis or even that rates of hydrolysis increased with increasing crystallinity.

This discrepancy in the effect of cellulose crystallinity on biomass recalcitrance can be attributed to a number of aspects mainly related to the complexity of (1) plant cell wall material and (2) biological systems designed for its deconstruction. In many studies, the effect of cellulose crystallinity on enzymatic hydrolysis and biomass recalcitrance is studied via various methods to systematically alter substrate crystallinity. However, the methods intended to vary and control crystallinity not only target the plant cell wall characteristic of interest (e.g., crystallinity) but also to invariably change a variety of other substrate properties. For example, gamma irradiation, hydrothermal treatments, and ball-milling can be used to prepare biomass samples of varying cellulose crystallinity. It has however been shown that these treatments simultaneously modify other crucial factors in enzymatic digestibility, e.g., cell wall chemistry, accessible surface area and cellulose degree of polymerization. Therefore, conclusions about the effect of cellulose crystallinity on recalcitrance could be over-simplified, particularly when the possible contributions of other substrate characteristics are not considered as well.

Another issue is related to the fact that different studies use different biomass substrates and enzymatic systems. Biomass substrates of the same genome have shown considerable variation in substrate phenotypes such as cellulose crystallinity, which depends on a number of growth and genetic factors.[7,8] Similarly, there are a vast number of cellulolytic enzymes isolated from different fungal and microbial sources, and depending on the exact enzyme or enzyme cocktail utilized enzymatic hydrolysis results can vary with respect to cellulose crystallinity.

An additional source of confusion is related to the fact that cellulose crystallinity can be measured by a wide variety of techniques, at times based on different operating principles and different physical characteristics, that generate different crystalline indices. Typically cellulose crystallinity is assessed as the ratio of two phases, defined as either: (1) crystalline versus amorphous, (2) solvent inaccessible versus accessible or, (3) ordered versus disordered domains. Techniques intended to distinguish among these phases and determine their relative ratio rely on: (1) physical methods, generally measuring disorder; (2) chemical swelling methods, in which reagents swell cellulose by disrupting hydrogen and inter-chain bonding and penetrate the cellulose molecular structure in the amorphous region by either reactive or non-reactive means, or (3) chemical non-swelling methods, which instead of disrupting the hydrogen and inter-chain bonding, only interact with external crystalline surfaces and detect the proportion of cellulose crystal surface along with surfaces of internal voids, capillaries and fibril structures. Usually physical evaluation of the relative fraction of crystalline cellulose can also be accomplished through IR or Raman spectroscopy, NMR spectroscopy along with X-ray diffraction (XRD) and electron diffraction/microscopic methodologies. Again, in most cases, these techniques performed on the same sample will produce differing results. This technique related variation in apparent crystallinity can be further compounded by the necessity of sample preparation (e.g., drying and/or isolation methods) and the effect sample preparation has on crystallinity.

Cellulose crystallinity may also affect the ability of cellulase enzyme modules, including cellulose binding modules (CBM) and catalytic domains, to adsorb or to function on cellulose. In general, it has been shown that increases in cellulose crystallinity negatively impact the adsorption of cellulase to cellulose. Moreover, purified cellulase components have shown preferential cellulose morphological attack and adsorption. For example, endoglucanase I (EGI) appears to have a greater capability to adsorb to crystalline cellulose and displays higher activity than cellobiohydrolase 1 (CBH1).[9]

Enzymatic hydrolysis of cellulose is actually a process that relies on a synergy of various cellulase components. For example, endocellulase depolymerizes linear cellulose molecules, producing reducing and non-reducing ends, while exocellulases only attack at reducing end. Cellobiose, a disaccharide consists of two glucose molecules, is a strong inhibitor to endocellulase and exocellulase activities. Consequently, another enzyme, β-glucosidase, is typically required to reduce this inhibition and allow continued cellulolytic activity. There is also evidence that the amount of crystalline cellulose alters the synergy between cellulase components. Studying CBHI and endoglucanase II (EGII) from *T. reesei*, a trend was found correlating increased cellulase component synergy with increased crystallinity.[10]

3.2. *Cellulose degree of polymerization*

Cellulose molecular weight or degree of polymerization (DP) is defined as the number of glucose monomer units in a cellulose chain. Cellulose

DP can greatly vary depending on substrate origin and preparation and, has been observed from <100 to> 15,000 units. Several studies have investigated the change in cellulose as a function of enzymatic hydrolysis with cellulase or purified cellulase components; however, similarly to cellulose crystallinity the role of cellulose DP on enzymatic hydrolysis is still not clearly defined. The reasons for this are also quite similar to the ones discussed above.

The recognition required for binding of cellulase components to cellulose chains is highly related to the chain length. A chain of at least seven to ten units is required for glucanase activity, whereas β-glucosidases can hydrolyze low-DP chains as well. Based on this, it would be reasonable to expect higher β-glucosidase activity on shorter chains or at least that exoglucanases display higher activity with a increasing number of reducing chain ends. When concentrating on changes in cellulose DP before and after pretreatment and the effect of DP on enzymatic deconstruction of cellulose, a major conclusion indicated shorter cellulose chains can be more readily hydrolyzed enzymatically.[11] However, other studies have shown that decreases in cellulose DP have less effect in accelerating enzymatic hydrolysis than changes in other substrate characteristics such as accessibility.[4] In research focused on the enzymatic hydrolysis of organosolv-pretreated softwood materials, it was shown that any single observed substrate characteristics such as fiber length, DP, and crystallinity have little effect on enzymatic hydrolysis.[12] Similarly to cellulose crystallinity, cellulose DP effects on cellulase adsorption and synergy must also be considered.

As in cellulose crystallinity, deconvoluting the effect of cellulose DP on enzymatic hydrolysis is difficult. Moreover, cellulose DP can be measured via multiple methods adding to possible variations or inconsistencies. The average molecular weight of a polydisperse distribution of polymers is typically expressed as an ordinary arithmetic mean and weighted mean denoted as the number average molar mass (M_n) and weight average molar mass (M_w), respectively. The relative molecular weight of cellulose, often only measuring either M_n or M_w, can be determined by techniques such as vapor pressure osmometry, reducing end concentration, electron microscopy, light scattering, sedimentation equilibrium, X-ray small angle scattering, intrinsic viscosity, and gel permeation chromatography.

Solution based techniques typically require cellulose isolation and solubilization. Intermolecular hydrogen bonding in cellulose is significant and as DP increases, its solubility dramatically decreases. Cellulose can be solubilized by two categories of solvents: non-derivatizing and derivatizing. Non-derivatizing systems of dissolving cellulose function by disrupting intermolecular bonding and included systems such as *N,N*-dimethylacetamide (DMA)/LiCl, ionic liquids, and dimethyl sulfoxide (DMSO)/SO_2/diethylamine. Derivatizing systems accomplish solubilization by formation of ether, ester, or acetyl derivatives. Typically, derivatization is required for molecular weight determination in a common solvent. Depending on the isolation and derivatization method used, the measured cellulose DP can vary inconsistently with the actual cellulose DP.

3.3. *Cellulose accessibility*

A requirement for enzymatic hydrolysis of insoluble cellulose is cellulase binding to the surface of biomass. Biomass is a complex three-dimensional structure, and the combination of cellulase size and shape with respect to the cell wall three-dimensional structure determines whether the bonds in cellulose are accessible to enzymatic attack. Some research fails to make the critical distinction among particle size, specific surface area, and enzyme accessible surface area. As mentioned earlier, particle size does little to account for the interior surface area in highly porous biomass. The internal surface area of biomass has been estimated as much as one to two orders of magnitude greater than the external surface area. External surface area is mainly described by particle size or diameter; however, specific surface area, is also governed by the size of the lumen and capillaries as well as voids and cracks localized in the lignocellulosic matrix, existing on a wide length-scale from nano to millimeter.

Enzyme accessible surface area can be also affected by the relative amount of exposed hydrophilic crystalline cellulose faces. Cellulase binding modules have binding mechanisms driven by both chemical and structural recognition to cellulose crystallite surfaces, demonstrated by certain cellulases which consistently attach to the two "hydrophobic" planar faces of the cellulose crystal.[13] An additional possible limiting factor in enzyme accessible surface area is related to the microstructure of the plant cell wall. Studies focused on correlations between the initial hydrolysis rate of pretreated biomass and pore size have determined pores with a minimal diameter of 51 Å are required for the accessibility of *T. reesei* cellulase.[14] It has been suggested pores of smaller dimensions cannot accommodate a complete cellulase enzyme; as a result, the cellulase modules cannot function with synergy and are not productive. This notion of accessibility has implications not only related to a minimal pore volume needed for enzymatic activity, but also suggests that pore tortuosity and inlet pore shape could be critical as well.

There is strong evidence that accessible surface area to cellulase is one of the most important factors for achieving high sugar yields from enzymatic deconstruction. Pretreatments have various mechanisms for reducing biomass recalcitrance. For example, soaking in aqueous ammonia (SAA) pretreatment is designed to remove lignin and to dissolve some hemicellulose at a moderate condition.[15] Whereas cellulose solvent- and organic solvent-based lignocellulose fractionation (COSLIF) focuses on disrupting inter- and intra-molecular bonds among lignocellulose components. The effect in both cases being increased cellulose accessibility. Comparing these two pretreatments, SAA pretreatments were found to remove large amounts of lignin and to increase cellulose accessibility mildly, yet displayed slower hydrolysis rates with respect to COSLIF pretreatment. Another pretreatment, ammonia fiber explosion (AFEX) does little to extract lignin and hemicellulose from the cell wall, however does cause significant modifications in microstructure, specifically increasing specific surface area.[16] AFEX pretreatments have been shown to enhance enzymatic hydrolysis yield by four-to-five-fold over that of untreated cell walls. Other proof of the importance of accessible surface area to biomass recalcitrance can be seen in studies that show enzymatic digestibility of biomass pulps decrease after drying.[17] This research compared the enzymatic digestibility of never-dried materials to dried materials varying drying method and drying duration. It was found drying effects sugar yields, and that effect was attributed to irreversible partial collapse and closure of pores.

There are several methods that can be used to determine enzyme accessible surface area. A direct method to measure accessibility is to monitor the adsorption of fluorescently tagged CBMs; however, the presence of lignin can skew results due to non-specific binding. Pore structure and accessibility can be otherwise assessed by solute exclusion (a technique based on the accessibility of different sized solute molecules), differential scanning calorimetry (based on the principle that water contained inside pores has a lower freezing point than that of bulk water) as well as closely related techniques such as cryoporometry NMR, NMR relaxometry, water or gas adsorption, small angle X-ray scattering, or mercury porosimetry.[18,19] Again depending on the techniques used, measures of accessibility differ and have further contributed to the confusion correlating the effect of cellulose accessibility to biomass recalcitrance.

4. Chemical Features

4.1. *Lignin*

Lignin is integrated into the cell wall microstructure between cellulose and hemicellulose, in which its critical biological functions include mechanical strength, defensive, and water transport. The presence, amount, and monomer distribution of lignin is an important aspect in biomass recalcitrance and shown to limit the extent of enzymatic hydrolysis in biomass. The effect of lignin has been in general attributed to both physical and chemical factors. Lignin physically reduces the accessibility of cellulose to enzymes by covering cellulose fibrils and by restricting fiber swelling. Lignin also causes recalcitrance via chemical means by irreversibly binding cellulase enzymes and by preventing further enzymatic activity on cellulose.

Like much of the literature regarding the effects of various substrate characteristics on biomass recalcitrance, the exact role of lignin in enzymatic hydrolysis is unclear. However, several factors such as the degree and type of cross-linkages between lignin and hemicellulose and the distribution of (1) monolignols, (2) lignin inter-unit linkages, and (3) terminal phenolic groups seem to have some effect on enzymatic digestibility. Some treatments such as alkaline, organosolv, and other pretreatments that mimic pulping processes, seem to increase enzymatic digestibility by causing hemicellulose and lignin removal. Removal of lignin would increase specific surface area and expose more cell wall voids for enzyme access, while also reducing the amount of non-specific adsorption. Cellulose fibrils may also be physically separated to a higher degree during pretreatments that target lignin removal, again affecting cellulose accessibility.

However, in many pretreatments, the relative amount of lignin does not change, however significant lignin redistribution occurs. In these cases, lignin depolymerization and repolymerization is suspected as the major mechanism for redistribution. Acid-catalyzed hydrothermal pretreatment should cause chain scission in lignin, primarily at β–O–4 ether bonds. However, the modest changes in the lignin molecular weight typically observed do not correspond to the significant linkage degradation observed spectrally.[20] This suggests that condensation reactions also occur during pretreatment. Lignin depolymerization and repolymerization provides a possible route to explain why lignin droplets deposit on cellulose after hydrothermal pretreatment.[21] Hot-water and dilute acid flow-through pretreatment systems have been developed as an alternative to conventional batch or concurrent systems. Due to the continuous-flow configuration, cell wall fragments have short residence times in the reactor. This short reactor residence time does not allow depolymerized lignin to react further and repolymerize. As a result, flow-through pretreatment produces more digestible lignocellulosic substrates, with (1) greater lignin removal, (2) no evidence of lignin droplet formation, and (3) increased cellulose accessibility. This research with flow-through pretreatment systems seems to suggest this repolymerized lignin could have significant negative impacts on cellulose enzymatic hydrolysis. Moreover, more recent studies show pseudo-lignin, which can be broadly defined as an aromatic material resulting from the dehydration of cellulose, forms as a result of severe hot water and acid pretreatment. Again, the presence and structure of pseudo-lignin has been correlated with decreases in enzymatic conversion.[22,23]

The ratio of monolignol units (S to G) has also been thought to affect biomass enzymatic digestibility. Lignin modifications in transgenic biomass have shown to increase sugar yields from enzymatic hydrolysis and fermentability to ethanol by both simultaneous saccharification and fermentation and separate hydrolysis and fermentation. Characterization of the transgenic biomass showed no change in cell wall polysaccharides with respect to the native control sample; yet, displayed lower lignin content and a shift in the ratio of S to G moieties.[24] The modification in the lignin structure was a result of the down-regulation of the caffeic acid O-methyl transferase gene. It has been suggested that monolignol distribution has some influence on the linearity of lignin and propensity to contain condensed inter-unit linkages, which in turn could affect biomass microstructure and accessibility.

A major mechanism in CBMs for cellulose binding relies on hydrophobic and electrostatic interactions. A consequence of this mechanism is spontaneous thermodynamically driven adsorption of cellulases onto lignin. This non-productive binding is so significant, it is described as irreversible. Experiments found that adding bovine serum albumin (BSA) to biomass prior to enzymatic hydrolysis increases the adsorption of cellulase from solution.[25] Delignification is therefore not only useful in removal of physical barriers to cellulase accessibility, but also chemical structures which cause non-productive binding.

4.2. *Hemicellulose*

Hemicelluloses include xylan, glucuronoxylan, arabinoxylan, glucomannan, and xyloglucan. These branched heteropolymers are polysaccharides, typically consisting of shorter chains relative to cellulose, of ~500–3,000 sugar monomers. Besides glucose and xylose, the major sugar monomers in hemicellulose include mannose, galactose, rhamnose, arabinose, mannuronic acid, and galacturonic acid. Much like lignin, hemicellulose is an integrated component of the cell wall composite, and physically limits the accessibility of cellulose enzymes. A variety of studies show hemicellulose removal by hydrothermal treatment significantly improves the enzymatic digestion of cellulose in biomass, suggesting that hemicellulose is a barrier to enzymatic hydrolysis of cellulose. Removal of hemicellulose is also associated with the breakdown of the cross-linked polysaccharides and LCCs. Although the role of hemicellulose in cellulose digestion is ambiguous, it is much easier to remove from the cell wall and its removal seems to have less effect on the accessibility of cellulose than lignin removal.

4.3. *Pectin*

Pectin is part of a complex family of polysaccharides, which are primarily found in primary walls that surround growing and dividing cells and in the middle lamella and cell corners. Pectin is also present in very small proportion in the secondary cell wall. Structurally, pectin is a galacturonic acid-rich polysaccharide, and part of a family of polymers including homogalacturonan (HG), rhamnogalacturonan I (RG-I), substituted galacturonans rhamnogalacturonan II (RG-II), and xylogalacturonan (XGA). There is evidence that indicates pectin is utilized by the plant for variety of functions including binding the cell wall composite together, plant growth, defense, cell–cell adhesion, cellular signaling, cell wall porosity, binding of ions, etc.

Targeting pectin's role in wood tissue integrity and enzymatic accessibility, transgenic biomass with reduced contents of de-methyl-esterified HG have been researched. Modified levels of HG were achieved through the expression *in planta* a fungal polygalacturonase and in a separate genetic construct, overexpressing plant inhibitors of endogenous pectin methylesterases.[26] In each case, the transgenics displayed improved enzymatic saccharification efficiency on substrates with and without dilute acid pretreatment. The results seem to suggest that the presence and structure of pectin are contributing factors in biomass recalcitrance, and that reduction of de-methyl-esterified HG can facilitate biomass saccharification. In another recent study utilizing a diverse antibody toolkit of over 155 anitbodies and carbohydrate-binding modules to monitor glycan-related epitopes as a result of hydrothermal pretreatment, the loss of lignin-polysaccharide interactions, arabinogalactans, and pectin was correlated to an increase in sugar yields from enzymatic hydrolysis.[27] Though far from conclusive, this research would suggest that even at the low-levels detected in the secondary cell wall of grasses and woody biomass, biomass recalcitrance is affected by the presence and structure of pectin. This effect may be due to a cell wall templating or cross-linking functions associated with pectin.

4.4. *Acetyl groups*

The plant cell wall is extensively acetylated, and is a functionality primarily found on xylan backbones and in lignin. Direct ester linkages between hemicellulose and lignin are mediated through carboxyl groups, also known as LCC. Acetyl groups can account for ~1–6% of the carbon in various biomass substrates. There are no direct correlations to define the role of acetyl groups in enzymatic hydrolysis; however, in vitro de-esterification with hydroxylamine solutions has been used to deacetylate biomass with little disruption of other cell wall components. This treatment can produce biomass where hemicellulose is much more digestible, which in turn renders a biomass substrate more accessible. Acetyl groups might inhibit enzymatic hydrolysis (1) by altering surface hydrophobicity and disrupting surface recognition sites critical to CBM productive binding or (2) by decreasing accessibility via increased sterics. Research has shown that cellulose accessibility is related to alkali swelling, which occurs in soaked aqueous biomass upon rapid increases in pH.[28] This is attributed to xylan acetyl groups that undergo alkali deacetylation, affecting water absorption and swelling, and thereby controlling enzyme accessibility.

References

1. M. Himmel, S. Ding, D. K. Johnson, W. Adney, M. Nimlos, *et al.* Biomass recalcitrance: Engineering plants and enzymes for biofuels production. *Science,* **315**, 804–807 (2007).
2. S. Mansfield, C. Mooney, J. Saddler. Substrate and enzyme characteristics that limit cellulose hydrolysis. *Biotechnol. Prog.,* **15**, 804–816 (1999).
3. W. Grous, A. Converse, H. Grethlein. Effect of steam explosion pretreatment on pore size and enzymatic hydrolysis of poplar. *Enzyme and Microbial Technology,* **8**, 274–280 (1986).
4. B. Yang, Z. Dai, S. Ding, C. E. Wyman. Enzymatic hydrolysis of cellulosic biomass *Biofuels,* **2**, 421–450 (2011).
5. Chundawat, Shishir PS, Giovanni Bellesia, Nirmal Uppugundla, Leonardo da Costa Sousa, Dahai Gao, Albert M. Cheh, Umesh P. Agarwal *et al.* (2011), Restructuring the crystalline cellulose hydrogen bond network enhances its depolymerization rate. JACS, 133(29), 11163–11174.
6. Y. Pu, C. Ziemer, A. J. Ragauskas. CP/MAS ^{13}C NMR analysis of cellulase treated bleached softwood kraft pulp. *Carbohydr. Res.,* **341**, 591–597 (2006).
7. Sannigrahi, P., Ragauskas, A. J., & Tuskan, G. A. (2010). Poplar as a feedstock for biofuels: a review of compositional characteristics, Biofuels, Bioproducts and Biorefining, 4(2), 209–226.
8. David, K., & Ragauskas, A. J. (2010). Switchgrass as an energy crop for biofuel production: A review of its ligno-cellulosic chemical properties. Energy & Environmental Science, 3(9), 1182–1190.
9. Kubicek, C. P. (1992). The eellulase proteins of Trichoderma reesei: structure, multiplicity, mode of action and regulation of formation. In Enzymes and Products from Bacteria Fungi and Plant Cells (pp. 1-27). Springer Berlin Heidelberg.
10. X. Zhao, L. Zhang, D. Liu. Biomass recalcitrance. Part I: The chemical compositions and physical structures affecting the enzymatic hydrolysis of lignocellulose. *Biofuels Bioprod. Bioref.,* **6**, 465–482 (2012).
11. M. Hall, P. Bansal, J. H. Lee, M. J. Realff, A. S. Bommarius. Cellulose crystallinity — a key predictor of the enzymatic hydrolysis rate. *FEBS,* **277**, 1571–1582 (2010).

12. L. F. Del Rio, R. P. Chandra, J. N. Saddler. The ease of enzymatic hydrolysis of organosolv-pretreated softwoods. *Biores. Technol.,* **107**, 235–242 (2010).
13. Y. Liu, J. Baker, Y. Zeng, M. Himmel, T. Hass, *et al.,* Cellobiohydrolase hydrolyzes crystalline cellulose on hydrophobic faces. *J. Biol. Chem.,* **296**, 11195–11201 (2011).
14. C. Ishizawa, M. Davis, D. Schell, *et al.,* Porosity and Its Effect on the digestibility of dilute sulfuric acid pretreated corn stover. *J. Agric. Food. Chem.,* **55**, 2575–2581 (2007).
15. J. A. Rollin, Z. Zhu, N. Sathitsuksanoh, Y. H. P. Zhang. Increasing cellulose accessibility is more important than removing lignin: A comparison of cellulose solvent-based lignocellulose fractionation and soaking in aqueous ammonia. *Biotechn. Bioengin.,* **108**, 22–30 (2011).
16. S. Chundawat, B. Venkatesh, B. Dale. Effect of particle size based separation of milled corn stover on AFEX pretreatment and enzymatic digestibility. *Biotechn. Bioengin.,* **96**, 219–231 (2007).
17. Luo, X., & Zhu, J. Y. (2011). Effects of drying-induced fiber hornification on enzymatic saccharification of lignocelluloses. Enzyme and microbial technology, 48(1), 92–99.
18. M. Foston, A. J. Ragauskas. Biomass characterization: Recent progress in understanding biomass recalcitrance. *Industrial Biotechnology,* **8**, 191–208 (2012).
19. S. Lin, Accessibility of cellulose: A critical review. *Fibre Science and Technology,* **5**, 303–314 (1972).
20. R. Samuel, Y. Pu, B. Raman, A. J. Ragauskas. Structural characterization and comparison of switchgrass ball-milled lignin before and after dilute acid pretreatment. *Appl. Biochem. Biotechnol.,* **162**, 62–74 (2010).
21. M. J. Selig, S. Viamajala, S. R. Decker, M. P. Tucker, M. E. Himmel, T. B. Vinzant, Deposition of lignin droplets produced during dilute acid pretreatment of maize stems retards enzymatic hydrolysis of cellulose. *Biotechnol. Prog.,* **23**, 1333–1339 (2007).
22. B. Yang, C. E. Wyman. Effect of xylan and lignin removal by batch and flowthrough pretreatment on the enzymatic digestibility of corn stover cellulose. *Biotechnol. Bioeng.,* **86**, 88–95 (2004).
23. Hu, F., Jung, S., & Ragauskas, A. (2012). Pseudo-lignin formation and its impact on enzymatic hydrolysis. Bioresource Technology.
24. L. Chen, *et al.,* Transgenic down-regulation of caffeic acid O-methyltransferase (COMT) led to improved digestibility in tall fescue (*Festuca arundinacea*). *Funct. Plant Biol.,* **31**, 235–245 (2004).
25. Y. Zhang, L. Lynd. Toward an aggregated understanding of enzymatic hydrolysis of cellulose: Noncomplexed cellulase systems. *Biotechnol. Bioengin.,* **88**, 797–824 (2004).
26. V. Lionetti, F. Francocci, S. Ferrari, C. Volpi, D. Bellincampi, *et al.* Engineering the cell wall by reducing de-methyl-esterified homogalacturonan improves saccharification of plant tissues for bioconversion. *PNAS,* **107**, 616–621 (2010).
27. J. D. DeMartini, S. Pattathil, Y. Avci, *et al.* Application of monoclonal antibodies to investigate plant cell wall deconstruction for biofuels production. *Energy Envir. Sci.,* **4**, 4332–4339 (2011).
28. F. Kong, C. Engler, E. Soltes. Effects of cell-wall acetate, xylan backbone, and lignin on enzymatic hydrolysis of aspen wood. *Appl. Biochem. Biotechnol.,* **34–35**, 23–35 (1992).

CHAPTER 3

REDUCTION OF BIOMASS RECALCITRANCE VIA WATER/ACID PRETREATMENTS

FAN HU

Department of Chemistry and Biochemistry, Institute of Paper Science and Technology, Georgia Institute of Technology
500 10th Street NW, Atlanta, GA 30332, USA
fhu6@gatech.edu

Lignocellulosic biomass is abundant and inexpensive feedstock that can be utilized to produce biofuels and bio-based chemicals. The yield of ethanol production from native lignocellulosic biomass is relatively low due to its native recalcitrance, which is attributed to, in part, lignin content/structure, hemicelluloses, cellulose crystallinity and other factors. Pretreatment of lignocellulosic biomass is required to overcome this recalcitrance. The goal of pretreatment is to alter the physical features and chemical composition/structure of lignocellulosic biomass, therefore making cellulose more accessible to enzymatic hydrolysis for sugar conversion. Dilute acid pretreatment (DAP) and hydrothermal pretreatment (HTP) are two of the most important pretreatment technologies developed to reduce recalcitrance and to increase sugar yield during the past two decades. This chapter examines the changes in lignocellulosic structure during these two major pretreatment technologies.

1. Introduction

Increasing global energy demand, unstable and expensive petroleum resources and concern over global climate changes lead to the development of renewable energy sources such as bioethanol that can supplement fossil fuels.[1] At present, bioethanol production largely represents the first-generation biofuel, which is produced from readily processable bioresources such as starch from corn and simple sugars from sugar cane.[2–4] However, as the demand for food resources increases, the search for renewable non-food resources to displace substantial amounts of non-renewable fossil fuels rests largely on low-cost and abundant lignocellulosic biomass.[5]

Lignocellulosic biomass is a complex composite mainly consisting of cellulose, hemicellulose and lignin. For the conversion of lignocellulosic biomass to ethanol, its polysaccharides (cellulose and hemicellulose) need to be broken down into the corresponding monosaccharides, which subsequently are fermented to ethanol by microorganisms. Although acid hydrolysis of lignocellulosics has been used to produce sugars suitable for ethanol production, utilization of enzymes is viewed as the most viable strategy, since enzymatic hydrolysis of lignocellulosics offers several advantages such as higher yield, lower byproduct formation and energy requirement, milder operation condition and environmental benign processing compared to chemical hydrolysis. However, native lignocellulosics are recalcitrant to decomposition from microbes and enzymes due to their physical features and chemical composition/structure.[1] This recalcitrance has been attributed to the presence and structure of hemicellulose, cellulose crystallinity/degree of polymerization (DP), lignin content/structure, as well as substrate accessibility to enzymes. Pretreatment is thus an essential step for overcoming the recalcitrance and increasing fermentable sugar yields from biological deconstruction step. It utilizes various technologies such as chemical treatment to alter the physicochemical, structural and compositional properties of lignocellulosic biomass, thereby making cellulose more accessible to enzymes during hydrolysis step.[6,7]

Although pretreatment has achieved various levels of success to overcome the recalcitrance of lignocellulosic biomass, it is still one of the most expensive processing steps. Among various pretreatment technologies developed during the past two decades, dilute acid pretreatment (DAP) and hydrothermal pretreatment (HTP) are two of the major chemical techniques. Understanding the chemistry behind these two leading pretreatment technologies is therefore essential since pretreatment chemistry is central of importance owing to its impacts on lignocellulosics processing and bioethanol conversion. The purpose of this chapter is to review the lignocellulosic structural changes associated with recalcitrance reduction during DAP and HTP. This will be done by introducing the technical process of DAP and HTP, followed by summarizing the hemicellulose hydrolysis, the change in cellulose crystallinity/DP and lignin behaviors during these two pretreatment processes. This fundamental insight is a key feature needed to develop more efficient/cost-effective pretreatments in the future.

2. Technical Process of DAP and HTP

2.1. *Dilute acid pretreatment*

Dilute acid pretreatment is usually performed over a temperature range of 120°C to 210°C, with acid concentration typically less than 4 wt%, and residence time from a few minutes to an hour in different types of reactors such as batch,[8] plug flow,[9] percolation,[10] countercurrent,[11] and shrinking-bed reactors.[11–13] Although a variety of acids such as hydrochloric acid, nitric acid, phosphoric acid and peracetic acid have been employed, sulfuric acid has been most widely used since it is inexpensive but effective.[14,15] DAP has been successfully applied to a wide range of feedstocks, including softwoods, hardwoods, herbaceous crops, and agricultural residues.[16–20] The combined severity (CS) factor is used for an easy comparison of pretreatment conditions and for facilitation of process control, which relates the experimental effects of temperature, residence time and acid concentration.[21] Table 1 summarizes recent DAP results for different substrates.

$$CS = \log\left\{t \exp\left[\left(T - T_{\mathrm{ref}}\right)/14.7\right]\right\} - \mathrm{pH},$$

- t is the pretreatment time (min)
- T is the pretreatment temperature (°C)
- T_{ref} is 100°C.

DAP is one of the most important chemical pretreatment technologies because of its high hemicellulose solubilization and recovery, and high yields in subsequent enzymatic deconstruction of cellulose. However, DAP is still among the most expensive steps in biomass conversion to

Table 1. Recent DAP results for different substrates.

Substrate	Pretreatment conditions	CS	Cellulose conversion yield (%)	Enzymes loadings	Reference
Corn stover	121°C, 2.0% H_3PO_4, 120 min	1.28	56.0 in 72 h	40 FPU/g for cellulase from Celluclast 1.5 L, and 8 CBU/g for β-glucosidase from Novozym 188	16
	121°C, 2.0% H_2SO_4, 120 min	2.01	75.6 in 72 h	40 FPU/g for cellulase from Celluclast 1.5 L, and 8 CBU/g for β-glucosidase from Novozym 188	16
Cotton stalk	121°C, 2.0% H_2SO_4, 60 min	1.71	23.9 in 72 h	40 FPU/g for cellulase from Celluclast 1.5 L, and 70 CBU/g for β-glucosidase from Novozym 188	22
Aspen	175°C, 0.25% H_2SO_4, 30 min	2.10	42.3 in 72 h	60 FPU/g for cellulase from Spezyme CP and 120 CBU/g for β-glucosidase from Novozyme 188	17
	170°C, 1.10% H_2SO_4, 30 min	2.60	88.0 in 72 h	7.5 FPU/g for cellulase from Celluclast 1.5 L and 11.25 CBU/g for β-glucosidase from Novozyme 188	18
Switchgrass	140°C, 1.0% H_2SO_4, 40 min	1.79	75.0 in 72 h	15 FPU/g for cellulase from Spezyme CP and 30 CBU/g for β-glucosidase from Novozyme 188	19
Sweet sorghum bagasse	180°C, 1.0% H_2SO_4, 20 min	2.67	53.9 in 60 h	20 FPU/g for cellulase	20

fuels,[23] primarily owing to the additional costs for acid, special reactor material and acid-neutralization step.

2.2. *Hydrothermal pretreatment*

Hydrothermal pretreatment refers to the use of water in the liquid or vapor phase to pretreat lignocellulosic biomass, which includes liquid hot water (LHW) and uncatalyzed steam explosion pretreatment. HTP is particularly promising because it provides several potential advantages including no requirement for catalysts or special reactor materials, or preliminary feedstock size reduction.[24]

During LHW pretreatment, lignocellulosic biomass undergoes high temperature cooking in water with high pressure. Pressure is utilized to maintain water in the liquid state at elevated temperatures of 160-240 °C.[25] LHW has been performed in co-current, countercurrent or flow-through reactors. In the co-current process, the water-biomass slurry is heated to pretreatment conditions and held for the required residence time. In the countercurrent process, the water and biomass flow in the opposite directions. In the flow-through process, hot water flows through a stationary bed of biomass.[26] LHW pretreatment without adding any catalyst is attractive due to its simplicity and significant reduction in the chemical and materials of construction costs compared to DAP. LHW pretreatment has lower hemicellulose sugar yields than DAP and it also has much lower cellouse degradation.

In steam explosion pretreatment, lignocellulosic particles are treated with high-pressure saturated steam for a period of time and then the pressure is swiftly released, resulting in an explosive decompression of the lignocellulosic material. Steam explosion is typically carried out at a temperature of 160–260°C (corresponding pressure 0.69–4.83 MPa) for a few minutes before the material is exposed to atmospheric pressure.[7] Addition of an acid catalyst such as SO_2 or preferably H_2SO_4 because of its inexpensive[27] to steam explosion can significantly increase its hemicellulose sugar yields.[28] Therefore the sugar yields are highly dependent on the pretreatment conditions. It is generally accepted that the chemistry in the acid catalyzed steam explosion pretreatment is similar to that in DAP. This technology combines both chemical and physical pretreatments into one step. The mechanical effects are caused by a sudden decompression so that fibers are separated. Similar to DAP, the severity factor is used to measure the treatment intensity in LHW and uncatalyzed steam explosion pretreatment, which is defined below:[29,30]

$$S_0 = \log\{t \exp[(T - T_{\text{ref}})/14.75]\}$$

- t is the pretreatment time (min)
- T is the pretreatment temperature (°C)
- T_{ref} is 100°C.

Table 2 summarizes recent HTP results for different substrates. In general, both LHW and uncatalyzed steam explosion pretreatments have the potential to reduce lignocellulosics recalcitrance and increase sugar yields from enzymatic hydrolysis of cellulose. It was reported that LHW pretreatment was comparable to uncatalyzed steam explosion pretreatment with respect to the cellulose to glucose conversion yield at equivalent pretreatment severity.[31] However, steam explosion pretreatment can be performed at higher solids concentrations, which often generates more inhibitors. It also suffers from lower hemicellulose recovery than LHW pretreatment.[32,33] Whereas LHW pretreatment produces large amounts of hemicellulose oligosaccharides that must be further hydrolyzed to fermentable monosaccharides.[34,35] In addition, the concentration of solubilized products is lower in LHW pretreatment compared to uncatalyzed steam explosion pretreatment.[36] This is probably caused by higher water input in LHW pretreatment than in steam explosion pretreatment.[25] However, higher water input in LHW pretreatment demands more energetic requirements, which is not favorable for developing at commercial scale.

3. Hemicelluloses Hydrolysis and Porosity during DAP and HTP

Hemicelluloses sheath cellulose microfibrils and their acetyl groups would sterically hinder enzyme attack.[26] Unbranched hemicelluloses form hydrogen bonds with cellulose fibrils surface; on the other hand, both unbranched and branched hemicelluloses may be covalently bonded to lignin to create enzyme-impenetrable cross-links namely

Table 2. HTP results for different substrates.

Substrate	Cellulose conversion yield (%)	Pretreatment conditions	Enzymes loadings	Reference
Wheat straw	95.8 in 72 h	200°C, 40 min	15 FPU/g for cellulase from Celluclast 1.5 L, and 15 IU/g for β-glucosidase from Novozyme 188	37
	87.5 in 72 h	195°C, 20 min	15 FPU/g for cellulase from Celluclast 1.5 L, and 15 IU/g for β-glucosidase from Novozyme 188	38
	81.2 in 72 h	195°C, 40 min	15 FPU/g for cellulase from Celluclast 1.5 L, and 15 IU/g for β-glucosidase from Novozyme 188	38
Corn stover	69.6 in 168 h	190°C, 15 min	15 FPU/g for cellulase from Spezyme CP, and 65 IU/g for β-glucosidase from Novozyme 188	39
Switchgrass	77.4 in 48 h	200°C, 10 min	49 FPU/g for cellulase from Celluclast 1.5 L, and 40 IU/g for β-glucosidase from Novozyme 188	40

lignin-carbohydrate complexes (LCCs),[41] as shown in Fig. 1. LCCs are thought to be one of the major impediments to enzyme access to cellulose.[41,42] Hydrolysis of hemicelluloses and cleavage of LCCs are important, due to the fact that they can open the plant cell wall structure. In addition, the mean pore size and the specific surface area of cellulose can also be enlarged.[43] These are influential structural features related to

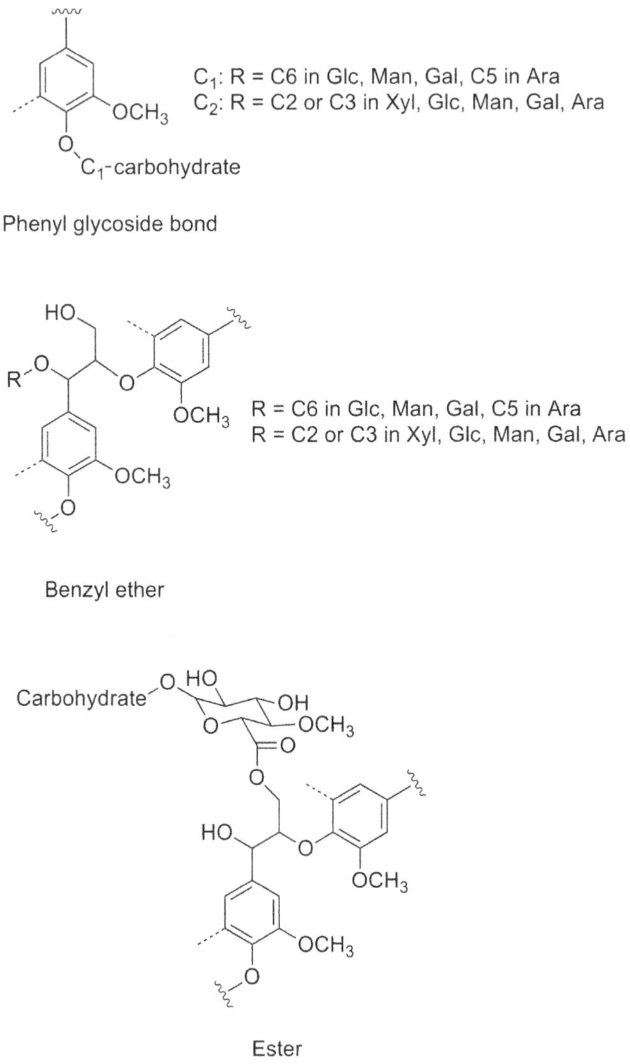

C_1: R = C6 in Glc, Man, Gal, C5 in Ara
C_2: R = C2 or C3 in Xyl, Glc, Man, Gal, Ara

OCH_3
C_1-carbohydrate

Phenyl glycoside bond

HO
R–O
O
OCH_3
R = C6 in Glc, Man, Gal, C5 in Ara
R = C2 or C3 in Xyl, Glc, Man, Gal, Ara
OCH_3
O

Benzyl ether

Carbohydrate–O HO
OH
O
OCH_3
O
O
HO
O
OCH_3
OCH_3
O

Ester

Fig. 1. Lignin-carbohydrate complex linkage structure.

noncomplexed cellulase adsorption on the cellulose surface and subsequent enzymatic deconstruction,[44] since cellulases must bind to the surface of cellulose before hydrolysis can take place. Zeng and coworkers[39] showed that an increased surface area resulted in more exposed cellulose, thereby increasing the initial enzymatic hydrolysis rate of cellulose.

One of the impacts of DAP and HTP is to hydrolyze hemicelluloses and to disrupt the lignin structure, thus the cellulose fraction of the treated biomass has enhanced enzymatic digestibility. Hemicellulose, mainly xylan, is hydrolyzed to xylose or xylo-oligomers during DAP or HTP respectively, as glucomannan is relatively stable in acid.[43] The hydronium ions released by the acid or water cause depolymerization of hemicellulose by selective hydrolysis of glycosidic linkages, liberating *O*-acetyl group and other acid moieties to form acetic and uronic acids. The release of these acids is thought to catalyze the hydrolysis of hemicelluloses and oligosaccharides from hemicelluloses, particularly in HTP.[32,45–47]

In general, the degree of xylan hydrolysis increases as the DAP or HTP severity increases.[48–51] Xylan is dissolved in the reaction media first as high M_w (DP > 25) material followed by cleavage of more and more bonds between xylose residues upon higher pretreatment severity.[48] The initially dissolved high DP xylo-oligomers have a high degree of acetylation since the acetyl groups increase xylan solubility.[52] The medium M_w (DP 9-25) xylo-oligomers are predominate in HTP, and their proportions decrease slightly with severity due to the increased decomposition.[53,54] On the other hand, most of the released xylan is accumulated in the liquors in the form of xylose at lower severity conditions for DAP.[49] The more severe the pretreatment, the more low Mw (DP < 9) xylo-oligomers and the less high M_w xylo-oligomers are detected in the reaction media for both DAP and HTP.[49,53,54] For DAP at higher severity conditions, the released xylan in the liquors is partially converted to furfural (Fig. 2),[48,49] which is a by-product inhibitory to the formation of ethanol during the fermentation process. According to the studies by Wyman Group,[55–57] hemicellulose hydrolysis is controlled by both chemical reaction (i.e., temperature and acid concentration) and mass transfer effect. In DAP where the acid concentration is relatively high, mass transfer is insignificant since the hydronium ions would rapidly hydrolyze long-chain

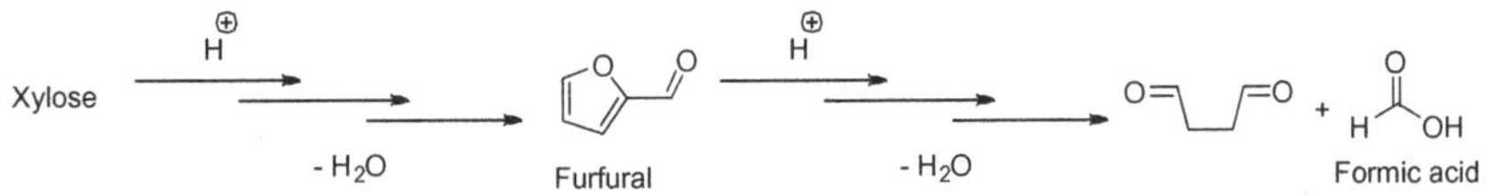

Fig. 2. Reaction pathways of inhibitory byproducts formation from xylose.

hemicellulose oligomers to more rapidly dissolving short-chain species and then to monomers. This is why hemicellulose solubilization is mainly controlled by temperature and acid concentration in DAP. On the other hand, mass transfer is believed to play a more important role in HTP, where most of hemicellulose sugars are released as oligomers.[55–59] This is owing to the long-chain hemicellulose oligomers and unreacted hemicellulose could form hydrogen bonds with water molecules to form an "ice-like" layer that slows the access of water to hemicellulose.[55] Increasing the flow of liquid would enhance the removal of less-soluble oligomers and disturb the "ice-like" layer, thereby increasing hemicellulose removal (Fig. 3).[55] In addition, lignin-hemicellulose-oligomers and their solubility are also postulated to significantly affect the rates and yields of hemicellulose solubilization.[57]

With respect to xylan hydrolysis, two fractions of xylan with different reactivities towards hydronium-catalyzed hydrolysis exist in feedstocks (two-fraction theory).[60,61] This is attributed to the difference in accessibility and in chemical structure between different zones of xylan,[60,62,63] leading to a fast and a slow reacting xylan. Although this two-fraction model has been applied to investigate the kinetics and mechanisms of xylan hydrolysis for two decades,[64–66] the fast and slow reacting xylans were not distinctly defined until recently, Shen and Wyman[67] defined that the fast reacting xylans were the portion that directly forms monomeric xyloses

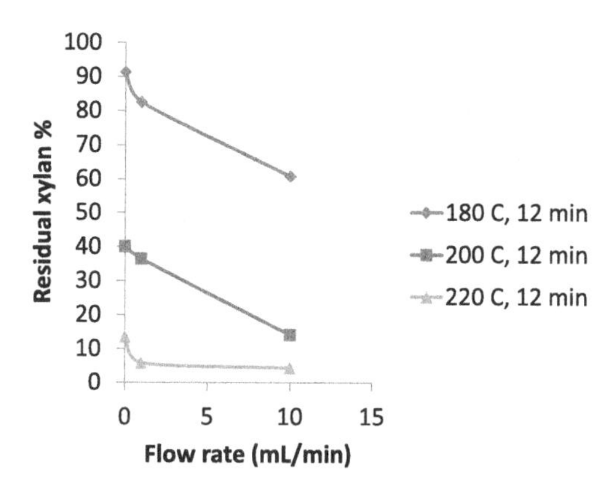

Fig. 3. The impact of flow rate on xylan hydrolysis for hydrothermal pretreatment of corn stover.

through an autohydrolysis mechanism. On the other hand, the slow reacting xylans were the portion that forms oligosaccharides by an autohydrolysis mechanism (or further forms monomeric xyloses by acid catalysis in DAP) plus the portion that remains unreacted in the solid xylan in biomass. Based on these definitions, Shen and Wyman[67] calculated that the percent of fast and slow reacting xylans in corn stover during DAP were 9.0% and 91.1% respectively. 10.6% of the slow reacting xylan (or 9.7% of total xylans) belonged to the unreacted fraction. Whereas these numbers became 9.0% and 91.8% for corn stover during HTP. 31.0% of the slow reacting xylan (or 28.5% of total xylans) belonged to the unreacted fraction.[67] In addition, the two-fraction theory can also be applied to deacetylation since the accessibility problems cause differences in the reactivity of both xylan and acetyl groups bound to xylan chains.[68] It was reported that the hydrolyzed acetyl groups became an in-situ source of acetic acids that could further catalyze xylan depolymerization, whereas another fraction of the acetyl esters remained covalently linked to the xylan backbone and were released from the residue together with the xylan as esterified xylo-oligomers. This latter fraction is believed to be a stronger inhibitor to cellulases than pure xylo-oligomers owing, in part, to the steric hindrances of the acetyl groups.

It was thought that the center of the cell wall would be the most protected from the acid hydrolysis, thus xylan hydrolysis during DAP would begin at the solvent exposed cell walls, and then work its way inward.[69] Contrary to this hypothesis, Brunecky *et al.*[69] and Jung *et al.*[70] observed the migration of xylan from the center cell wall to the lumen and middle lamella prior to being hydrolyzed into soluble oligomers during DAP especially at high temperature. This is attributed to, in part, the solubility of the large xylo-oligomers drops sufficiently for them to precipitate from the solution when the reaction medium cools at the end of pretreatment. The large xylo-oligomers could then redeposit on the outer surfaces of the cell walls.[69] Furthermore, the association of a fraction of the xylan with lignin would also cause xylan redistribution.[69] Lignin is well known to migrate and form spherical droplets on the outer surfaces of cell walls, particularly on the lumen and in the middle lamella during DAP at high temperature.[71,72] The hydrophobic nature of lignin could hinder the access of the acid in the hydrolyzate to the ether linkages of the xylo-oligomers, which explains why this fraction of xylan appears to be more difficult to hydrolyze during DAP.[73]

Several studies have indicated that the breakdown and loosen of the lignocellulosic structure by DAP and/or HTP increases the pore volume and the pore size or porosity of the biomass.[39, 74–80] Hsu *et al.*[75] suggested that this was not only caused by hemicellulose removal but also by rearrangement of the lignin structure. A further study by Foston and Ragasukas[76] revealed that the increase in pore size during DAP was due to existing pores within the system expanding. In addition, Chen *et al.*[81] observed that the small holes merged into the large holes, rendering the swelling behavior of the biomass during DAP. They figured out later that the lignocellulosic structure of biomass simultaneously underwent fragmentation and swelling during DAP.[77] The fragmentation produced more small particles, thereby enlarging the specific surface area of the biomass. On the other hand, the swelling behavior of the biomass would reduce rather than increase the specific surface area.[77]

4. Cellulose Crystallinity and Degree of Polymerization during DAP and HTP

The effect of cellulose crystallinity on the overall cellulose-to-glucose conversion has been an issue for extensive studies. The degree of cellulose crystallinity is expressed in terms of the crystallinity index (CrI), which can be measured by either X-ray diffraction (XRD) technique or solid-state NMR analysis.[82, 83]

The initial enzymatic hydrolysis rate of different cellulose allomorphs decreases in the following order: amorphous > III_I > II > I, which has been attributed to their enhanced specific surface areas compared to cellulose I.[84] The slower hydrolysis rate of native crystalline cellulose (cellulose I) is due to the presence of strong interchain hydrogen bonding between adjacent chains in a cellulose sheet and weaker hydrophobic interactions between cellulose sheets, resulting in the stability of crystalline cellulose nanofibers that strongly resist chemically or biologically catalyzed degradation.[85–87] In addition, Weimer *et al.*[84] pointed out that the lower crystallinity, lower packing density and higher distances between hydrophobic surfaces of cellulose III_I as compared to cellulose I were responsible for its increased enzymatic hydrolysis rate.

Several recent studies suggested the degree or rate of enzymatic hydrolysis of cellulose declined with increasing cellulose crystallinity,[88–94] whereas some researchers did not observe this strong correlation.[95–97] Furthermore, differences in the adsorption properties of cellulases on crystalline and amorphous cellulose are also believed to be related to the reactivity difference between crystalline and amorphous cellulose.[98,99] Hall *et al.*[94] observed that the adsorbed enzymes on the cellulose with low degrees of crystallinity were more active at the same overall concentration, due to a more open cellulose structure that hindered enzymes from residing on neighboring chains from hindering one another.[100] Exo-cellulases may also locate a chain end faster with an open structure so that hydrolysis may occur quicker. On the other hand, the hydrolysis rate was limited to the high degrees of crystallinity, since the internal surface of highly crystalline cellulose was poorly accessible to enzymes, leading to low enzyme adsorption. Nevertheless, based on the conflicting results from the literature, further studies are needed to determine whether cellulose crystallinity provides a clear indication of enzymatic digestibility of cellulose.

The degree of polymerization of cellulose has also been postulated to play a role in its susceptibility to enzymatic deconstruction of cellulose. During enzymatic hydrolysis, endo-cellulases are involved in cleaving internal $\beta(1\rightarrow4)$ linkages of cellulose chains decreasing the DP of cellulose and exposing reactive ends that can be attacked by exo-cellulases.[101,102] Exo-cellulases have a marked preference for substrates with lower DP,[88,103] thus reduction in the DP of cellulose would produce more chain ends with higher enzyme accessibility to cellulose. Furthermore, it was reported that cellulose with shorter chains was more amenable to enzymatic deconstruction due to the absence of strong hydrogen bonding.[104–106] One early study focusing on effect of the cellulose DP on enzymatic saccharification suggested that the DP of cellulose may play an important role in the initial rate of enzymatic hydrolysis.[95] However, Sinistyn and co-workers[107] showed that reduction in the DP of cotton linters while keeping crystallinity index constant, had negligible impact on the enzymatic hydrolysis rate. Zhang and Lynd[108] observed that a decrease in cellulose DP was less effective in accelerating enzymatic hydrolysis than an increase in accessibility of β-glycosidic bonds. Such results indicate that the understanding of the impact of cellulose chain length on enzymatic hydrolysis is still under developing.

During DAP and HTP, the degradation of cellulose is an acid catalyzed and thermally accelerated chain scission reaction, which occurs within the fibril structure from either a crystalline or amorphous region of cellulose.[109] Two major stages are involved in this process; an initial rapid hydrolysis of the more solvent accessible amorphous region, accounting for the increase in crystalline and para-crystalline percentages; and a latter much slower hydrolytic attack of the crystalline portion.[110,111] Yu and Wu[112] investigated the hydrolysis behavior of amorphous and crystalline cellulose in HTP. The minimal temperature required to rupture the glycosidic bonds in the chain segments within the amorphous portion of cellulose appears to be approximately 150°C whereas for crystalline portion of cellulose it is 180°C. Furthermore, low-DP glucose oligomers are produced at 180°C, whereas large-DP glucose oligomers are released at temperatures above 200°C. Clearly, this difference in the hydrolysis behavior between amorphous and crystalline cellulose is due to the significant ultrastructural differences in the amorphous and crystalline portions of cellulose. Based on the study focusing on changes in the cellulose ultrastructure during DAP, it was observed that the cellulose I_α form was susceptible to either selective degradation by acidic hydrolyzation and/or transformation to other crystal allomorphs such as cellulose I_β during pretreatment.[109] This transformation is attributed to the metastable properties of the triclinic one-chain crystal structure of cellulose I_α.[111,113,114] It was suggested that during DAP cellulose I_α was primarily converted to para-crystalline cellulose, followed by conversion to cellulose I_β while simultaneously a small fraction of para-crystalline cellulose slowly transformed into crystalline cellulose.[109] In addition, it has been reported that steam explosion can selectively hydrolyze amorphous portion of cellulose,[115,116] and transfer some amorphous portion of cellulose to crystalline portion.[113,117–119] Therefore, the relative proportion of both the crystalline and para-crystalline forms of cellulose can be affected by ultrastructural transformation mechanisms and/or hydrolyzation at crystalline surfaces during DAP and HTP. It is also expected that the crystallinity of cellulose would

Table 3. Crystallinity index (CrI) before and after DAP or HTP for different substrate.

Substrate	Pretreatment conditions	CrI (%) before pretreatment	CrI (%) after pretreatment	Reference
Corn stover	160 °C, 0.5% H_2SO_4, 20 min	50.3	52.5	120
Poplar	160 °C, 1.0% H_2SO_4, 5 min	49.9	70.5	109
	200 °C, 10 min	49.9	54.0	120
Loblolly pine	180 °C, 0.5% H_2SO_4, 10 min	62.5	69.9	121
Tamarix	160 °C, 8 min	41.0	50.6	122
Coastal bermuda grass	170 °C, 60 min	50.2	69.4	123

Table 4. DP of cellulose after DAP or HTP for different substrates.[120]

Substrate	Pretreatment conditions	DP
Corn stover	160 °C, 0.5% H_2SO_4, 20 min	2700
	190 °C, 15 min	5700
	190 °C, 3% SO_2-steam explosion, 5 min	3000
Poplar	200 °C, 10 min	1800
	200 °C, 3% SO_2-steam explosion, 5 min	500

increase after the pretreatments as shown in Table 3. Nevertheless, the enzymatic digestibility of lignocellulosic biomass increases after the pretreatment despite the fact that the CrI also increases.

Both of DAP and HTP result in the reduction in the DP of cellulose especially at high severity conditions. This is favorable since the reduction in cellulose DP can provide more reducing ends thereby increasing enzymatic digestibility of cellulose. Table 4 shows the DP of cellulose of different substrates after DAP or HTP. During the acid hydrolysis of cellulose, the DP of cellulose from different substrates decreases gradually until reaching a nominal value namely the leveling-off degree of polymerization (LODP).[124,125] The initial faster DP reduction phase is believed to represent the hydrolysis of the reactive amorphous region of cellulose; whereas the slower plateau rate phase corresponds to the hydrolysis of the slowly reacting crystalline fraction of cellulose.[124,125]

5. Lignin Behavior during DAP and HTP

The chemistry and physiological functions of lignin are distinct from those of either hemicelluloses or cellulose. Lignin is a polyphenolic, highly cross-linked macromolecular polymer, which confers integrity and structural rigidity on the plant cell wall. It consists of three phenylpropane units called guaiacyl (G), syringyl (S) and p-hydroxyphenyl (H) units, and their respective precursors are three aromatic alcohols (monolignols) namely coniferyl, sinapyl, and p-coumaryl alcohols (Fig. 4).[126] During the lignification process, these monolignols produce a complex three-dimensional amorphous lignin polymer via β–*O*–4, α–*O*–4, β–5, β–1, 5–5, and β–β linkages, which lacks the regular and ordered repeating units found in other polymers such as cellulose (Fig. 5).[127] Lignin acts as a physical barrier to prevent enzyme access to the carbohydrate fraction of biomass. Although the detailed mechanism that explains the protective

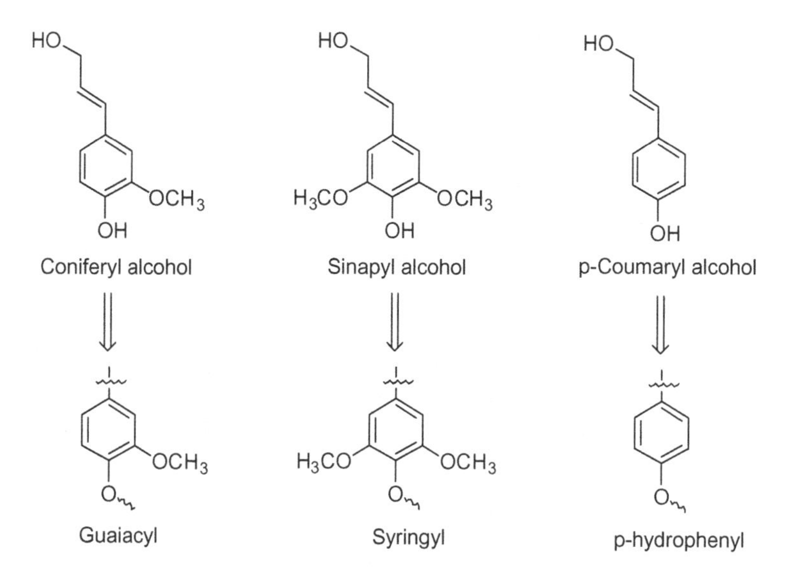

Fig. 4. Three building blocks of lignin.

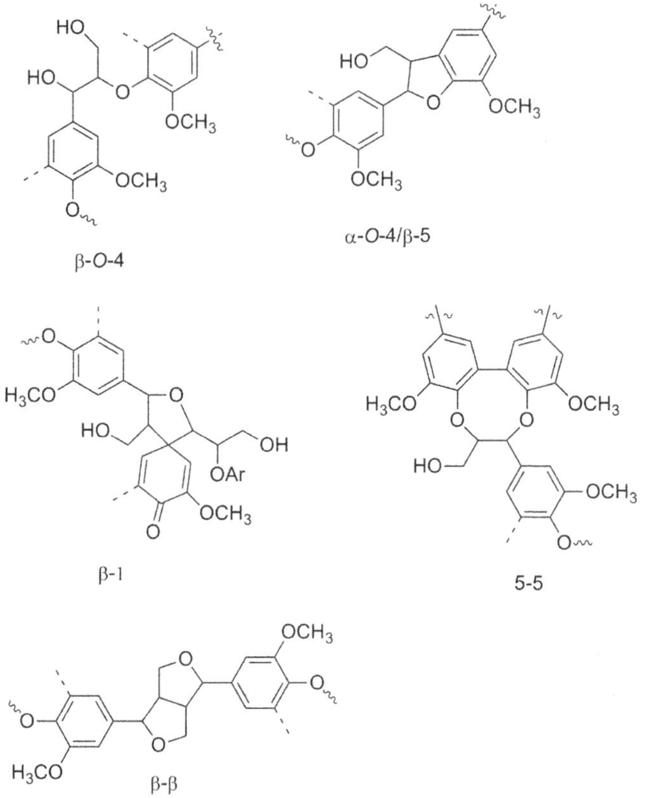

Fig. 5. Different lignin linkages.

effect of lignin against enzymatic hydrolysis is still unclear, the degree and type of cross-linkage between lignin and carbohydrate, the structure of lignin and its distribution through the cell wall are important.[128] In addition, during enzymatic hydrolysis, enzymes tend to non-productively bind to lignin through hydrophobic interactions that causes a loss in their activities.[129] Such non-productive binding of enzymes to lignin has been suggested to be responsible for the requirement of high enzyme loadings.[130–134] Yoshida and coworkers[91] reported that lignin was the most significant resistance factor against enzymatic hydrolysis, particularly in the initial hydrolysis phase.

One of the impacts of DAP and HTP is the translocation and redistribution of lignin, resulting in formation of lignin droplets of various morphologies on biomass cell walls.[71,72,135] The pretreatments that exceed the critical phase transition temperature of lignin allow it to expand and migrate to large voids, where lignin is shaped by the aqueous environment and forms spherical droplets.[71] It was also suggested that lignin aggregated prior to the formation of larger lignin droplets, which was attributed to its self-aggregation property.[135] The lignin droplets can leave the cell wall matrix, move into the solution and redeposit on the pretreated biomass surface during cooling. The redistribution of lignin is believed to open up the cell wall structure, which is almost as important as lignin removal in increasing the accessibility of cellulose microfibrils to enzymes.

The quantitative NMR studies revealed that the generation of unsaturated carbons and the accumulation of lignin occur during the DAP and HTP process.[136–138] Indeed, the acid-insoluble (Klason) lignin content of biomass after the pretreatment was often found higher than that of the starting material (Table 5). DAP and HTP thus do not lead to significant delignification, instead, they alter the lignin structure. The cleavage of β-*O*-4 linkage is the primary structural change of lignin occurring with

Table 5. Klason lignin content before and after DAP or HTP for different substrates.

Substrate	Pretreatment conditions	Klason lignin % before pretreatment	Klason lignin % after pretreatment	Reference
Corn stover	160°C, 0.5% H_2SO_4, 20 min	17.2	26.8	137
Loblolly pine	200°C, 1.0% H_2SO_4, 2 min	29.4	46.2	121
Poplar	175°C, 2.0% H_2SO_4, 10 min	23.2	38.5	70
Switchgrass	160°C, 1% H_2SO_4, 10 min	31.2	44.3	135
Tamarix	180°C, 9min	15.1	29.3	122
Wheat straw	195°C, 6 min	22.6	25.5	145
Coastal Bermuda grass	170°C, 60 min	17.6	21.9	123

DAP and HTP.[137–141] A decrease of β-β, β-5 and/or 5-5 linkages of the switchgrass lignin after the pretreatment was also reported.[138,140] In addition, DAP and HTP lead to the change in the amounts of aromatic methoxy groups.[136,137,139] A decrease in the amounts of aliphatic OH groups and an increase in the total phenolic OH group amounts have also been noticed after DAP and HTP.[139,140] The increase in the phenolics during HTP process has been attributed to the hemolytic cleavage of β-*O*-aryl ethers followed by the elimination of primary and secondary aliphatic hydroxyl groups (Fig. 6).[142] Although the degradation of lignin during the pretreatment is confirmed by the NMR analysis, the molecular weights of lignin have been reported to be unchanged or slightly increased.[142,143] This is due

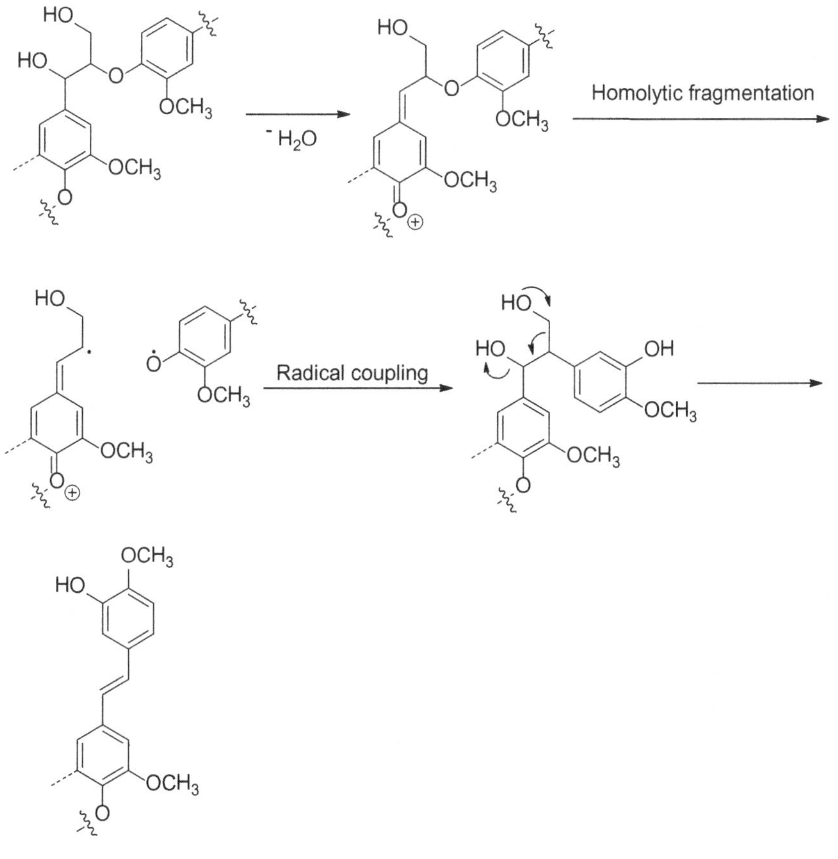

Fig. 6. The homolytic cleavage of lignin during HTP.

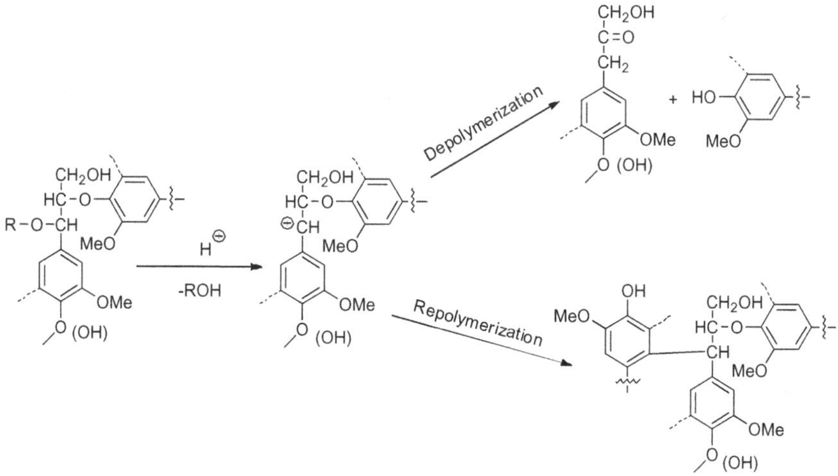

Fig. 7. Lignin depolymerization and repolymerization through carbonium ion intermediate.

to the fact that lignin is depolymerized and repolymerized through carbenium ion intermediate during the DAP and HTP process (Fig. 7).[144] The lignin condensation reaction increases the hydrophobicity of lignin particularly at high pretreatment severity, which increases the non-productive binding of cellulases to lignin. The mechanism implies an application of a scavenger of carbonium ion may suppress the lignin condensation reaction, thereby increasing the enzymatic digestibility of pretreated biomass.

6. Pseudo-lignin Formation

Although the increased Klason lignin content of biomass after DAP or HTP might be explained by the removal of significant amounts of hemicelluloses and certain amounts of cellulose while retaining most of lignin, some researchers hypothesized this is due to the repolymerization of polysaccharides degradation products and/or polymerization with lignin (Fig. 8), which form a lignin-like material termed pseudo-lignin.[144,146] Recently, Sannigrahi and co-workers[147] demonstrated that pseudo-lignin can be generated from carbohydrates without significant contribution

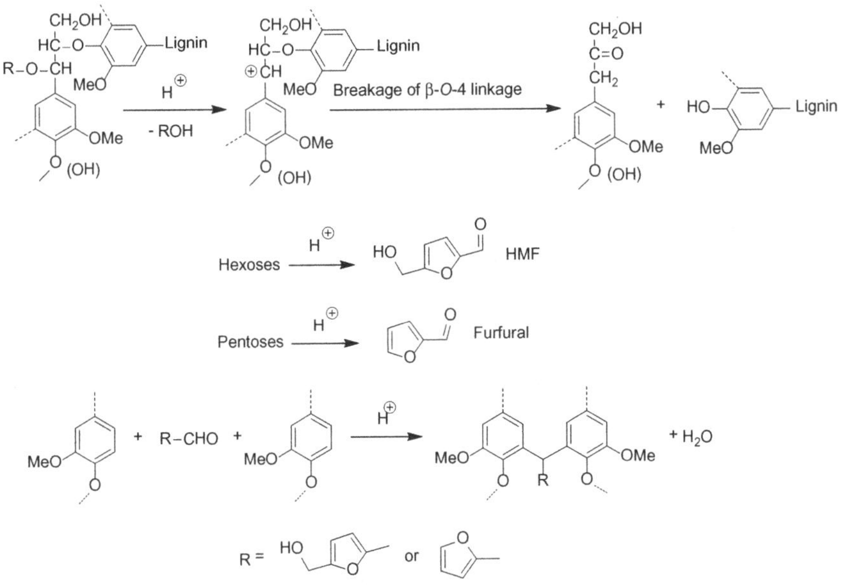

Fig. 8. Repolymerization of HMF and/or furfural with lignin during DAP.

from lignin during DAP especially under high severity pretreatment conditions. Scanning electron microscope (SEM) studies on dilute acid pretreated holocellulose revealed the presence of pseudo-lignin spherical droplets on the surface of dilute acid pretreated holocellulose. ^{13}C CP/MAS analysis of pretreated holocellulose indicated significant peaks from carbonyl, aromatic, methoxy and aliphatic structure. These peaks were attributed to the structure of pseudo-lignin, which accounted for the additional lignin detected by wet chemical analysis. Furthermore, the intensities of these peaks increased as pretreatment severity increased, suggesting an acid-catalyzed disproportionation mechanism of formation for pseudo-lignin. Furthermore, Hu *et al.*[148] isolated and characterized pseudo-lignin produced from dilute acid pretreated poplar holocellulose and α-cellulose. They showed that pseudo-lignin was polymeric with a $M_w \sim 5000$ g/mol and contained carbonyl, carboxylic, aromatic methoxy and aliphatic structures, which were produced from both dilute acid pretreated cellulose and hemicellulose. In addition, it was postulated that during DAP, the hydrolysis of polysaccharides, which leads to some release of monosaccharides, and their subsequent dehydration reactions to form furfural and 5-hydromethylfurfural (HMF) takes place. Further rearrangements of furfural and/or HMF may produce aromatic compounds, which undergo further polymerization and/or polycondensation reactions to form pseudo-lignin (Fig. 9).[148] Equally important, these studies indicated that the presence of pseudo-lignin on the surface of pretreated biomass can significantly inhibit enzymatic hydrolysis of cellulose. It is well known that lignin has non-productive association with cellulases due to its hydrophobic structural features including hydrogen-bonding, methoxy groups and polyaromatic structures.[149] The structural functionality of methoxy and polyaromaticity of pseudo-lignin revealed by the FT-IR and ^{13}C NMR analysis suggests its hydrophobicity. Indeed, pseudo-lignin is insoluble in water. The hydrophobic structural functionality of pseudo-lignin is speculated to account for its non-productive association with cellulases, resulting in the inhibition effects to enzymatic hydrolysis of cellulose. A follow-up study by Hu *et al.*[150] revealed that although pseudo-lignin that is not derived from native lignin is even more detrimental to enzymatic hydrolysis of cellulose than dilute acid pretreated lignin, thus its formation should be avoided.

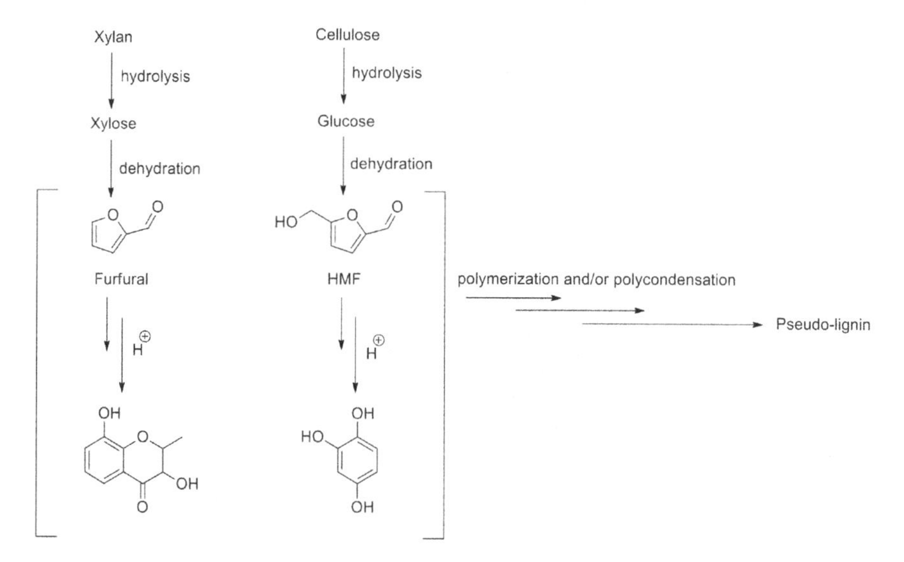

Fig. 9. Hypothesized reaction pathways for pseudo-lignin formation during DAP.

7. Conclusions and Outlook

The biodegradability of lignocellulosic biomass is limited by several factors such as lignin content/structure, specific surface area and cellulose CrI/DP. How DAP and HTP address these factors to reduce the recalcitrance and increase the enzymatic digestibility of lignocellulosic biomass have been discussed herein. Although researchers have put substantial amount of effort into understanding the chemistry and the plant cell wall structure changes during DAP, HTP and other pretreatment technologies, the insufficient knowledge of plant cell wall structure and ultrastructure still limits the economics and effectiveness of pretreatment. Plants are composed of at least 35 different cell types that are distinct in composition, structure, and ultrastructure, resulting in their very complex biological and chemical properties.[151] For instance, the detailed protection mechanism of LCCs towards chemical and biological deconstruction of cell walls is poorly understood.[41] The fundamental science behind the optimizations of pretreatment effectiveness is still not fully understood, due to the lack of mechanistic understanding of the ultrastructural and physicochemical changes occurring within the cell wall at the molecular level and the cellular/tissue scale during various pretreatment technologies.[41] It is thus important to understand the pretreatment effects on plant cell walls at a more fundamental level, in order to develop a more cost-effective pretreatment technology.

Acknowledgments

The authors are grateful for financial support from the Paper Science & Engineering (PSE) Fellowship program at Institute of Paper Science & Technology (IPST) and the School of Chemistry and Biochemistry at Georgia Institute of Technology.

References

1. M. E. Himmel, S. Y. Ding, D. K. Johnson, W. S. Adney, M. R. Nimlos, J. W. Brady and T. D. Foust, *Science*, 804 (2007).
2. R. J. Bothast and M. A. Schlicher, *Applied Microbiology and Biotechnology*, 19 (2005).

3. K. A. Gray, *International Sugar Journal*, 145 (2007).
4. J. M. Marchetti, V. U. Miguel and A. F. Errazu, *Renewable & Sustainable Energy Reviews*, 1300 (2007).
5. R. Perlack, L. L. Wright, A. F. Turhollow, R. L. Graham, B. J. Stokes and D. C. Erbach, (2005) Biomass as feedstock for bioenergy and bioproducts industry: The technical feasibility of a billion-ton annual supply. http://feedstockreview.ornl.gov/pdf/billion_ton_vision.pdf. Accessed 04 May 2012
6. N. Mosier, C. Wyman, B. Dale, R. Elander, Y. Y. Lee, M. Holtzapple and M. Ladisch, *Bioresource Technology*, 673 (2005).
7. S. Y. C. J, *Bioresource Technology*, 1 (2002).
8. J. Shi, Y. Pu, B. Yang, A. Ragauskas and C. E. Wyman, *Bioresource Technology*, 5952 (2011).
9. G. Ucar, *Wood Science and Technology*, 171 (1990).
10. Y. M. Zhu, Y. Y. Lee and R. T. Elander, *Applied Biochemistry and Biotechnology*, 1045 (2005).
11. Y. Y. Lee, Z. W. Wu and R. W. Torget, *Bioresource Technology*, 29 (2000).
12. M. J. Taherzadeh and K. Karimi, *Bioresources*, 707 (2007).
13. R. F. Chen, Z. W. Wu and Y. Y. Lee, *Applied Biochemistry and Biotechnology*, 37 (1998).
14. M. J. Taherzadeh and K. Karimi, *Bioresources*, 742 (2007).
15. P. Sannigrahi, D. H. Kim, S. Jung and A. Ragauskas, *Energy & Environmental Science*, 1306 (2011).
16. B. H. Um, M. N. Karim and L. L. Henk, *Applied Biochemistry and Biotechnology*, 115 (2003).
17. J. R. Jensen, J. E. Morinelly, K. R. Gossen, M. J. Brodeur-Campbell and D. R. Shonnard, *Bioresource Technology*, 2317 (2010).
18. S. Tian, W. Zhu, R. Gleisner, X. J. Pan and J. Y. Zhu, *Biotechnology Progress*, 419 (2011).
19. C. E. Wyman, V. Balan, B. E. Dale, R. T. Elander, M. Falls, B. Hames, M. T. Holtzapple, M. R. Ladisch, Y. Y. Lee, N. Mosier, V. R. Pallapolu, J. Shi, S. R. Thomas and R. E. Warner, *Bioresource Technology*, 11052 (2011).
20. J. Zhang, X. Ma, J. Yu, X. Zhang and T. Tan, *Bioresource Technology*, 4585 (2011).
21. H. L. Chum, S. K. Black, D. K. Johnson, K. V. Sarkanen, D. Robert, *Clean Technol. Environ. Policy*, 187 (1999).
22. R. A. Silverstein, Y. Chen, R. R. Sharma-Shivappa, M. D. Boyette and J. Osborne, *Bioresource Technology*, 3000 (2007).
23. C. G. Liu and C. E. Wyman, *Industrial & Engineering Chemistry Research*, 2781 (2004).
24. G. P. VanWalsum, S. G. Allen, M. J. Spencer, M. S. Laser, M. J. Antal and L. R. Lynd, *Applied Biochemistry and Biotechnology*, 157 (1996).
25. P. Alvira, E. Tomas-Pejo, M. Ballesteros and M. J. Negro, *Bioresource Technology*, 4851 (2010).

26. R. Sierra, A. Smith, C. Granda, M. T. Holtzapple, *Chemical Engineering Progress*, S10 (2008).
27. T. A. Hsu, in Wyman CE (Ed.) *Handbook on Bioethanol Production and Utilization*, Ed. C. E. Wyman (Taylor & Francis, Washington DC, 1996), p. 179.
28. C. E. Wyman, B. E. Dale, R. T. Elander, M. Holtzapple, M. R. Ladisch and Y. Y. Lee, *Bioresource Technology*, 1959 (2005).
29. R. P. Overend and E. Chornet, *Philosophical Transactions of the Royal Society of London Series a-Mathematical Physical and Engineering Sciences*, 523 (1987).
30. F. Alfani, A. Gallifuoco, A. Saporosi, A. Spera and M. Cantarella, *Journal of Industrial Microbiology & Biotechnology*, 184 (2000).
31. J. Bouchard, T. S. Nguyen, E. Chornet and R. P. Overend, *Bioresource Technology*, 121 (1991).
32. M. Laser, D. Schulman, S. G. Allen, J. Lichwa, M. J. Antal and L. R. Lynd, *Bioresource Technology*, 33 (2002).
33. N. Mosier, R. Hendrickson, N. Ho, M. Sedlak and M. R. Ladisch, *Bioresource Technology*, 1986 (2005).
34. C. Liu and C. E. Wyman, *Bioresource Technology*, 1978 (2005).
35. Y. Kim, R. Hendrickson, N. Mosier and M. R. Ladisch, *Energy & Fuels*, 2189 (2005).
36. O. Bobleter, *Progress in Polymer Science*, 797 (1994).
37. J. A. Pérez, A. González, J. M. Oliva, I. Ballesteros, P. Manzanares, *Journal of Chemical Technology & Biotechnology*, 929 (2007).
38. J. Perez *et al.*, *Fuel*, 3640 (2008).
39. M. Zeng, N. S. Mosier, C. P. Huang, D. M. Sherman, M. R. Ladisch, *Biotechnol. Bioeng.*, 265 (2007).
40. Z. Hu, A. J. Ragauskas, *Industrial & Engineering Chemistry Research*, 4225 (2011).
41. S. P. S. Chundawat, G. T. Beckham, M. E. Himmel, B. E. Dale, *Annual Review of Chemical and Biomolecular Engineering*, 121 (2011).
42. R. D. Hatfield, J. Ralph, J. H. Grabber, *Journal of the Science of Food and Agriculture*, 403 (1999).
43. A. T. Hendriks, G. Zeeman, *Bioresource Technology*, 10 (2009).
44. B. Yang, C. E. Wyman, *Biotechnology and Bioengineering*, 611 (2006).
45. S. G. Allen *et al.*, *Industrial & Engineering Chemistry Research*, 2934 (2001).
46. G. Garrote *et al.*, *Journal of Agricultural and Food Chemistry*, 9006 (2007).
47. R. Vegas, M. Kabel, H. A. Schols, J. L. Alonso, J. C. Parajó, *Journal of Chemical Technology & Biotechnology*, 965 (2008).
48. T. Lloyd, C. E. Wyman, *Applied Biochemistry and Biotechnology*, 53 (2003).
49. M. A. Kabel, G. Bos, J. Zeevalking, A. G. Voragen, H. A. Schols, *Bioresource Technology*, 2034 (2007).
50. Y. Sun, J. J. Cheng, *Bioresource Technology*, 1599 (2005).
51. A. M. Adel, Z. H. Abd El-Wahab, A. A. Ibrahim, M. T. Al-Shemy, *Bioresource Technology*, 4446 (2010).

52. X. W. Chen, M. Lawoko, A. van Heiningen, *Bioresource Technology*, 7812 (2010).
53. G. Garrote *et al.*, *Journal of Agricultural and Food Chemistry*, 9006 (2007).
54. R. Vegas, M. Kabel, H. A. Schols, J. L. Alonso, J. C. Parajó, *Journal of Chemical Technology & Biotechnology*, 965 (2008).
55. C. G. Liu, C. E. Wyman, *Industrial & Engineering Chemistry Research*, 5409 (2003).
56. C. G. Liu, C. E. Wyman, *Industrial & Engineering Chemistry Research*, 2781 (2004).
57. C. Liu, C. E. Wyman, *Bioresour Technol.*, 1978 (2005).
58. Y. Kim, R. Hendrickson, N. Mosier, M. R. Ladisch, *Energy & Fuels*, 2189 (2005).
59. M. C. Gray, A. O. Converse, C. E. *Industrial & Engineering Chemistry Research*, 2383 (2007).
60. A. H. Conner, *Wood and Fiber Science*, 268 (1984).
61. A. H. Conner, L. F. Lorenz, *Wood and Fiber Science*, 248 (1986).
62. F. Carrasco, C. Roy, *Wood Science and Technology*, 189 (1992).
63. K. Belkacemi, N. Abatzoglou, R. P. Overend, E. Chornet, *Industrial & Engineering Chemistry Research*, 2416 (1991).
64. D. J. Schell, J. Farmer, M. Newman, J. D. McMillan, *Applied Biochemistry and Biotechnology*, 69 (2003).
65. S. E. Jacobsen, C. E. Wyman, *Industrial & Engineering Chemistry Research*, 1454 (2002).
66. Y. Lu, N. S. Mosier, *Biotechnology and Bioengineering*, 1170 (2008).
67. J. Shen, C. E. Wyman, *Bioresource Technology*, 9111 (2011).
68. G. Garrote, H. Dominguez, J. C. Parajo, *Process Biochemistry*, 1067 (2002).
69. R. Brunecky *et al.*, *Biotechnol. Bioeng.*, 1537 (2009).
70. S. Jung, M. Foston, M. C. Sullards, A. J. Ragauskas, *Energy & Fuels*, 1347 (2010).
71. B. S. Donohoe, S. R. Decker, M. P. Tucker, M. E. Himmel, T. B. Vinzant, *Biotechnol. Bioeng.*, 913 (2008).
72. M. J. Selig *et al.*, *Biotechnology Progress*, 1333 (2007).
73. F. R. Kong, C. R. Engler, E. J. Soltes, *Applied Biochemistry Biotechnology*, 23 (1992)
74. M. Zeng *et al.*, *Biotechnol. Bioeng.*, 398 (2012).
75. T. C. Hsu, G. L. Guo, W. H. Chen, W. S. Hwang, *Bioresource Technology*, 4907 (2010).
76. M. Foston, A. J. Ragauskas, *Energy & Fuels*, 5677 (2010).
77. W. H. Chen, Y. J. Tu and H. K. Sheen, *Applied Energy*, 2726 (2011).
78. C. I. Ishizawa, M. F. Davis, D. F. Schell and D. K. Johnson, *Journal of Agricultural and Food Chemistry*, 2575 (2007).
79. C. T. Yu, W. H. Chen, L. C. Men, W. S. *Industrial Crops and Products*, 308 (2009).
80. J. Zhang, X. Ma, J. Yu, X. Zhang, T. Tan, *Bioresource Technology*, 4585 (2011).
81. W.-H. Chen, Y.-J. Tu, H.-K. Sheen, *International Journal of Energy Research*, 265 (2010).
82. L. Segal, J. J. Creely, A. E. Martin Jr. and C. M. Conrad, *Textile Research Journal*, 786 (1959).

83. P. T. Larsson, U. Westermark, T. Iversen, *Carbohydrate Research*, 339 (1995).
84. P. J. Weimer, A. D. French, T. A. Calamari, *Applied and Environmental Microbiology*, 3101 (1991).
85. Y. Nishiyama, P. Langan, H. Chanzy, *Journal of the American Chemical Society*, 9074 (2002).
86. S. Y. Ding, M. E. Himmel, *Journal of Agricultural and Food Chemistry*, 597 (2006).
87. S. P. Chundawat *et al.*, *J. Am. Chem. Soc.*, 11163 (2011).
88. Y. H. P. Zhang, L. R. Lynd, *Biotechnology and Bioengineering*, 797 (2004).
89. P. Sannigrahi, S. J. Miller, A. J. Ragauskas, *Carbohydr. Res.*, 965 (2010).
90. L. Zhu, J. P. O'Dwyer, V. S. Chang, C. B. Granda, M. T. Holtzapple, *Bioresource Technology*, 3817 (2008).
91. M. Yoshida *et al.*, *Bioscience, Biotechnology, and Biochemistry*, 805 (2008).
92. A. Mittal, R. Katahira, M. E. Himmel, D. K. Johnson, *Biotechnol Biofuels*, 41 (2011).
93. M. Ioelovich, E. Morag, *Bioresources*, 2818 (2011).
94. M. Hall, P. Bansal, J. H. Lee, M. J. Realff, A. S. Bommarius, *FEBS J*, 1571 (2010).
95. V. P. Puri, *Biotechnology and Bioengineering*, 1219 (1984).
96. H. E. Grethlein, *Bio-Technology*, 155 (1985).
97. D. N. Thompson, H. C. Chen, H. E. Grethlein, *Bioresource Technology*, 155 (1992).
98. A. A. Klyosov, O. V. Mitkevich, A. P. Sinitsyn, *Biochemistry*, 540 (1986).
99. J. Hong, X. H. Ye, Y. H. P. Zhang, *Langmuir*, 12535 (2007).
100. P. Va ljamae, V. Sild, G. Pettersson, G. Johansson, *Eur. J. Biochem.*, 469 (1998).
101. K. M. Kleman-Leyer, N. R. Gilkes, R. C. Miller Jr., T. K. Kirk, *Biochem. J.*, 463 (1994).
102. M. Srisodsuk, K. Kleman-Leyer, S. Keranen, T. K. Kirk, T. T. Teeri, *European Journal of Biochemistry*, 885 (1998).
103. T. M. Wood, *Biotechnology and Bioengineering*, 111 (1975).
104. X. Pan, D. Xie, K. Y. Kang, S. L. Yoon, J. N. Saddler, *Applied Biochemistry and Biotechnology*, 367 (2007).
105. X. J. Pan, D. Xie, R. W. Yu, J. N. Saddler, *Biotechnology and Bioengineering*, 39 (2008).
106. B. B. Hallac *et al.*, *Industrial & Engineering Chemistry Research*, 1467 (2010).
107. A. P. Sinitsyn, A. V. Gusakov, E. Y. Vlasenko, *Applied Biochemistry and Biotechnology*, 43 (1991).
108. Y. H. P. Zhang, L. R. Lynd, *Biotechnology and Bioengineering*, 888 (2006).
109. M. Foston, A. J. Ragauskas, *Biomass and Bioenergy*, 1885 (2010).
110. C. Stephens, P. Whitmore, H. Morris, M. Bier, *Biomacromolecules*, 1093 (2008).
111. A. Emelsy, R. Heywood, *Cellulose*, 1 (1997).
112. Y. Yu, H. W. Wu, *Industrial & Engineering Chemistry Research*, 3902 (2010).
113. M. Rinaudo, J. P. Merle, *Euro. Polym.*, 41 (1970).
114. A. O'Sullivan, *Cellulose*, 173 (1997).
115. K. Wang, J. X. Jiang, F. Xu, R. C. Sun, *Bioresource Technology*, 5288 (2009).

116. D. Y. Corredor *et al.*, *Applied Biochemistry and Biotechnology*, 164 (2009).
117. T. Yamashiki, T. Matsui, M. Saitoh, K. Okajima, K. Kamide, *British Polymer Journal*, 121 (1990).
118. J. Fernandez-Bolanos, B. Felizon, A. Heredia, R. Guillen, A. Jimenez, *Bioresource Technology*, 121 (1999).
119. S. Deguchi, K. Tsujii, K. Horikoshi, *Green Chemistry*, 191 (2008).
120. R. Kumar, G. Mago, V. Balan, C. E. Wyman, *Bioresource Technology*, 3948 (2009).
121. P. Sannigrahi, A. J. Ragauskas, S. J. Miller, *Bio Energy Research*, 205 (2008).
122. L. P. Xiao, Z. J. Sun, Z. J. Shi, F. Xu, R. C. Sun, *Bioresources*, 1576 (2011).
123. J. M. Lee, H. Jameel, R. A. Venditti, *Bioresource Technology*, 5449 (2010).
124. J. M. Martinez *et al.*, *Industrial & Engineering Chemistry Research*, 688 (1997).
125. H. Håkansson, P. Ahlgren and U. Germgård. The Degree of Disorder in Hardwood Kraft Pulps Studied by Means of LODP. *Cellulose*, 12:327–335 (2005).
126. Y. Pu, D. Zhang, P. M. Singh, A. J. Ragauskas, *Biofuels, Bioproducts and Biorefining*, 58 (2008).
127. A. Buranov, G. Mazza, *Industrial Crops and Products*, 237 (2008).
128. L. Laureano-Perez, F. Teymouri, H. Alizadeh, B. E. Dale, *Applied Biochemistry and Biotechnology*, 1081 (2005).
129. V. Balan *et al.*, *Biotechnology Progress*, 365 (2009).
130. L. R. Lynd, *Annual Review of Energy and the Environment*, 403 (1996).
131. L. R. Lynd, R. T. Elander, C. E. Wyman, *Applied Biochemistry and Biotechnology*, 741 (1996).
132. C. A. Mooney, S. D. Mansfield, M. G. Touhy, J. N. Saddler. *Bioresource Technology*, 113 (1998).
133. J. B. Kristensen, J. Borjesson, M. H. Bruun, F. Tjerneld, H. Jorgensen, *Enzyme and Microbial Technology*, 888 (2007).
134. B. Yang, C. E. Wyman, *Biotechnology and Bioengineering*, 611 (2006).
135. S. V. Pingali *et al.*, *Biomacromolecules*, 2329 (2010).
136. C. Sievers *et al.*, *Bioresource Technology*, 4758 (2009).
137. J. D. Mao, K. M. Holtman, D. Franqui-Villanueva, *J. Agric. Food Chem.*, 11680 (2010).
138. R. Samuel, M. Foston, N. Jiang, L. Allison, A. J. Ragauskas, *Polymer Degradation and Stability*, 2002 (2011).
139. J. Li, G. Gellerstedt, K. Toven, *Bioresource Technology*, 2556 (2009).
140. R. Samuel, Y. Pu, B. Raman, A. J. Ragauskas, *Applied Biochemistry and Biotechnology*, 62 (2010).
141. S. Nakagame, R. P. Chandra, J. F. Kadla, J. N. Saddler, *Bioresource Technology*, 4507 (2011).
142. M. Leschinsky, G. Zuckerstätter, H. K. Weber, R. Patt, H. Sixta, *Holzforschung*, 645 (2008).
143. M. Borrega, K. Nieminen, H. Sixta, *Bioresource*, 1890 (2011).
144. J. B. Li, G. Henriksson, G. Gellerstedt, *Bioresource Technology*, 3061 (2007).

145. J. B. Kristensen, L. G. Thygesen, C. Felby, H. Jorgensen, T. Elder, *Biotechnol. Biofuels*, 5 (2008).
146. J. B. Li, G. Henriksson, G. Gellerstedt, *Applied Biochemistry and Biotechnology*, 175 (2005).
147. P. Sannigrahi, D. H. Kim, S. Jung, A. Ragauskas, *Energy & Environmental Science*, 1306 (2011).
148. F. Hu, S. Jung, A. Ragauskas, *Bioresource Technology*, 7 (2012).
149. T. Vu, A. Chaffee, I. Yarovsky, *Molecular Simulation*, 981 (2002).
150. F. Hu, S. Jung, A. Ragauskas, *ACS Sustainable Chemistry & Engineering*, submitted.
151. D. J. Cosgrove, *Nature Reviews Molecular Cell Biology*, 850 (2005).

CHAPTER 4

REDUCTION OF BIOMASS RECALCITRANCE VIA ORGANOSOLV PRETREATMENTS

XIANZHI MENG

Department of Chemistry and Biochemistry, Institute of Paper Science and Technology, Georgia Institute of Technology
500 10th Street NW, Atlanta, GA 30332, USA
xmeng30@gatech.edu

Bioconversion of lignocellulosic biomass to ethanol is significantly hindered by the structural complexity of biomass, therefore overcome the resistance of plant cell walls to deconstruction of lignocellulosic biomass is the key step in the production of fuels. To overcome this native biomass recalcitrance, pretreatment is always necessary to separate the cellulose from the matrix polymers therefore make it more accessible for enzymatic hydrolysis. Various promising pretreatment methods have been developed during the past few decades. Besides dilute acid and hydrothermal pretreatments, organosolv pretreatment is another most commonly used technique. This chapter reviews this pretreatment technology along with the changes in lignocellulosic structure during the pretreatment.

1. Introduction

Nowadays, shortages of fossil energy, economic and environmental concerns, together with increasing global energy demand have reinforced the need of using renewable energy resources to replace the petroleum based materials.[1] Lignocellulosic biomass, the most abundant organic materials in nature, have been considered as one of the most important energy feedstock with great potential to be used in the production of bioenergy, because of its low cost and large scale availability.[2–5] In particular, the combustion of lignocellulosic ethanol produces almost no net carbon dioxide into the atmosphere because it comes from plants; therefore lignocellulosic biomass is also a carbon-neutral source of energy.[6–8]

In terms of conversion of lignocellulosic biomass to ethanol, one of the most important steps is to depolymerize the cellulose to glucose by using appropriate enzyme. Unfortunately, this process is significantly hindered by the natural resistance of plant cell walls to enzymatic deconstruction, known as "biomass recalcitrance".[9–12] To overcome this recalcitrance, pretreatment is always necessary before the enzymatic hydrolysis. The goal of the pretreatment is to try to disrupt the crystalline structure and reduce the degree of polymerization of cellulose, and increase cellulose accessibility by removing/redistributing the outside shield formed by lignin and hemicellulose. The pretreatment step has been considered the most expensive processing step in the conversion of lignocellulosic biomass to ethanol, representing about 20% of the total cost. In addition, pretreatment also has strong impacts on other operations in the overall conversion scheme, including feedstock choice, size reduction prior to pretreatment, enzymatic hydrolysis and fermentation.[13,14] Over the past few decades, many technologies have been developed, which allow this bioconversion process to occur, but only few of them seem to be promising.[15] Obviously, the problem is how to make the technologies cost-competitive. These commonly used pretreatments include dilute acid pretreatment, steam explosion, ammonia fiber expansion, organosolv pretreatment and ammonia recycle percolation.[13,16–18] Each pretreatment has its own advantages and disadvantages as well as own mechanisms. For example, organosolv pretreatment is more expensive than other leading pretreatment techniques, but it can provide some valuable by-products.[19] The objective of this chapter is to give an overview of organosolv pretreatment, including the technical process, mechanisms, lignocellulosic structural changes associated with reduction of biomass recalcitrance, and some perspectives for further development. Understanding the mechanisms of organosolv pretreatment, especially how it alters chemical composition and physically change cell wall structure, will aid in better understanding of biomass recalcitrance and therefore help in developing novel pretreatment technologies.

2. Overview of Organosolv Pretreatment

The organosolv pretreatment representing an alternative to traditional pulping method, is originated from the organosolv pulping process that extracts lignin from lignocellulosic feedstocks with organic solvents, but does not require the equivalent degree of delignification. It removes extensive lignin and hemicelluloses and increases accessible surface area and pore volume. In the early 1990s, the organosolv process has been already applied specifically as a biomass pretreatment method in ethanol production.[20] A flowchart summarizing ethanol organosolv pretreatment is shown in Fig. 1.

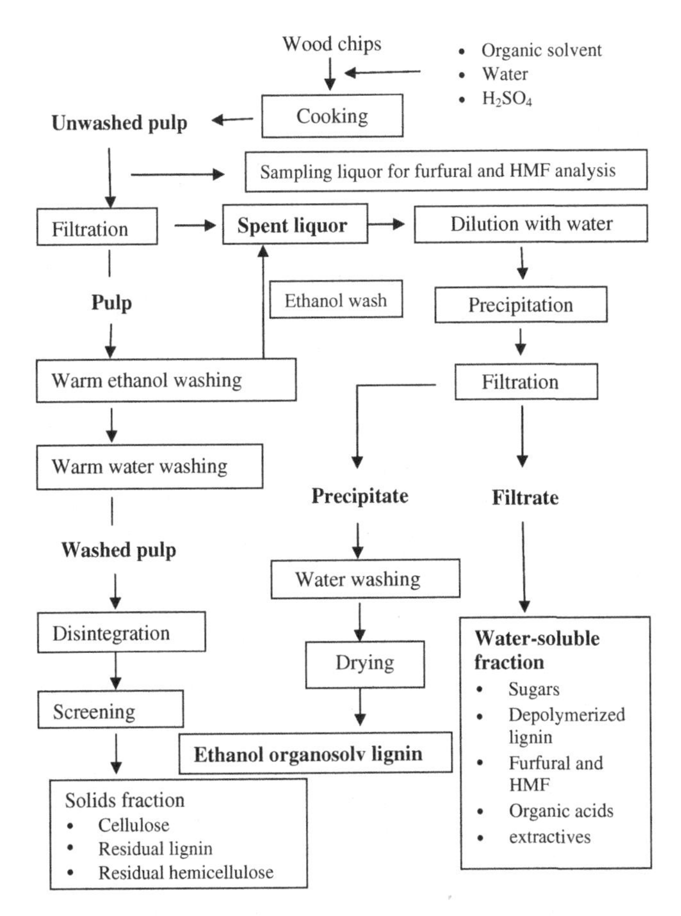

Fig. 1. Flowchart of the laboratory-scale organosolv pretreatment (Reproduced from Ref. 24 with permission).

In organosolv pretreatment, the lignocellulosic biomass is treated in an organic or aqueous organic solvent mixture with or without an acid catalyst, such as H_2SO_4. The acid catalyst is used to increase the solubilization of hemicellulose and enhance the digestibility of substrate.[15] Both organic acids (i.e., formic acid, acetylsalicylic and salicylic acid) and mineral acid (i.e., hydrochloric acid, sulfuric acid and phosphoric acid) are considered as good acid catalysts to accelerate delignification and xylan degradation. The operation temperature is usually in the range of 100–250°C, it is reported that the addition of catalyst was unnecessary for satisfactory delignification if the temperature is higher than 185°C, because organic acids released from the biomass can act as the catalyst for the rupture of the lignin-carbohydrate complex.[21,22] However, the presence of acid catalyst can increase the rate of delignification and generate higher yields of xylose. Several organic solvents can be used in the organosolv pretreatment, including solvents with low boiling points such as methanol and ethanol as well as a variety of alcohols such as ethylene glycol, triethylene glycol, tetrahydrofurfuryl alcohol, glycerol, phenol, and aqueous *n*-butanol which have much higher boiling point. Among the solvents listed above, lower molecular weight and boiling point alcohol such as methanol and ethanol are widely used for the economic purpose.[23] Since most organic solvents are usually expensive, it should be recovered and recycled at the end to reduce the operation costs. In addition, if the solvent is not removed from the system, it may become inhibitors to the growth of organisms, enzymatic hydrolysis and fermentation. Table 1 summarizes recent organosolv pretreatments for different substrates, including organic solvents used, pretreatment conditions, enzyme loadings and the enzymatic hydrolysis results.

3. Mechanism of Organosolv Pretreatment for Reduction of Recalcitrance

It has been reported that the biomass recalcitrance is attributed to several chemical and physical factors, including cellulose crystallinity, degree of polymerization, lignin and hemicellulose structure/content, accessible surface area and pore volume.[9,11,12,32,33] The presence of hemicellulose and lignin in biomass physically block the access of enzyme to the cellulose

Table 1. Recent organosolv pretreatment results for different substrate.

Substrate	Pretreatment conditions	Enzymes loadings	Cellulose conversion (%)	Ref.
Poplar	180°C, 1.25% H_2SO_4, 50% ethanol, 60 min	20 FPU cellulase and 40 IU β-glucosidase (Novozym 188)/g cellulose	93 in 24 h 97 in 48 h	24
Switchgrass	180°C, 3:1 (v/v) ethanol/water ratio, 0.99% (w/w) H_2SO_4, 60 min	20 FPU cellulase and 40 IU β-glucosidase (Novozym 188)/g cellulose	~70 in 8 h ~98 in 72 h	25
Miscanthus	170°C, 0.8 ethanol/water ratio, 0.5% (w/w) H_2SO_4, 60 min	20 FPU cellulase and 40 IU β-glucosidase (Novozym 188)/g cellulose	98 in 48 h	26
Loblolly pine	170°C, 65% ethanol, 1.1% H_2SO_4, 60 min	8 FPU cellulase and 16 IU β-glucosidase (Novozym 188)/g cellulose	70 in 80 h	27
Eucalyptus globulus tension wood	60% ethanol, 6:1 liquid/wood ratio, with H factor 3900	20 FPU cellulase 1.5 L and 20 CBU β-glucosidase (Novozym 188)/g pulp	45 ± 2 in 24 h 69 ± 1 in 72 h	28
	60% ethanol, 6:1 liquid/wood ratio, with H factor 12500	20 FPU cellulase 1.5 L and 20 CBU β-glucosidase (Novozym 188)/g pulp	55 ± 2 in 24 h 73 ± 2 in 72 h	28

(Continued)

Table 1. *(Continued)*

Substrate	Pretreatment conditions	Enzymes loadings	Cellulose conversion (%)	Ref.
Eucalyptus globulus oppsite wood	60% ethanol, 6:1 liquid/ wood ratio, with H factor 3900	20 FPU cellulase 1.5 L and 20 CBU β-glucosidase (Novozym 188)/g pulp	41 ± 2 in 24 h 65 ± 1 in 72 h	28
	60% ethanol, 6:1 liquid / wood ratio, with H factor 12500	20 FPU cellulase 1.5 L and 20 CBU β-glucosidase (Novozym 188)/g pulp	59 ± 1 in 24 h 77 ± 2 in 72 h	28
Olive tree	180°C, 43% ethanol, 15 min	15 FPU cellulase and 15 IU β-glucosidase (Novozym 188)/g substrate	89.6 in 72 h	29
Sugarcane bagasse	195°C, 30% (v/v) ethanol, 60 min	15 FPU cellulase and 15 IU β-glucosidase (Novozym 188)/g substrate	29.1 ± 0.4 g glucose/ 100 g initial raw material	30
Rye straw	440 K, 3.5 MPa pressure, 50% (w/w) ethanol, 0.5 N H_2SO_4. 35 min	5:1 volumetric ratio of celluclast 1.5 L to Novozyme 188	76 ± 2 in 48 h	31

surface, therefore an efficient pretreatment should be able to remove nearly complete lignin and hemicellulose so that the accessible surface area can be increased. In this respect, the organosolv pretreatment is a promising pretreatment strategy, because it can break the internal lignin and hemicellulose bond, remove most of the lignin and hemicelluloses from the solid substrate therefore yielding "purified" cellulose for the enzymatic deconstruction.[34,35] Basically, this process separates the lignocellulosic biomass into three major fractions: solid cellulose-rich pulp, liquid fraction that is rich in hemicellulose sugars and organosolv lignin precipitated from liquid fraction. Substrates pretreated by organosolv have been reported to have superior enzymatic digestibility over those pretreated by the other pretreatments.[24,25,29] It is well known that the reduction of cellulose crystallinity can increase the initial hydrolysis rate because the hydrogen bonds in crystalline region make the cellulose more recalcitrant to enzyme.[36,37] The change of cellulose crystallinity during organosolv pretreatment is still not clear yet and need further investigation. In addition, the cellulose degree of polymerization has been found decreased during the organosolv pretreatment.[25,38] All the details about lignin degradation/dissolution, cellulose crystallinity and DP, as well as chemistry reactions associated with organosolv pretreatment will be discussed in the next few sections.

4. Cellulose Behavior during Organosolv Pretreatment

Cellulose crystallinity and degree of polymerization are two important cellulose structure-relevant factors that may affect the accessibility of biomass cellulose. Therefore, analyzing the structure changes of cellulose before and after pretreatment is essential for overcoming the biomass recalcitrance.

Cellulose in biomass exists in both amorphous and crystalline forms, and the crystalline cellulose comprises the major proportion of cellulose. Native cellulose contains three types of crystalline allomorphs, i.e., cellulose I_α, cellulose I_β, and *para*-crystalline cellulose.[39] Cellulose I_α is a one-chain triclinic unit cell and I_β is a monoclinic two-chain unit cell, which is dominant in wood. In these crystalline regions, cellulose exists in the form of microfibrils that are paracrystalline assemblies of 1,4 β-D-glucan chains hydrogen bonded to one another along their length. These hydrogen bonds make the crystalline region of cellulose more recalcitrant to the enzyme attachment than amorphous region.[12,40–42] One of the most controversial theories is the influence of cellulose crystallinity on the enzymatic hydrolysis. It was found that the initial degree of crystallinity of cellulose played an important role as a rate determinant in the enzymatic hydrolysis reaction.[37] It is generally believed that cellulose crystallinity is increased during the pretreatment primarily due to localized hydrolyzation and removal of cellulose in the amorphous regions.[43] However, the effect of organosolv pretreatment on cellulose crystallinity is still not fully explored. Table 2 summarized the cellulose crystallinity index (CrI) before and after organosolv pretreatment for different lignocellulosic substrates. Some studies show that the cellulose crystallinity increases with increasing severity of the organosolv pretreatment conditions, which is attributed to the preferential degradation of the less structured forms of cellulose under harsher pretreatment conditions, while other studies observed that cellulose crystallinity remained approximately constant after organosolv pretreatment.[25,44] In addition, some other researchers also found that organosolv pretreatment was capable of disrupting the crystalline regions of cellulose, with resulting decrease of cellulose crystallinity after pretreatment.[37,38,45] Therefore, how organosolv pretreatment affects the cellulose crystallinity still needs further investigation. But it has been clear that pretreatment variables such as temperature, ratio of wood to liquor, ethanol concentration and holding time on the process can all significantly affect the cellulose crystallinity index.[46] Hallac *et al.*[38] found after organosolv pretreatment for Buddleja *davidii*, the amount of cellulose I_α and I_β decreased, whereas the relative intensity of para-crystalline and amorphous cellulose increased. This result indicates there was decrystallization of cellulose during organosolv pretreatment, and cellulose I_α and I_β allomorphs are converted to *para*-crystalline cellulose because of the depolymerization of the glucan chains.

The number of glucose units that make up one polymer molecule is referred to its degree of polymerization. During enzymatic hydrolysis, the endocellulase opens up the linear cellulose molecules, producing reducing ends that can be attacked by exocellulases. So it is generally believed that the reduction in cellulose degree of polymerization increases the

Table 2. Cellulose crystallinity index (CrI) before and after organosolv pretreatment.

Substrate	Pretreatment conditions	CrI (%) before pretreatment	CrI (%) after pretreatment	Ref.
Hybrid Poplar	165°C, 1.0 (w/w) H_2SO_4 dosage, 65% (v/v) ethanol, 40 min	N/A	59	47
	195°C, 1.0 (w/w) H_2SO_4 dosage, 65% (v/v) ethanol, 40 min	N/A	78	47
	165°C, 1.5 (w/w) H_2SO_4 dosage, 65% (v/v) ethanol, 80 min	N/A	80	47
Loblolly Pine	170°C, 65% ethanol, 1.1% H_2SO_4, 60 min	63	53	27
Lodgepole Pine	170°C, 0.76 (w/w) H_2SO_4 dosage, 65% (v/v) ethanol, 60 min	N/A	73	44
	170°C, 1.10 (w/w) H_2SO_4 dosage, 65% (v/v) ethanol, 60 min	N/A	75	44
	187°C, 1.10 (w/w) H_2SO_4 dosage, 65% (v/v) ethanol, 60 min	N/A	83	44
Switchgrass	180°C, 3:1 (v/v) ethanol/ water ratio, 0.99% (w/w) H_2SO_4, 60 min	50	52	25
Buddleja davidii	180°C, 1.25 (w/w) H2SO4, 50% (v/v) ethanol, 60 min	55	53	38
	180°C, 1.25 (w/w) H_2SO_4, 50% (v/v) ethanol, 60 min	55	53	38
	180°C, 1.25 (w/w) H_2SO_4, 50% (v/v) ethanol, 60 min	55	49	38

enzymatic digestibility of cellulose because it can provide more reducing ends for the cellulase attacking.[41,47,48] In addition, researchers also found that the shorter the cellulose chain is, the faster the cellulose is hydrolyzed, which is attributed to the absence of strong hydrogen bonds.[49] However, biomass recalcitrance does not come from a signal structural factor, and unfortunately it is impossible to remove one of the compositions without changing the others. These interactions make the mechanism for the effects of the reduction of biomass recalcitrance via pretreatment become very complicated. Similar to cellulose crystallinity, the degree of polymerization is also not an independent factor because its alteration is usually accompanied by change of crystallinity.[50] The two most commonly used techniques to measure the DP of cellulose are the gel-permeation chromatography (GPC) and viscometry methods.[51–55] In order to analyze the DP, cellulose must be first purified and isolated from its native source. Organosolv pretreatment is a good technique to get fully purified cellulose. However, a complete delignification process can result in excessive loss of polysaccharides, which may significantly reduce the chain length of cellulose.[54] Hubbell and Ragauskas found that lignin-free filter paper showed a DP reduction of nearly 35%, while a reduction of 17% was observed for the samples that only contained 1% lignin.[56] Therefore, small amount of lignin left can minimize the loss of cellulose DP during a delignification process. Although the effect of degree of polymerization on enzymatic hydrolysis is still undefined, it has been clear that cellulose undergoing depolymerization during organosolv pretreatment leads to the decrease of cellulose degree of polymerization. Table 3 summarizes the change in DP of cellulose after various organosolv pretreatment.

Apparently, for organosolv pretreatment, the cellulose DP varied across the different organic solvents used. Ethanol organosolv pretreatment caused more depolymerization of cellulose than methanol, formic acid and acetic acid. In addition, the increase amount of organic solvent preserved the reduction in DP of cellulose.

5. Lignin Behavior during Organosolv Pretreatment

Lignin is a complex, large molecular structure containing cross-linked polymers of phenolic monomers. Its structure, distribution and content are

Table 3. Cellulose DP after organosolv pretreatment for different lignocellulosic substrates.

Substrate	Solvents used	Solvents ratio (% v/v)	DP	Ref.
Switchgrass	Ethanol-water	75/25	2412	25
Buddleja davidii	Ethanol-water	50/50	970	38
	Ethanol-water	65/35	420	38
Wheatstraw	Ethanol-water	60/40	1356	57
	Methanol-water	60/40	1519	57
	Formic acid-acetic acid-water	30/60/10	2289	57
	Formic acid-acetic acid-water	20/60/20	2182	57
	Acetic acid-water	90/10	1952	57
	Acetic acid-water	80/20	1763	57
	Acetic acid-water	65/35	1594	57

considered very important factors responsible for the recalcitrance of lignocellulosic to enzymatic degradation.[58–60] The biosynthesis of lignin is generally considered to stem from the polymerization of three types of phenylpropane units as monolignols: conifer, sinapyl, and *p*-coumaryl alcohol, which give rise, respectively, to the so-called guaiacyl (G), syringyl (S), and *p*-hydroxyphenyl (H) lignin units.[61,62] These monolignols can undergo an *in situ* radical polarization, yielding some inter-unit linkages, including β-*O*-aryl ether, resinol, phenylcoumarane, biphenyl and 1,2-diaryl propane, which are all shown in Fig. 2.

Several studies have shown that lignin is an important factor limiting the extent and rate of enzymatic hydrolysis of biomass and the remove of lignin from biomass greatly enhances the cellulose digestibility. First of all, lignin acts as a physical barrier to restrict the access of enzyme to cellulose. Although how lignin protects the polysaccharides from enzymatic degradation is still not quite clear due to the lack of research, it has been clear that some factors such as degree and type of cross-linkage of lignin to polysaccharides, the distribution of phenolic polymers through the cell wall, and the diversity of structures found in the lignin composition are all playing important role in this protective effect.[63] Besides this direct physical effect, lignin has also been found to prevent the cellulase enzyme action on cellulose by adsorbing it irreversibly.[64–66] Tu and co-workers

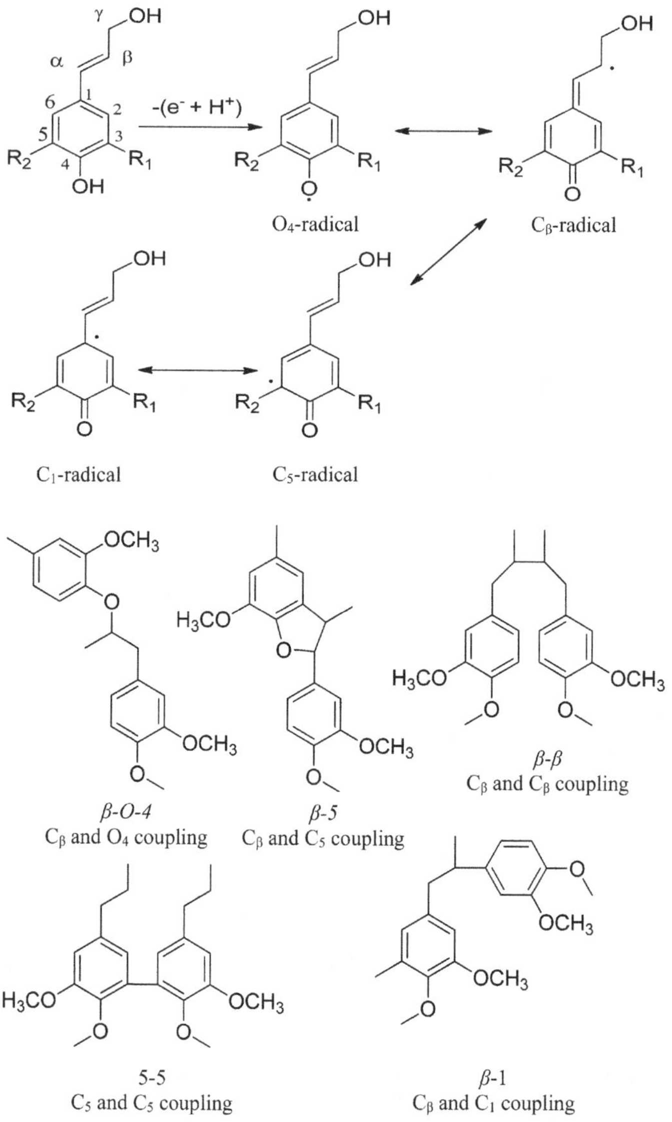

Fig. 2. Common linkages found in lignin formed via radical coupling.[72] Figure was reproduced from Ref. 72 with permission.

reported that cellulase enzymes were effectively recycled on ethanol-pretreated Lodgepole pine, but not on steam-exploded Lodgepole pine. The adsorption ability of lignin to cellulases depends on the structure of the lignin, which is moreover strongly dependent on the conditions and methods of the biomass pretreatment.[67] It was found that the organosolv lignins seemed to be more effective to absorb the cellulase than steam explosion lignin mainly because of the higher content of phenolic hydroxyl group in organosolv lignin.[68]

Delignification is a process that not only can destroy the physical barrier caused by lignin, but also can eliminate the adsorption of enzyme cellulases by lignin. As mentioned before, one of the great advantages of organosolv pretreatment is its ability to remove nearly complete lignin. After pretreatment, a large amount of lignin with high hydrophobicities, low glass transition temperatures, low polydispersity and high functionalization are generated. These lignins have great promising applications in the fields of biodegradable polymers and adhesives.[69] Gellerstedt *et al.*[70] reported the possibility of converting technical lignins to bio-oil by pyrolysis in the presence of formic acid and alcohol. In addition, a conversion process of ethanol organosolv lignin to a potential fuel precursor for green gasoline via pyrolysis and catalytic hydrogenolysis has also been demonstrated.[71]

In terms of structural characterization of lignin, it is well known that the nuclear magnetic resonance (NMR) spectroscopy is one of the most widely used techniques. The combination of quantitative ^{13}C and two-dimensional heteronuclear single quantum coherence (2D HSQC) NMR is frequently used to determine the amount of several lignin interlinkages, providing a comprehensive structural information on lignin before and after pretreatment.[73–76] In general, the organosolv pretreatment resulted in a decrease of the content of aliphatic hydroxyl groups and an increase of the content of phenolic and carboxylic hydroxyl groups. The decrease of the aliphatic hydroxyl groups is attributed to the loss of the γ-methylol group, and the increase of the phenolic OH groups indicates the occurrence of the acid-catalyzed cleavage of aryl ether bonds during the ethanol organosolv pretreatment.[38] Hu *et al.*[77] reported that the ethanol organosolv pretreatment resulted in an increase in all three types of lignin phenols (guaiacyl, syringyl, and *p*-hydroxyphenyl OH groups), and the content of carboxyl OH groups was found to increase after organosolv pretreatment from 0.07 mmol/g to 0.33 mmol/g for switchgrass ethanol organosolv pretreated lignin. Hallac and co-workers reported that the syringyl/guaiacyl ratio (s/g) remained relatively constant after ethanol organosolv pretreatment for *Buddleja davidii*, whereas Villaverde *et al.* found that the s/g of the lignin increased after the pretreatment for *Miscanthus x giganteus*.[72,78] However, it is worth mentioning that both s and g lignin relative content decreased after the pretreatment, which is in part due to the condensation reactions occurring on C_2, C_5, or C_6 of the aromatic ring of lignin.

Referring to the chemical processes, literature suggests the cleavage of ether linkages (α and β aryl ether) is the major mechanism of lignin degradation and dissolution during organosolv pretreatment.[79–84] Furthermore, α-aryl-ether bonds are broken more easily than β-aryl-ether bonds, especially when they occur in a lignin structural unit containing a free phenolic hydroxyl group in the para-position.[35] Figure 3 shows various solvolytic cleavage of α-aryl-ether linkage under acid condition. The likelihood of

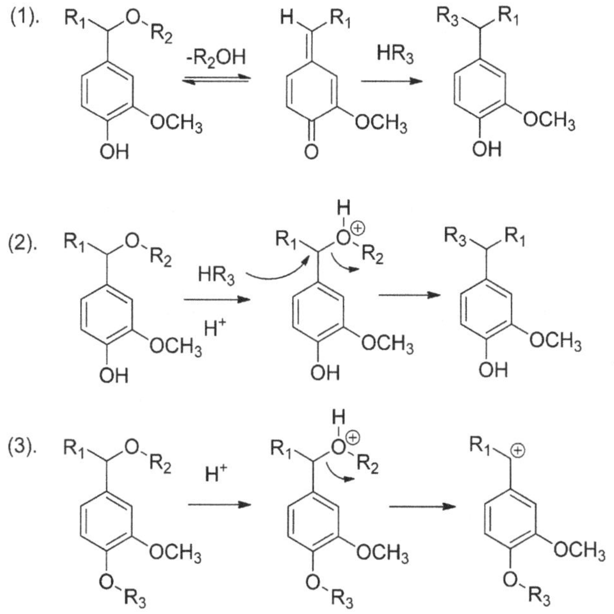

Fig. 3. Solvolytic cleavage of α-aryl-ether linkage in lignin under acidic: (1) Cleavage via a quinone methide intermediate. R_3 = OH, OCH_3 etc. (2) Cleavage via nucleophilic substitution. (3) Formation of a benzyl carbocation.

β-aryl-ether cleavage is greater in more strongly acidic systems, due to the fact that stronger acidic systems bear a greater resemblance to the classical acidolysis systems in which the breakage of *β*-ethers has been established.[85] In a word, the exact lignin breakdown process taking place during an organosolv pretreatment is very complex and can be determined by several factors, including physical and chemical properties of solvent as well as the solution pH. A proposed scheme for the depolymerization by fragmentation of *β*-O-4-linked structures and re-polymerization by acid-catalyzed condensation between the aromatic C_6 and a carbonium ion, which have been reported as one of the predominant reactions in organosolv pretreatment, is given in Fig. 4.[82] Hage *et al.*[86] reported that an increase in the severity of the organosolv pretreatment for *Miscanthus* can enhance the dehydration reactions on the side chain and the condensation of lignin, can decrease the concentration of phenol groups and the molecular mass of lignin fragments. In addition, at higher severity conditions, the FTIR analysis reveals the presence of a carbonyl peak at 1705 cm^{-1}, which

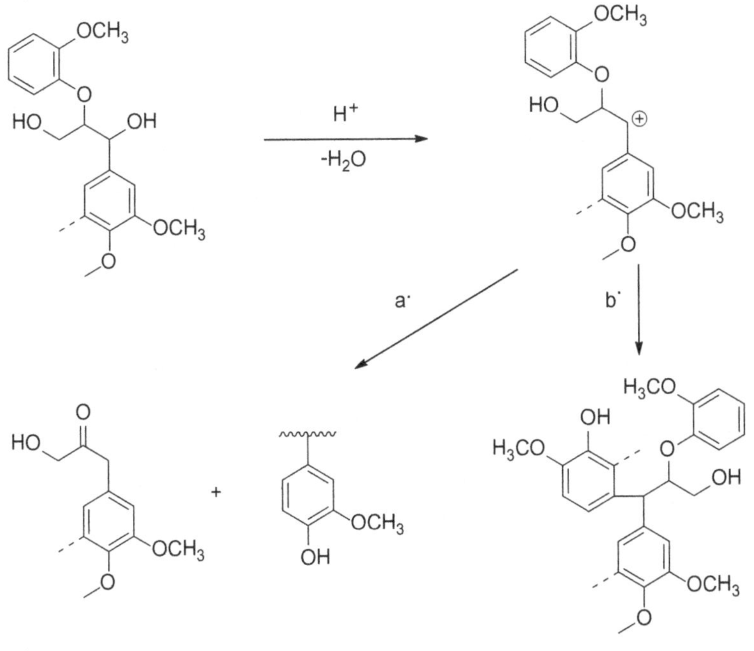

Fig. 4. Reaction scheme showing (a) Depoymerization (b) Repolymerization reactions in lignin under acid conditions (Reproduced from Ref. 82 with permission).

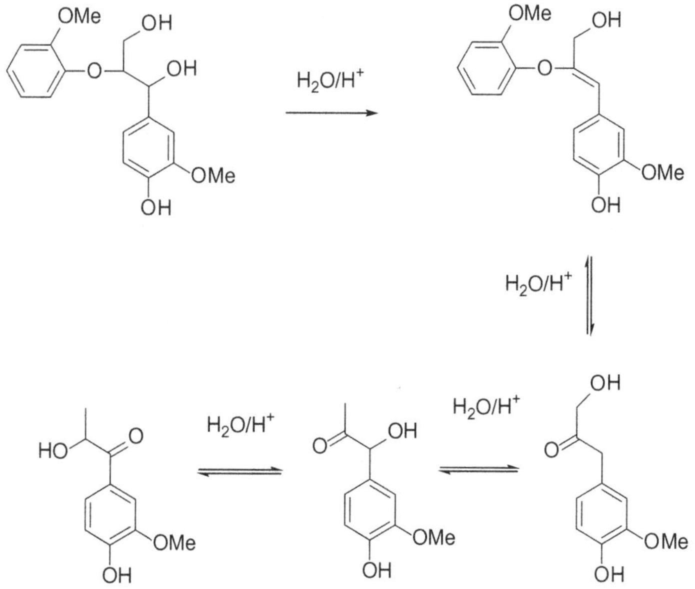

Fig. 5. Mechanism of the formation of Hibbert's ketones (Reproduced from Ref. 82 with permission).

could be assigned to ketone groups arising from dehydration reactions occurred in the lateral chain of lignin. A possible mechanism of the formation of Hibbert's ketones involving a β-O-4 cleavage is shown in Fig. 5.

6. Conclusions and Outlook

Lignocellulosic biomass has evolved complex structural and chemical mechanisms for resisting assault on its sugars, and factors such as lignin and hemicellulose content, cellulose crystallinity, degree of polymerization, and accessible surface area all contribute to the resistance of biomass to enzymatic hydrolysis. Various pretreatment technologies have been developed but only a few of them have been tested on a demonstration scale, and each of these pretreatments changes at least two of the above factors. This book chapter documented the important physical and chemical changes in cellulose and lignin structure caused by the organosolv pretreatment as well as its mechanism for the reduction of biomass recalcitrance. However, due to the complex chemical and biological properties of lignocellulosic biomass, the substrates features are still not quite controllable and adjustable during the pretreatment. Therefore, a better fundamental understanding of both the physical and chemical mechanisms that occur during organosolv pretreatment is required for the development of advanced effectiveness pretreatment technologies with minimum energy input and maximum fermentable sugar recovery.

Acknowledgments

The authors are grateful for the financial support from the School of Chemistry and Biochemistry at Georgia Institute of Technology. I would also like to thank those who provided relevant material that has been used in this article.

References

1. A. J. Ragauskas, C. K. Williams, B. H. Davison, G. Britovsek, J. Cairney, C. A. Eckert, W. J. Fredcrick, J. P. Hallett, D. J. Leak, C. L. Liotta, J. R. Mielenz, R. Murphy, R. Templer and T. Tschaplinski, *Science*, 484 (2006).
2. Y. Pu, D. Zhang, P. M. Singh and A. J. Ragauskas, *Biofuels Bioprod Biorefin.*, 58 (2008).
3. A. J. Ragauskas, M. Nagy, D. H. Kim, C. A. Eckert, J. P. Hallett and C. L. Liotta, *Ind. Biotechnol.*, 2 (2006).
4. R. C. Saxena, D. K. Adhikari and H. B. Goyal, *Renew. Sust Energ. Rev.*, 167 (2009).
5. M. Galbe and G. Zacchi, *Appl. Microbiol. Biotech.*, 59 (2002).
6. P. A. M. Claassen, J. B. van Lier, A. M. L. Contreras, E. W. J. van Niel, L. Sijtsma, A. J. M. Stams, S. S. de Vries and R. A. Weusthuis, *Appl Microbiol. Biotech.*, 52 (1999).
7. L. R. Lynd, J. H. Cushman, R. J. Nichols and C. E. Wyman, *Science*, 1318 (1991).
8. A. Demirbas, *Energy Convers. Manag.*, 2239 (2009).
9. M. E. Himmel, S. Y. Ding, D. K. Johnson, W. S. Adney, M. R. Nimlos, J. W. Brady and T. D. Foust, *Science*, 804 (2007).
10. B. Yang, Z. Dai, S. Ding and C. E. Wyamn, *Biofuels,* 421 (2011).
11. S. Mansfield, C. Mooney and J. Saddler, *Biotechnol. Prog.*, 804 (1999).
12. L. Fan, Y. Lee and D. Beardmore, *Biotechnol. Bioeng.*, 177 (1980).
13. B. Yang and C. E. Wyman, *Biofuels. Bioprod. Biorefin.*, 26 (2007).
14. C. Brodeur, E. Yau, K. Badal, J. Collier, K. B. Ramachandram and S. Ramakrishnan, *Enzyme Research*, 1 (2011).
15. Y. Sun and J. Cheng, *Bioresour Technol.*, 1 (2002).

16. N. Mosier, C. E. Wyman, B. Dale, R. Elander, Y. Y Lee, M. Holtzapple and M. Ladisch, *Bioresour. Technol.,* 673 (2005).
17. M. J. Taherzadeh and K. Karimi, *Int. J. Mol Sci.,* 1621 (2008).
18. J. Y. Zhu, X. Pan and R. S. Zalesny, *Appl Microbiol. Biotech.*, 847 (2010).
19. H. J. Chum, D. K. Johnson and S. K. Black, *Ind Eng Chem Res.,* 156 (1990).
20. E. K. Pye and J. H. Lora, *TAPPI J.* 113 (1991).
21. S. Aziz and K. Sarkanen, *TAPPI J.* 169 (1989).
22. S. J. B. Duff and W. D. Murray, *Bioresource Technol.,* 1 (1996).
23. D. Sidiras and E. Koukios, *Bioresource Technol.,* 91 (2004).
24. X. Pan, N. Gikes, J. Kadla, K. Pye, S. Saka, D. Gregg, K. Ehara, D. Xie, D. Lam and J. Saddler, *Biotechnol. Bioeng.,* 851 (2006).
25. C. Cateto, G. Hu and A. J. Ragauskas, *Energy Environ. Sci.,* 1516 (2011).
26. N. Brosse, P. Sannigrahi and A. J. Ragauskas, *Ind. Eng. Chem. Res.,* 8328 (2009).
27. P. Sannigrahi, S. J. Miller and A. J. Ragauskas, *Carbohydr. Res.,* 965 (2010).
28. C. Munoz, J. Baeza, J. Freer and R. T. Mendonca, *J. Ind. Microbiol. Biotechnol.,* 1861 (2011).
29. M. J. Diaz, W. J. J. Huijgen, R. R. van der Laan, J. H. Reith, C. Cara and E. Castro, *Holzforschung.,* 177 (2011).
30. L. Meas, E. Gonzalez, C. Cara, M. Gonzalez, E. Castro and S. I. Mussatto, *Chem. Eng. J.* 1157 (2011).
31. T. Ingram, K. Wormeyer, J. C. I. Lima, V. Bockemuhl, G. Antranikian, G. Brunner and I. Smirnova, *Bioresource Technol.,* 5221 (2011).
32. R. Kumar, S. Singh and O. Singh. *J. Ind. Microbiol. Biotechnol.,* 377 (2008).
33. S. K. Ritter, *Plant Biochem.,* 15 (2008).
34. A. Johansson, O. Aaltonen and P. Ylinen, *Biomass.,* 45 (1987).
35. T. J. McDonough, *TAPPI J.* 186 (1993).
36. M. Foston, C. A. Hubbell, M. Davis and A. J. Ragauskas, *Bio energy Res.,* 193 (2009).
37. M. Hall, P. Bansal, J. H. Lee, M. J. Realff and A. S. Bommarius, *FEBS J.* 1571 (2010).
38. B. B. Hallac, P. Sannigrahi, Y. Pu, M. Ray, R. J. Murphy and A. J. Ragauskas, *Ind. Eng. Chem. Res.,* 1467 (2010).
39. R. H. Atalla and D. L. VanderHart, *Science,* 283 (1984).
40. P. Beguin and J. P. Aubert, *FEMS Microbiol. Rev.,* 25 (1994).
41. Y. H. P. Zhang and L. R. Lynd, *Biotechnol. Bioeng.,* 797 (2004).
42. L. Fan, Y. Lee and D. R. Beardmore, *Biotechnol. Bioeng.,* 419 (1981).
43. P. Kumar, D. M. Barrett, M. J. Delwiche and P. Stroeve, *Ind. Eng. Chem. Res.,* 3713 (2009).
44. X. Pan, D. Xie, R. W. Yu and J. N. Saddler, *Biotechnol. Bioeng.,* 39 (2008).
45. J. Baeza, A. M. Fernandez, J. Freer, A. Pedreros, E. Schmidt and N. Duran, *Appl. Biochem. Biotechnol.,* 273 (1991).
46. N. Kang, Z. Liu, L. Hui, C. Si, L. Cui, T. Zhao and S. Mao, *Biores.,* 578 (2012).
47. X. Pan, D. Xie, K. Kang, S. Yong and J. N. Saddler, *Appl. Microbiol. Biotechnol.,* 367 (2007).

48. X. Pan, D. Xie, R. W. Yu and J. N.Saddler, *Ind. Eng. Chem. Res.*, 2609 (2007).
49. R. Gupta and Y. Y. Lee, *Biotechnol. Bioeng.*, 1570 (2009).
50. B. B. Hallac and A. J. Ragauskas, *Biofuels. Bioprod Biorefin.*, 215 (2011).
51. V. P. Puri, *Biotechnol. Bioeng.*, 1219 (1984).
52. T. E. Timell, *Pulp. Pap. Mag. Can.*, 104 (1955).
53. J. L. Snyder and T. E. Timell, *Sven. papperstidn.*, 851 (1955).
54. R. Kumar, G. Mago, V. Balan and C. E. Wymand, *Bioresour. Technol.*, 3948 (2009).
55. M. S. Sweet and J. E. Winandy, *Holzforschung.*, 311 (1999).
56. C. A. Hubbell and A. J. Ragauskas, *Bioresource Technol.*, 7410 (2010).
57. X. F. Sun, R. C. Sun, P. Fowler and M. S. Baird, *Carbohydr. Polym.*, 379 (2004).
58. K. M. Draude, C. B. Kurniawan and S. T. B. Duff, *Bioresource Technol.*, 113 (2001).
59. X. Zhao, L. Wang and D. Liu, *J. Chem. Technol. Biotechnol.*, 1115 (2007).
60. M. Yoshida, Y. Liu, S. Uchida, K. Kawarada, Y. Ukagami, H. Ichinose, S. Kaneko and K. Fukuda, *Biosci. Biotechnol. Bioch.*, 805 (2008).
61. W. Boerjan, J. Ralph and M. Baucher, *Annu. Rev. Plant Biol.*, 519 (2003).
62. L. B. Davin and N. G. Lewis, *Curr. Opin. Biotechnol.*, 407 (2005).
63. L. Laureano-Perez, F. Teymouri, H. Alizadeh and B. E. Dale, *Appl. Biochem. Biotechnol.*, 1081 (2005).
64. M. Tu, X. Pan and J. N. Saddler, *J. Agric. Food Chem.*, 7771 (2009).
65. P. Y. Lu, B. Yang, D. Gregg, J. N. Saddler and S. D. Mansfield, *Appl Biochem Biotechnol.*, 641 (2002).
66. M. Tu, R. P. Chandra and J. N. Saddler, *Biotechnol. Prog.*, 398 (2007).
67. R. Sutcliffe and J. N. Saddler, *Biotechnol. Bioeng. Symp.*, 749 (1986).
68. V. J. H. Sewalt, W. G. Glasser and K. A. Beauchemin, *J. Agric. Food Chem.*, 1823 (1997).
69. N. Brosse, M. N. M. Ibrahim and A. A. Rahim, *ISRN Mater. Sci.*, 1 (2011).
70. G. Gellerstedt, J. Li, I. Eide, M. Kleinert and T. Barth, *Energy Fuels,* 4240 (2008).
71. M. Nagy, K. David, G. J. P. Britovsek and A. J. Ragauskas, *Holzforschung.*, 513 (2009).
72. B. B. Hallac, Y. Pu and A. J. Ragauskas, *Energy Fuels,* 2723 (2010).
73. E. A. Capanema, M. Y. Balakshin and J. F. Kadla, *J. Agric. Food Chem.*, 1850 (2004).
74. Z. Xia, L. G. Akim and D. S. Argyropoulos, *J. Agric. Food Chem.*, 3573 (2001).
75. L. Zhang and G. Gellerstedt, *Magn. Reson. Chem.*, 37 (2007).
76. K. H. Holtman, H. M. Chang, H. Jammel and J. F. Kaddla, *J. Wood Chem. Technol.*, 21 (2006).
77. G. Hu, C. Cateto, Y. Pu, R. Samuel and A. J. Ragausaks, *Energy Fuels,* 740 (2011).
78. J. J. Villaverde, J. Li, M. Ek, P. Ligero and A. D. J. Vega, *Agric. Food Chem.*, 6262 (2009).
79. M. Meshgini and K. V. Sarkanen, *Holzforschung.*, 239 (1989).
80. F. Xu, K. Cheng and D. Liu, *Ind. Eng. Chem. Res.*,180 (2006).
81. X. Zhao, K. Cheng and D. Liu, *Appl. Microbiol. Biotechnol.*, 815 (2009).
82. P. Sannigrahi, A. J. Ragauskas and S. Miller, *Energy Fuels,* 683 (2010).

83. R. E. Hage, N. Brosse, L. Chrusciel, C. Sanchez, P. Sannigrahi and A. J. Ragauskas, *Polym. Degrad. Stab.*, 1632 (2009).
84. R. Sun, Q. Lu and X. F. Sun, *Polym. Degrad. Stab.*, 229 (2001).
85. T. J. Mcdonough, *TAPPI J.* 1 (1992).
86. R. E. Hage, N. Brosse, P. Sannigrahi and A. J. Ragauskas, *Polym. Degrad. Stab.*, 997 (2010).

CHAPTER 5

REDUCTION OF BIOMASS RECALCITRANCE VIA IONIC LIQUID PRETREATMENTS

ALISTAIR W. T. KING*, HAIBO XIE,[‡,§] JUHA FISKARI,[†]
ILKKA KILPELÄINEN[†]

*,[†] *Laboratory of Organic Chemistry, Department of Chemistry, University of Helsinki, A. I. Virtasen Aukio 1, 00014, Helsinki, Finland*
alistair.king@helsinki.fi
[‡] *Division of Bioenergy Research, Dalian National Laboratory for Clean Energy, 457 Zhongshan Rd. Dalian,116023, China*
[§] *Division of Biotechnology, Dalian Institute of Chemical Physics, CAS, 457 Zhongshan Rd. Dalian,116023, China;*
hbxie@dicp.ac.cn

Ionic liquids are known to be effective solvents for lignin, cellulose, hemicelluloses and even intact wood. This opens the door to new methods of fractionation and depolymerisation for the purposes of production of bio-based materials, chemicals and fuels. The efficiency of future processes entirely depends on the capabilities of ionic liquids, as direct dissolution solvents for biomass. While fractionation efficiency for wood is generally poor with ionic liquids, more harsh conditions have allowed for further reduction in molecular weights, more efficient separation of degraded polysaccharide and lignin and reduction in crystallinity of the cellulosic material. This can afford a more easily digestible material. This chapter describes the solubility and chemistry of wood in ionic liquids. It also describes the treatment of lignocellulosics with ionic liquids for the purposes of production of biofuels in general. As ionic liquids have been found to be highly effective at reducing the recalcitrance of biomass, the chapter focuses on how ionic liquid pretreatment increases the efficiency of digestion by enzymes or microorganisms, with bioethanol production used as the reference process. Recent results in this area are reviewed. As a drawback of the use of ionic liquids in bioprocessing, the inhibition of enzymes and microorganisms by ionic liquids is also reviewed. As little is known about the recyclability of ionic liquids for biomass processing, this chapter will also give an overview of recent strategies into ionic liquid recycling, followed by a summary of the future outlook and challenges for the area.

1. Introduction

1.1. *What are ionic liquids?*

Ionic liquids (ILs) are molten salts, typically ion pairs containing one or more organic counter ions. The general distinction between molten salts and ILs is that the melting point (m.p.) of ILs should be less than 100 °C. This is an arbitrary distinction in the context of technical processing, as is the presence of traces of molecular solvent (in less than molar equivalents) making the distinction between ILs and organic electrolyte solutions. The earliest known ILs are the protic ILs ethanolammonium nitrate (1888, m.p. 52–55 °C), described by Gabriel and Weiner,[1] and ethylammonium nitrate (1914, m.p. 12 °C), described by Walden.[2] Although, these were not described as 'ionic liquids' at the time. The first useful molten salts, for cellulose dissolution and processing, were the *N*-alkylpyridinium chlorides (1933, m.p.s typically greater than 100 °C), patented by Graenacher.[3] They do not come under the common definition of an IL as the examples described in the patent have m.p.s equal to or greater than 100 °C. Never the less, they were technically useful. Swatlowski *et al.*[4,5] published and patented the first confirmed ILs, such as 1-butyl-3-methylimidazolium chloride ([bmim]Cl, 2002, m.p. 64–66 °C), for cellulose dissolution. This was quickly followed by the development of 1-allyl-3-methylimidazolium chloride ([amim]Cl, 2004, m.p. 55 °C), which is commonly described as the first room temperature ionic liquid (RTIL, molten salts with m.p.s less than room temperature) for cellulose dissolution. This is not strictly correct as the m.p. is greater than room temperature but the IL remains in a super-cooled glassy state unless highly purified. This is practically quite important as it makes the IL easy to dispense by needle for direct dissolution of small lignocellulose samples. This, combined with the fact that the IL allows for efficient esterification reactions on wood and cellulose,[6,7] makes [amim]Cl highly suitable for direct solvation/dissolution, homogeneous derivatisation and liquid-phase analysis of previously poorly soluble samples.[8–11] The first RTILs for cellulose dissolution were reported by Fukaya *et al.*[12] These were 1-alkyl-3-methylimidazolium formates, however not all m.p.s were determined for these compounds. These were followed by the 1-alkyl-3-methylimidazolium phosphates and phosphonates,[13] which were easy to prepare in a pure state and allowed for rapid dissolution of cellulose, even at room temperature. BASF, who purchased the IL-cellulose dissolution intellectual property from the Rogers-Alabama group, determined that 1-ethyl-3-methylimidazolium acetate ([emim][OAc], originally discovered in 1992, m.p. –45 °C)[14] was a highly effective IL for cellulose dissolution. It remains so to this day although as we will see later there are some obvious drawbacks to the use of this IL that require further understanding for the development of sustainable processes.

Spange *et al.*[15] determined, using Kamlet-Taft parameterization,[16] that it was the H-bond basicity of lithium chloride/*N,N*-dimethylacetamide (LiCl/DMA) mixtures that was important for the direct dissolution of cellulose in this now common solvating mixture. This was confirmed to be true for ILs also by Fukaya and Ohno *et al.*,[12,13] using Kamlet-Taft parameterization of formate, phosphate and phosphonate-based ILs. It is a common feature for all direct dissolution cellulose solvents (in the absence of derivatization or depolymerisation during dissolution). Hauru *et al.*[17] have compared this this for a series of ILs and more traditional direct dissolution solvents such as LiCl/DMA and *N*-methylmorpholine-*N*-oxide hydrate (NMMO•xH_2O). The need for the effective H-bond basicity is to break H-bonds between adjacent cellulose layers in the crystalline structure of cellulose. The relevant bonding interactions in [amim]Cl, as an exemplary direct dissolution solvent, are shown in Fig. 1. Here the anion is the basic species with only a minor H-bond donation interaction from the $[amim]^+$ C2, C4 and C5. In the aprotic environment of pure cellulose-dissolving ILs, there is an enthalpy gain in formation of H-bonds between the IL anions and cellulose hydroxyl functionalities.

When it comes to wood dissolution and processing, cellulose is the major polymeric component in wood. It is also regarded as the hardest to dissolve, due to its crystalline nature. Therefore, it follows that most ILs

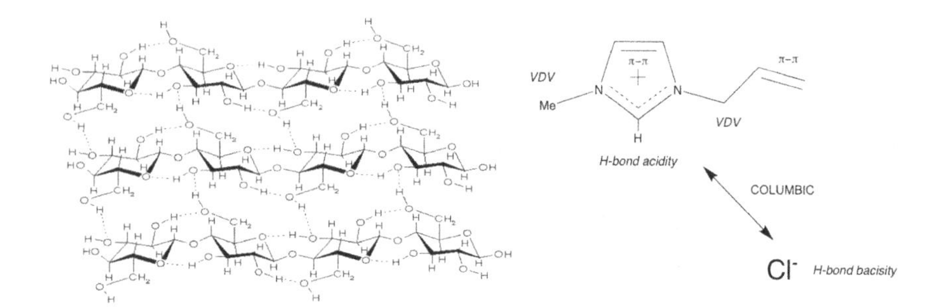

Fig. 1. H-bonding interaction potential between crystalline cellulose (left) and [amim]Cl (right).

that are capable of dissolving cellulose are also highly effective at solvating wood. This is important as reduction of the recalcitrance of wood obviously partly relies on the ability of the solvent to effectively solvate cellulose crystallites. The first publication describing the complete dissolution of wood was by Kilpeläinen *et al.*,[18] where intact wood was dissolved and esterified using [amim]Cl. Beyond this publication, a better understanding of wood solubility has been achieved and will be detailed later in the chapter.

1.2. *What biofuels are possible from IL pretreatments*

Lignocellulosic biomass, primarily being a complex mixture of cellulose, hemicellulose and lignin, is naturally resistant to breakdown by pests, disease, and weather. This inherent recalcitrance makes the production of monosugars or other valuable chemicals from lignocellulose expensive and inefficient. The full dissolution of lignocellulose in ILs is typically accompanied by the reduction of the cellulose crystallinity by breaking the inter- and intra-molecular H-bonding network. To achieve full dissolution this should also be accompanied by breakage of some covalent bonds between polysaccharides and lignin. This results in an easier separation of lignin to give cellulose-rich products. All of the above factors are beneficial for further chemical or biological conversion of carbohydrates into monosugars and chemicals. Currently, substantial effort has been put into this area with the aim of producing biofuels from lignocellulose. Well-investigated pathways are summarized in Fig. 1.

Although acid catalysis in IL media will allow for further depolymerisation to a mixture of oligomers and monomers, further upgrading to purified energy-rich materials should be best performed after isolation of polymeric material from the pretreatment media. This is due to the difficulty in separating low molecular weight (MW) materials from IL media. Enzymes and microorganisms are also inhibited in these media. Despite these issues, process efficiency relies on the ability of ILs to reduce the recalcitrance of, and fractionate, biomass. After the dissolution pretreatment with ILs, two divergent processing strategies can typically be followed:

1. Extended Chemical Route. The *in-situ* catalytic conversion of cellulose and hemicellulose in lignocellulose into 5-hydroxylmethyl furfural, and furfural by acidic catalysts can be performed. The furfurals can be further upgraded into alkanes and levulinic acidic esters, which have high energy density, comparing with that of carbohydrates, and are potential sustainable biofuels for the future.[19] An example of an advanced process for downstream production of biofuels from polysaccharide is the Sylvan process.[20]

2. Extended Biological Route. This includes sugar platform bioenergy. Currently, three types of approach have been carried out based on the ILs platform, to produce sugars for bioenergy production. The first is enzymatic hydrolysis of cellulosic materials, regenerated from ILs with addition of an anti-solvent. The second pathway is the *in situ* enzymatic hydrolysis of cellulose in biocompatible ILs. The third pathway is to use acidic catalysts to hydrolyze the dissolved carbohydrates in ILs. The obtained sugars can be further fermented into bioethanol and biobutanol[21] and microbial lipids. The microbial lipids can be further converted into fatty acid esters, which are currently widely used as biofuels.[22–23] In terms of bioethanol production, The Joint Bioenergy Institute (JBEI), as a Department of Energy funded project in the USA has made considerable inroads in the development the area, specifically using ILs as a pretreatment method.[24] At the present moment, the work is still academic but highly motivated and subject to continuous technoeconomical process modeling.

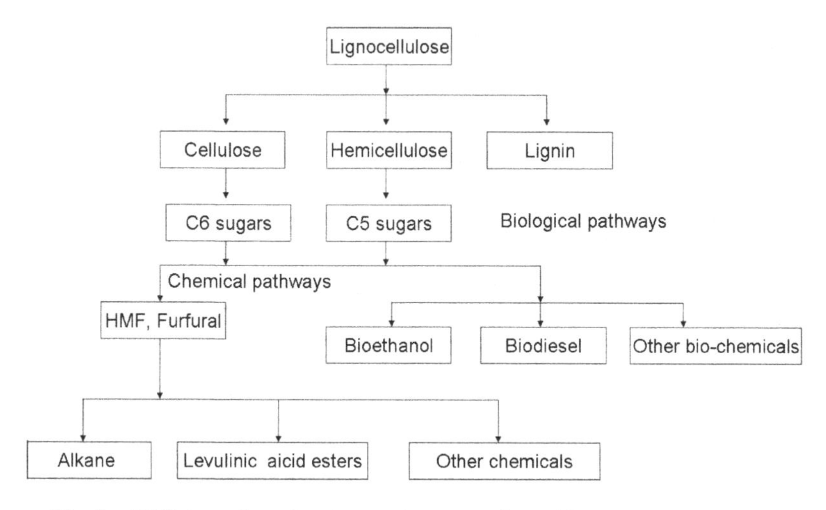

Fig. 2. Well-investigated pathways for conversion of lignocellulose to fuels.

2. Biomass Solubility

2.1. *Solubility and stability of wood and wood biopolymers in ILs*

Since the Kilpeläinen publication,[18] and a previous patent by Myllymäki and Aksela,[25] it has been assumed that native wood is fully soluble in ILs under mild treatment conditions. This has been reflected in almost all publications describing wood fractionation with ILs. Using the dissolution/regeneration strategy for fractionation; prior to the fractionation the authors describe the wood as being 'dissolved' into IL before selective precipitation of one or other components. This is rather missinformative and requires clarification of what dissolution is and under what conditions it occurs for different ILs. While it is possible to dissolve wood in some IL and IL-electrolyte solutions, mild treatments (roughly below 100 °C) often result in formation of 'wood suspensions', where some of the components are in solution and much of the mass is suspended in the IL media. This does not fit the simple definition of a solution, which is a 'homogeneous' media. The solubility of untreated Norway spruce sawdust, under mild dissolution conditions ([amim]Cl, 80 °C, 18 h), has been charted using a novel ^{31}P derivatization and solution-phase NMR analysis technique.[9] In these experiments, the spruce sawdust was milled in a planetary mill to different degrees (0–96 h). After the dissolution stage in

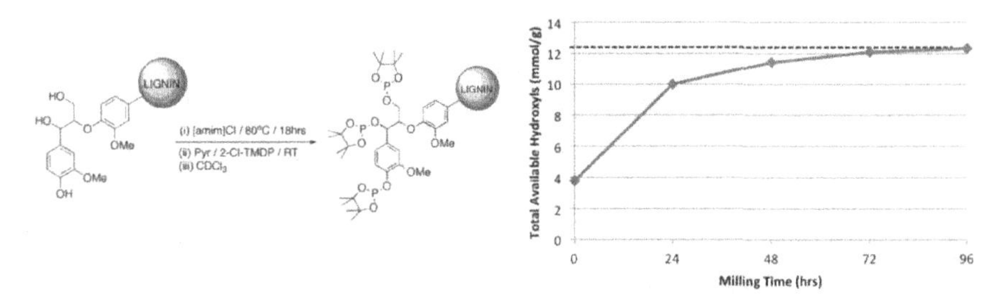

Fig. 3. Increasing solubility of wood, as a function of mechanical treatment (right), as characterized by pre-solvation with [amim]Cl, ^{31}P derivatisation and solution-phase NMR analysis (reaction shown on lignin to the left).[9]

[amim]Cl, those polymers which were effectively solvated are functionalized at their free hydroxyl groups, which are then quantifiable as phosphite esters, by the solution-phase NMR analysis. It was demonstrated that only after extensive pulverization that the wood became fully soluble, with a maximum number of available hydroxyls at approximately 12 mmol/g of wood (Fig. 3). This was repeated for both spruce and birch in a later publication demonstrating a similar effect for birch.[26] There could be a multitude of possible reasons for this 'recalcitrance'. The most likely explanation is that the high crystallinity of cellulose prevents dissolution and certainly, in regard to further chemical or enzymatic modification, this is a major issue. We do, however, have solvents that can dissolve purified cellulose and reduce its crystallinity, therefore this is logically not the main reason for the recalcitrance of wood. Another obvious reason is the high MW of the wood biopolymers. It is fundamentally understood that the higher MW polymers are, the harder it is to dissolve them in directly solvating media.

Moreover, as their concentrations increase in solution, viscosity drastically increases and mass transport drastically decreases. This gives limits to how much material can be practically dissolved in the already viscous ILs. High MW is to some extent preventing complete dissolution of wood to high concentration, which is a practical disadvantage to any process as high loadings per processing step are techno-economically preferable. However, even at low loadings of sawdust, or even wood chips, particles do not swell significantly, so do the purified lignocellulose polymers during dissolution into ILs. It seems that the recalcitrance is caused by additional factors, rather than the high MW of the native polymers. Some evidence shows that it is actually the extended lignin-carbohydrate complex (LCC) matrix that prevents the dissolution of wood. Recently, in the work carried out by Leskinen *et al.*,[26,27] it has been demonstrated the possibility to extract cellulose from spruce sawdust and it is actually the LCC matrix, with effectively infinite MW, is what prevents extensive swelling and eventually dissolution. This is achieved by using a relatively mild treatment with [amim]Cl and fractional precipitation with acetonitrile. The recovered fractions were characterized by Klason lignin content analysis, IR and NMR. The insoluble fraction, which was isolated first, was found to contain mainly hemicelluloses and lignin, whereas after the filtration a second cellulose-rich and lignin-poor fraction was precipitated after a second addition of acetonitrile. This is consistent with a report by Lawoko *et al.*[28] which has demonstrated experimentally that LCCs in spruce are mainly of the hemi-lignin-hemi type, with very little LCC linkages between cellulose and lignin. A publication by Hauru *et al.*[29] has demonstrated that it can be practically overcome by using a pretreatment, prior to fractionation. In this report hot water autohydrolysis[30] was used, which was known to deacetylate hemicelluloses and potentially cleave labile LCC linkages. After the autohydrolysis pretreatment, the resulting wood was found to be more soluble in [emim][OAc] than the untreated wood. Moreover a more efficient fractionation of lignin from polysaccharide was achieved using a modified method originally laid out by Sun *et al.*[31] These results of course require further understanding but have demonstrated a better understanding of the mechanisms required for efficient wood fractionation and recalcitrance reduction, which are both necessary for providing pure fractions for subsequent biofuel production steps.

Despite the poor overall solubility of wood in ILs, under mild dissolution conditions, higher temperatures can of course be used to dissolve wood. This typically results in breakage of LCCs and depolymerisation of biopolymers, which greatly enhances the solubility of wood. The obvious culprit to catalyse these reactions is the presence of trace acids in the media. In the aprotic and highly polarized IL media, acidity is greatly enhanced and rapid reduction of biopolymer MW ensues. A practical demonstration of this is an article by Rinaldi *et al.*[32] where cellulose is rapidly depolymerized in [bmim]Cl, in the presence of solid and homogeneous acid catalysts. The same depolymerisation rates are not observed in aqueous media. This phenomena is further discussed in an article by Mihichuk *et al.*[33] If the reduction in MW of wood biopolymers is one step on the way to synthesis of biofuels, this depolymerisation in itself may be advantageous providing the resulting enriched fractions can be efficiently isolated from the IL, without significant contamination of the non-volatile ILs. Other more selective side-reactions are now becoming apparent, however, that may or may not consume the IL during the pretreatment. The first of these was discovered by by Liebert *et al.*[34] and further characterized by Ebner *et al.*[35] They describe the potential for reaction of the imidazolium C2 with C1 reducing end groups in cellulose. This is likely dependent on the basicity of the anion in solution, catalysing the reaction *via* an initial deprotonation step at C2-H. A publication by Çetinkol *et al.*[36] describes the deacetylation of xylan residues and simultaneous acetylation of lignin residues when Wiley milled *Eucalyptus globulus* wood was treated with [emim][OAc] at 120 °C for 3 h. Hauru *et al.*[37] also described the deacetylation of residues in birch after treatment with [emim][OAc] at 100°C for 30 h. This selective reactivity may occur *via* a hydrolysis mechanism or it could also be considered to occur *via* a transacetylation mechanism, involving an acetic anhydride intermediate or acetylated carbene intermediate (Fig. 4). Further work is required to prove this mechanism. However, much evidence exists indicating that *N*-heterocyclic carbenes (NHCs), including those derived from imidazoliums are capable of catalysing this type of carbonyl chemistry. Typical reactions are the Benzoin condensation and the Stetter reaction. More recently persistent carbenes, derived from dialkylimidazolium ILs, have been shown to catalyse transesterification reactions,[38] which is exactly the proposed mechanism that may be occurring during the deacetylation/acetylation of wood, after IL treatments. Two recent publications by Rodríguez *et al.*[39] and Gurau *et al.*[40] even demonstrate the reactivity of the carbene at 25 °C with the electrophiles CO_2 and elemental sulphur (S_8). The reactivity in question for all these publications suggests the potential involvement of the acetate anion in a nucleophilic or basic capacity.[41] So far this reactivity seems to be limited to [emim][OAc] as the most basic and most effective IL for cellulose dissolution. If the carbene-based mechanism for reaction

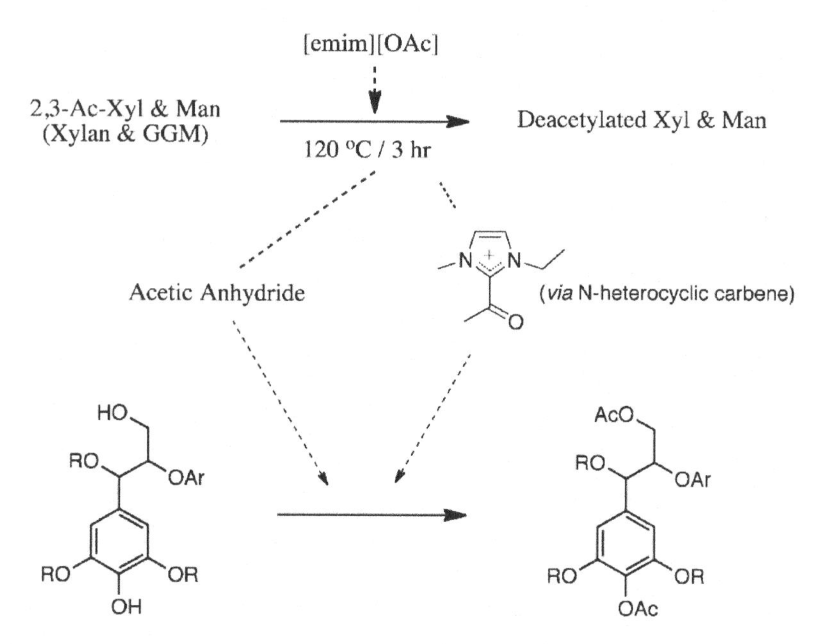

Fig. 4. Potential mechanism for deacetylation of hemicellulose and acetylation of lignin during wood treatment with [emim][OAc].

with wood functionalities is proven this implicates the IL in a catalytic role and as such, does not mean that the IL is consumed during the treatment. This is of course advantageous due to the high cost of the IL. The exception to this is in the case of reactivity with reducing end groups of cellulose. This is an actual covalent linkage between the imidazolium moiety and the cellulose backbone. It is possible, however, that this linkage may be cleaved during an aqueous workup treatment, as they are typically quite labile. With regard to dissolution of untreated wood, the increasing abundance of carbene at higher temperatures (greater than 100 °C), particularly with [emim][OAc] may facilitate the breakage of LCCs and hence enhanced dissolution, in comparison to mild dissolution temperatures or less basic ILs.

In addition to reaction with wood-based polysaccharides and potentially with LCCs, lignin itself has been shown to undergo changes upon wood pretreatment with ILs.[42] These changes include depolymerisation, demethylation and condensation reactions. Due to the complexity of lignin, in comparison to polysaccharides, it is difficult to say what factors cause these reactions. Temperature may play a big part here, in particular to the condensation of lignin at higher temperatures (greater than 140 °C). This condensation ultimately reduces the downstream processability of lignin. As we have seen earlier, higher temperatures could also lead to IL losses due to instability. It is therefore preferable to keep temperatures within a respectable range, in the neat ILs. Co-solvents can afford some additional stability to the ILs, in particular protic solvents (e.g., water and alcohols).

2.2. *Solubility of wood in IL-based organic electrolytes*

Since the 2002 report of cellulose dissolution into [bmim]Cl,[4] significant progress has been made, aimed at the efficient utilization of biomass. However, the traditional IL processes require high purity ILs and typically dry lignocellulose for an efficient dissolution and pretreatment process. Pretreatment reaction or dissolution kinetics are also rather slow due to high IL viscosities. With the typical literature procedures for neat ILs, this is overall rather an intensive process. Furthermore, the high price of ILs and energy requirements for IL recycling are highly prohibitive toward cost-competitive monosugar production based on IL pretreatment technologies, from a practical point of view. Fortunately, there are many recent reports that have demonstrated that IL-based electrolyte mixtures can effectively be used as solvents for biomass pretreatment. Good saccharification kinetics and monosugar yields, by enzymatic hydrolysis of regenerated materials, are still achievable, which will decrease the overall cost of IL-based pretreatment technologies. This is achieved by decreasing the operational costs through lower reagent cost, and potentially lower energy requirements through optimized processing conditions. However, water removal and IL purification are still problematic. In 2010, Rinaldi reported that organic electrolyte solutions, which contained just a small molar fraction of ILs ($X_{ILs} = 0.4$), dissolve 'instantaneously' 10 wt% of cellulose at 100 °C.[43] This demonstrates a novel dissolution process, by using less ILs to achieve dissolution and at much faster rates. Recently Xie *et al.*[44] demonstrated. that the mixed *N*-methylpyrrolidone (NMP) and [emim][OAc] system could completely dissolve corn stover at 140 °C in under 60 min.

Table 1. The solubility of different lignocellulosic biomass[a] in different organic electrolytes.[45]

Biomass Species	Organic electrolytes[b]	Temp. (°C)	Time (min)	Dissolution State[c]
Cover stover	NMP/[emim][OAc]	140	30	±
Cover stover	NMP/[emim][OAc]	140	60	+
Spruce TMP	NMP/[emim][OAc]	140	60	±
Spruce TMP	NMP/[emim][OAc]	140	120	+
Spruce TMP	DMSO/[emim][OAc]	140	120	+
Spruce TMP	DMF/[emim][OAc]	140	120	+
Spruce TMP	DMA/[emim][OAc]	140	120	+
Spruce TMP	DMI/[emim][OAc]	140	120	+
Peanut Straw	NMP/[emim][OAc]	140	120	+
Wheat Straw	NMP/[emim][OAc]	140	120	+

[a] Experimental conditions: solvent (20.0 g), biomass (2.0 g).
[b] Ionic liquids molar fraction is 0.2.
[c] ± partial dissolution; + dissolution.

The dissolution behavior was monitored by common optical and fluorescence microscopy, showing a homogeneous dispersion after complete dissolution. The results in Table 1 demonstrate that 10 wt% of corn stover can be partially dissolved in 30 min at 140 °C, in a mole fraction of 0.2 of [emim][OAc] in NMP. By prolonging the dissolution time to 60 min, it is possible to achieve complete dissolution. However, it was found that the spruce TMP was harder to dissolve, and it took 120 min to get complete dissolution under identical conditions. Other polar aprotic solvents, such as dimethylsulfoxide (DMSO), DMA, *N,N*-dimethylformamide (DMF), and *N,N*-dimethyl-2-imidazolidinone (DMI) are also good co-solvents for the dissolution of spruce TMP under these conditions. The system has been extended into other agriculture wastes, such as peanut straw and wheat straw.[45]

Lignin is one of the major components in lignocellulose and is a complex heteropolymer with auto-fluorescence in the visible as well as far-IR region, due to its phenolic structural units. It is known that the distribution of lignin varies over the different cell components in biomass, however the complete dissolution of lignocellulose in ILs will result in a homogeneous dispersion of lignin in the solution. Therefore it can be assumed that the dispersion degree of lignin in the solution is correlated with the dissolution behavior of biomass and can be evaluated by tracking the auto-fluorescent of lignin in ILs, during the dissolution process. This can provide significant insight in terms of the dissolution behavior of lignocellulose in ILs. During dissolution a series of comparative fluorescent microscopy pictures have been obtained and are shown in Fig. 5. At the beginning of dissolution, there is strong fluorescence tracked on the spruce TMP fibers with no fluorescence in the ILs, which indicates that the lignin is embedded in the fibre. When the mixture was heated up at 140 °C for 60 min, under mechanical stirring, almost all spruce TMP fibre disappeared. A solution with only a small portion of highly swollen aggregates of lignocellulose was observed; correspondingly, strong and homogeneous fluorescence was observed in the IL, except for the small portion of aggregation regions. The aggregates have stronger fluorescence than that of the solution, which implies that there are still some small fragments of lignocellulose remaining undissolved. By prolonging the dissolution time from 60 min to 120 min, a

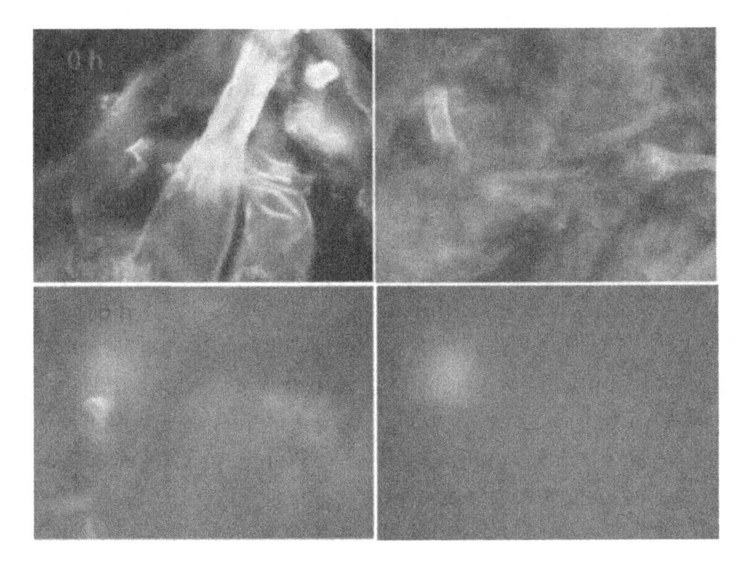

Fig. 5. Monitoring the dissolution behavior of 10 wt% spruce TMP in DMI/[emim][OAc] (X_{ILs} = 0.2) at 140 °C, by fluorescence microscopy.[45]

homogenous solution was achieved without any fluorescent aggregates observed. In effect, the wood was dissolved and lignin was dispersed throughout the media. As with the ^{31}P NMR analysis procedure, this fluorescence microscopy technique provides a rapid and descriptive method to analyse the solution state of wood in ILs.

3. IL-Aided Fractionation as a Pretreatment for Saccharification

As saccharification is a good descriptor of biomass reclalcitrance reduction, comparative assessment of this on fractionated materials is a good indicator of the ability of an IL to reduce biomass recalcitrance. To enhance the hydrolysis of cellulose by enzymes, a pretreatment is necessary on the lignocellulosic biomass to decrease the native crystallinity of cellulose and to partially remove the lignin and hemicellulose. On one hand, regeneration of cellulose from IL typically yields a pulp which has higher amorphous and cellulose II crystalline polymorph contents. However, recent studies have demonstrated that cellulose II is actually more readily digested than the cellulose I crystalline polymorph. This is due to the stronger van der Waals interactions between H-bonded sheets in cellulose I than those in cellulose II. It is these interactions that act as the main factor preventing the hydrolysis of cellulose by cellulases.[46,47] In addition to the modification or removal of crystallinity, the partial removal of lignin and hemicellulose can enhance the subsequent saccharification. This seems to be as a result of the increase in the cellulase accessible surface area of the cellulosic pulp.[48] Therefore enhancing the actual fractionation of wood into components during the pretreatment, avoiding precipitation of the dissolved components onto the cellulosic pulp surface, could have significant results.

In 2007, Kilpeläinen *et al.*[18] reported the dissolution of different woody lignocellulosic materials in [amim]Cl. It was found that under unoptimized conditions, about 60% of the theoretical amount of glucose in spruce wood was enzymatically released when predissolved in [amim] Cl and regenerated by the addition of water. Since then, to develop an economic and efficient IL-based pretreatment technology, much effort has been devoted to optimizing this new technology. This is achieved mainly by varying treatment conditions, modifying the structure of the ILs, using different lignocellulosic feedstocks and simplifying the sample regeneration, and IL recycling processes. Previous reviews have presented a comprehensive summary of the progress in this area, with the aim of developing IL-based pretreatment technologies.[49,50] Herein, with the limitation of the length of this chapter, some recent progress is highlighted.

Considering the good solubility of cellulose and lignocellulose in 1-ethyl-3-methylimidazolium diethylphosphate ([emim][Et_2PO_4]), Li *et al.* reported that the enzymatic hydrolysis of wheat straw pretreated with [emim][Et_2PO_4] at 130 °C for 30 min is enhanced significantly, and the yield of reducing sugars reached 54.8%.[51] In 2010, Nguyen demonstrated that the integration of aqueous ammonia and [emim][OAc], for rice straw pretreatment, could reduce the cost of the pure IL pretreatment technology.[52] This may make this emerging technology more adaptable to industrial application in enzymatic hydrolysis of different biomass feedstocks. It was found that the combined treatment exhibited a synergistic effect for rice straw, with 82% of the cellulose recoved and 97% of the enzymatic glucose conversion on this material. This cooperative effect showed over 90% of the glucose conversion even with a reduced enzyme usage and incubation time. Furthermore, the ILs could be recycled more than 20 times. Compared with the conventional pretreatments of ILs, this combined method for lignocellulosic biomass pretreatment was more economical and eco-friendly. As biomass usually contains much water (sometimes even more than 50% in industrial wood chips), an aqueous pretreatment system is preferred from a practical point of view. By using partial aqueous-based systems the pretreatment system can both minimize the use of ILs and avoid the costly biomass and IL drying steps. This can decrease the energy and equipment requirements of the process significantly. Recent studies using IL-water mixtures have stimulated substantial interest. In 2011, Brandt *et al.* reported that ground miscanthus (*Miscanthus giganteus*), pine (Pinus sylvestris) and willow (*Salix viminalis*) could be effectively pretreated using water mixedwith 1-butyl-3-methylimidazolium methylsulfate ([bmim][$MeSO_4$]) and 1-butyl-3-methylimidazolium hydrogensulfate ([bmim][$MeSO_4$]).[53] The optimum water content was found to be 10–40%. The recovered cellulose enriched solid fraction was subjected to enzymatic hydrolysis. Up to 90% of the glucose and 25% of the hemicellulose contained in the original biomass were released. After the pretreatment, the IL liquor contained the majority of the lignin and the hemicellulose. The lignin portion was partially precipitated from the liquor upon dilution with water. This was obtained relatively free from saccharide. The amount of hemicellulose monomers in the IL liquor and their conversion into furfurals was also examined and demonstrated that most hemicellulose is retained in the IL liquor and separated from the lignin by simple precipitation. This is a highly effective process for yielding digestable polysaccharide and pure lignin. However, recovery of the residual hemicellulose and furfurals may be problematic for full recyclability of the system.

Articles by Fu and Mazza[54,55] have demonstrated that mixtures of [emim][OAc] and water (10–30%) are also effective for pretreatment of lignocellulosic biomass, as evidenced by the removal of lignin and a reduction in cellulose crystallinity. A higher fermentable sugar yield (81%) was obtained more than that from the pure IL pretreatment (67%), under the same conditions. Cellulose and xylan digestibility generally increased with increasing temperature, time and IL concentration up to a certain level. The optimum pretreatment conditions were 158 °C at an IL concentration of 49.5 wt% and duration of 3.6 h.

A recent study by Zhang *et al.* demonstrated that pretreatment of sugarcane bagasse at 130 °C for 30 min by aqueous [bmim]Cl solution containing 1.2% HCl resulted in a glucan digestibility of 94–100% after 72 h of enzymatic hydrolysis.[56] HCl was found to be a more effective catalyst than H_2SO_4 or $FeCl_3$. Increasing the acid concentration from 0.4% to 1.2% and reaction temperature from 90 to 130 °C increased the glucan digestibility. The optimum water content was found to be 20%, in the same range as the Brandt and Fu publications. The addition of water to ILs reduced viscosity and allowed for higher biomass loading. It also reduced the lignin and hemicellulose recovery, likely due to inhibition of acid-catalysed degradation of these materials. One obvious disadvantage of the acid-catalysed degradation is that the formation of oligomers and monomers makes it difficult to purify the IL. Despite this the glucan digestibility of the solid residue, obtained with the acidified [bmim]Cl solution re-used three times, was greater than 97%.

In 2012, Xie *et al.*[44] reported a novel NMP/[emim][OAc] mixed solvent (0.2 mole fraction of IL) that can dissolve up to 10 wt% corn stover

Table 2. Composition comparison of raw corn stover and regenerated corn stover from NMP/[emim][OAc] (X_{ILs} = 0.2) solvent by ethanol.[44]

Materials	Cellulose (%)	Hemicellulose (%)				Lignin(%)		Ash (%)
Raw corn Stover	38.0	22.4				20.9		2.5
		Xylose	arabinose	galactose	mannose	AL[a]	AIL[b]	
		22.8	2.6	1.0	0.3	1.8	19.1	
Regenerated sample	52.8	28.4				9.8		2.2
		Xylose	arabinose	galactose	mannose	AL[a]	AIL[b]	
		28.9	3.3	0.8	0.4	2.1	7.7	

[a]acid soluble lignin;
[b]acid insoluble lignin.

at 140 °C in 60 min. With addition of ethanol, the dissolved biomass can be regenerated. Composition analysis (Table 2) of the regenerated biomass shows that they are rich in polysaccharide (cellulose 52.8%; hemicellulose 28.4%) and the lignin content significantly decreases from 20.9% to 9.8%. Enzymatic saccharification of the regenerated corn stover afforded an 82.9% total reducing sugars yield and a 60.8% glucose yield within 24 h. These results are comparative with those from the pure [emim][OAc] pretreatment process under similar conditions.[57]

Acetic and formic acid pulping of biomass is regarded as a novel fractionation method that allows for processing of non-woody biomass. The methods can tolerate high silicate contents of common grass species and typical products formed are cellulosic pulps, lignin, hemicelluloses and hydrolysates that can be further saccharified. The CIMV process is a recent commercial addition to these processes.[58] Strong acids, and H_2O_2, can also assist in delignification and saccharification. Acid catalysis in ILs is also a useful technique for biomass pretreatment. Due to an enhancement of acidity in IL media even acetic acid, a weak acid in aqueous solutions, has potential in biomass pretreatment. Use of this acid can minimize the formation of inhibitors, degradation of sugars and improve overall enzymatic hydrolysis. Recently, Xie *et al.*[59,60] found that carboxylic acid functionalized ILs, such as 1-carboxymethyl-3-methylimidazolium chloride

([HO_2CCH_2mim]Cl) in aqueous solutions could be used as a pretreatment reagent for corn stover, under mild conditions. Results revealed that the pretreatment of corn stover by [HO_2CCH_2mim]Cl/H_2O could selectively remove hemicellulose through hydrolysis, resulting in cellulose and lignin-rich fractions with enhanced saccharification behavior. For example, when the material was pretreated by 40 wt% [HO_2CCH_2mim]Cl/H_2O solution at 120 °C for 4 h, the contents of cellulose and lignin in the recovered biomass increased from 37.9%, and 21.1% to 55.6% and 26.5%, respectively. The hemicellulose contents decreased from 24.4% to 9.34%; further enzymatic hydrolysis of the pretreated corn stover resulted in 0.32 g/g and 0.45 g/g of glucose and total reducing sugars yields, respectively.

4. Tolerance of Enzymes/Microorganisms to IL Systems

With increasing interest in the IL-based pretreatment technology, it is necessary to develop systems that are stable and active in the presence of small amounts of ILs co-precipitated with recovered lignocellulose. However, ILs are potent inhibitors of the subsequent enzymatic hydrolysis and biological conversion to ethanol, or other processes like these. The identification of IL-tolerant enzymes that could be produced as a cellulase cocktail would reduce the costs and water use requirements of the IL pretreatment process. Similarly, the development or engineering of IL-tolerant organisms are necessary for all processes with IL-based pretreatments.

In 2010, Datta, *et al.*[61] investigated the stability of novel hyperthermophilic enzymes in the presence of [emim][OAc] and compared it to the industrial benchmark *Trichoderma viride* cellulase. The endoglucanase from a hyperthermophilic bacterium, *Thermatoga maritima*, and a hyperthermophilic archaeon, *Pyrococcus horikoshii*, were over expressed in *E. coli* and purified. Under their optimum conditions, both hyperthermophilic enzymes showed significantly higher [emim][OAc] tolerance than *T. viride* cellulase. By using differential scanning calorimetry, the authors determined the effect of [emim][OAc] on protein stability and the data indicated that higher concentrations of IL correlated with lowered protein stability. Both hyperthermophilic enzymes were active on [emim][OAc] pretreated microcrystalline cellulose and corn stover. Furthermore, these enzymes can be recovered with little loss in activity after exposure to 15% [emim][OAc] for 15 h. These results demonstrate the potential of using IL-tolerant extremophile-based cellulases for the hydrolysis of IL-pretreated lignocellulosic biomass, for biofuel production.

Nancharaiah *et al.*[62] investigated the effect of [emim][OAc], [emim][Et_2PO_4], and 1-methyl-3-methylimidazolium dimethylphosphate ([mmim][Me_2PO_4]) on the growth and glucose fermentation in *Clostridium sp.* It was found that [mmim][Me_2PO_4] was less toxic to the species. Growth of *Clostridium sp.* was not inhibited up to 2.5, 4 and 4 g/L of [emim][OAc], [emim][Et_2PO_4] and [mmim][Me_2PO_4], respectively. [emim][OAc] at less than 2.5 g/L showed a hormetic effect (positive effect at low concentrations) and stimulated the growth and fermentation by modulating medium pH. Total organic acid production increased in the presence of 2.5 and 2 g/L of [emim][OAc] and [mmim][Me_2PO_4]. ILs had no significant influence on alcohol production at a concentraion less than 2.5 g/L. Total gas production was affected by ILs at greater than 2.5 g/L and varied between ILs. Overall, the results show that the growth and fermentative metabolism of *Clostridium sp.* are not impacted by ILs at concentrations below 2.5 g/L.

To evaluate how the residual ILs in the fermentable sugars affect the subsequent ethanol fermentation process, Zhu *et al.*[63] investigated the effects of [bmim]Cl on the morphological structure, growth and ethanol fermentation of the yeast *Saccharomyces cerevisiae* AY92022. The morphology study showed that its single cell morphology remained unchanged at all [bmim]Cl concentrations, but its reproduction rate by budding decreased with increasing [bmim]Cl concentration. However, the results indicated that [bmim]Cl inhibited the yeast growth. Its specific growth rate during the log phase and bacterial concentration during the stationary phase all decreased with an increase in [bmim]Cl concentration. They found that [bmim]Cl had a negative effect on ethanol production. The increase in concentration of [bmim]Cl resulted in a decrease in final ethanol yield, although it was found that the ethanol-specific formation rate in the stationary phase remained unchanged for all [bmim]Cl concentrations. It was also observed that when the concentration of [bmim]Cl in the medium was 10^{-3} g/L, the ethanol fermentation process was no different from that of the control. This suggests that the [bmim]Cl in the fermentable sugars should be controlled below 10^{-3} g/L, thus it would not affect the subsequent ethanol fermentation process.

For the discovery of novel bacterial enzymes, with elevated stability in ILs, Ilmberger *et al.*[64] screened metagenomic libraries from three different hydrolytic communities (i.e., an enrichment culture inoculated with an extract of the shipworm *Teredo navalis*, a biogas plant sample and a sample from elephant faeces). 14 cellulolytic clones were identified and subsequently assayed in the presence of six different ILs. The most promising enzymes, CelA2, CelA3 (both derived from the biogas plant) and CelA84 (derived from elephant faeces), showed high activities (up to 6.4 U/mg) in the presence of 30 vol% ILs. As these enzymes were moderately thermophilic and halotolerant, they retained 40% to 80% relative activity after 34 days in 4 M NaCl, and they were benchmarked with two thermostable enzymes, CelA from *Thermotoga maritima* and Cel5K from a metagenome library derived from the Avachinsky volcanic crater in Kamchatka, Russia. These enzymes also exhibited high activity (up to 11.1 U/mg) in aqueous IL solutions (30 vol%). Furthermore some of the enzymes exhibited remarkable stability in 60 vol% IL. After 4 days, CelA3 and Cel5K retained up to 79% and 100% of their activity, respectively. Altogether, the results suggest that IL tolerance appears to correlate with thermophilicity and halotolerance.

Singer *et al.*[65] conducted enrichment experiments that demonstrated that IL-tolerant Aspergillus isolates were capable of growing in a radial plate growth assay, in the presence of 10% [emim][OAc]. When an [emim][OAc]-tolerant *Aspergillus fumigatus* strain was grown in the presence of switchgrass, endoglucanases and xylanases were secreted that retained residual enzymatic activity in the presence of 20% [emim][OAc].

Huang *et al.*[66] investigated the effects of three typical ILs, capable of dissolving cellulose, on lipid production by the oleaginous yeast *Rhodosporidium toruloides* AS 2.1389. When cultures were maintained at pH 6.0 in the presence of 30 mM of the ILs [emim]Cl, [emim][Et_2PO_4], or [emim][OAc], minor inhibition effects were observed. When cultures were performed in the presence of 60 mM ILs, or without pH control, inhibition was largely dependent on the structure of the ILs. Detailed analysis indicated that the anion of [emim][OAc] was consumed, concurrent to an alkaline-pH shift, leading to enhanced inhibition on cell growth and lipid production.

Aside from engineering isolating resistant strains of microorganism and enzymes that are not significantly inhibited by IL, JBEI is developing a new methodology to improve the resistance of microorganisms to IL inhibition. This involves the expression of genes for 'efflux pumps' in existing strains of microorganism that allows for the conversion of saccharides to biofuels.[67] Efflux pumps are mechanisms that may allow for the transport of toxic materials (e.g., ionic liquid, ethanol or butanol) through the cell wall and out of the cell, preventing the inhibition of conversion to biofuel, by lowering the concentration of toxic materials in the cells. Only initial studies have been published but this methodology may offer increased tolerance of micro-organisms to residual IL, left in the pulps after a pretreatment. Overall a combination of strategies (the use of efflux pumps and tolerant strains or enzymes) will likely be used to allow for suitable downstream processing conditions.

5. Ionic Liquid Recyclability and Recycling Strategies

A significant effort has been made into process optimization and the topic is becoming quite well understood, from a mechanistic point of view. However, most studies have focused on the use of common bioprocessing-related ILs such as [emim][OAc] due to its superior efficiency in dissolving cellulose. There are, however, some major drawbacks with the use of basic ILs.[41] As mentioned previously, they are known to react with lignocellulosic functionalities. Acetate or catalytic dialkylimidazol-2-ylidene carbene in solution may also cleave LCCs and enhance solubility of woody biomass. These may or may not reduce the viability of the process by consumption of the IL. The main reason being, that, ILs are very expensive at present, requiring at some estimates greater than 99% recovery per cycle. Another drawback of the use of archetypical cellulose-dissolving ILs arises from their non-volatility, almost a defining property of ILs. The non-volatility is one of the properties of ILs that allows them to be considered as potentially 'green' process media. The problem with this unfortunately is that when biomass is degraded and fractionated in ILs, low MW or oligomeric components (e.g., extractives, hemicellulose or lignin oligomers) and inorganic contaminants (e.g., metals, silicates) are soluble in the media, even upon addition of large amounts of non-solvents.

The non-solvents only remove polymeric materials. After many process cycles these contaminants may build up to such a level as to reduce process efficiency and will require a more advanced IL recycling. There are several potential ways to recycle ILs, including distillation of co-solvents and IL, phase-separation, extraction, complexation of components, membrane filtration and other membrance methods. In relation to bioprocessing, three have been studied in detail, i.e., IL distillation, phase-separation and extraction, excluding co-solvent evaporation, which is common to all processes.

BASF have patented a method of distillation of ILs such as [emim][OAc].[68] Further studies have shown however that [emim][OAc] irreversibly decomposes at around 175 °C.[41] By using high enough vacuum it is possible to distill [emim][OAc] below these temperatures but both these reports describe the use of high vacuum (typically less than 0.05 mbar). This can be achieved by using short-path distillation, with specialized apparatus to maximize surface area, which is still quite energy intensive due to the need for such low vacuums. As a consequence of this, King *et al.*[69] have described the development of a new class of 'distillable' ILs which are capable of dissolving cellulose, as a major component of woody biomass. Initial structures were based on the 1:1 combination of 1,1,3,3-tetramethylguanidine (TMG) with short-chain carboxylic acids, such as acetic and propionic acid. At low temperatures, the conjugated acid and base form ILs. If the samples are heated above 120 °C then the acid-base equilibrium shifts towards the neutral species affording a significant vapour pressure and thus 'distillability' (Fig. 6). As an example of

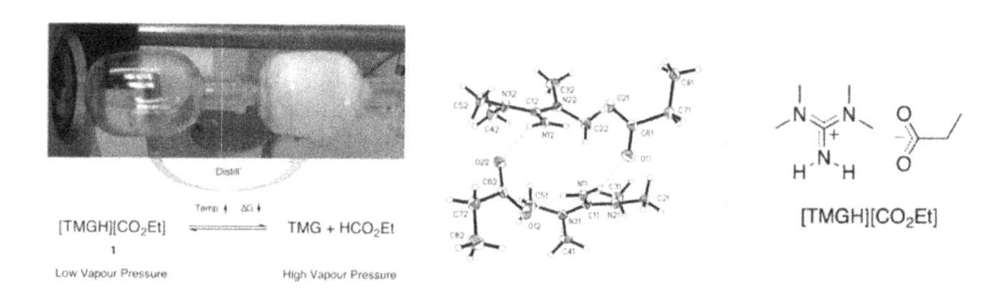

Fig. 6. Distillation concept, X-ray structure and Fischer projection for [TMGH][CO_2Et], as a distillable IL for biomass processing.[69]

their improved distillability over [emim][OAc] distillation of 1,1,3,3-tetramethylguanidinium propionate ([TMGH][CO_2Et]) was possible at 130 °C and 5 mbar, *versus* 130 °C and 0.05 for [emim][OAc]. Disadvantages of these initial TMG-based structures were however that they had quite high m.p.s and viscosities, which limited the practical cellulose solubility to less than 10% w/w. Since then other low viscosity structures have been developed.[70] It has been found that structures based on 1,5-diazabicyclo (4.3.0)non-5-ene (DBN) afford ILs that are now almost as low viscosity as [emim][OAc]. The result is that they can dissolve much larger amounts of cellulose, similar to [emim][OAc]. The full utility of these ILs in respect to biomass processing will be assessed in the coming years, although like [emim][OAc], where side reactions with lignocellulosics are known, they will also have to be assessed for their stabilities and reactivities with lignocellulosic functionalities.

The second suitable method for purifying these ILs is phase-separation. This may allow for a reduction in the energy required for removal of non-solvents, such as water. It may also allow for preferential separation of water-soluble components into an aqueous phase and separation of the IL to be reintroduced into the process. In a publication by Gutowski *et al.*,[71] it was demonstrated that the water miscibility of ILs such as [bmim]Cl, one of the initial ILs described for cellulose dissolution,[4] could be altered by addition of tripotassium phosphate (K_3PO_4) salt. This was implemented in an article by Shill *et al.*[72] where they used K_3PO_4 salt to regenerate [emim][OAc] after a biomass pretreatment step. It was found that the IL could be recovered with much lower water contents, affording significant energy reductions. They also found that the inclusion of the salt solution allowed for better separation of the polysaccharide from the lignin, which in turn afforded improved enzymatic hydrolysis kinetics. One problem, however, is that after several process cycles the separated lignin starts to accumulate in the IL, which requires additional purification techniques. Singh *et al.*[73] recently reported that sodium hydroxide solutions could also have the same phase-separation properties for [emim][OAc]. This effect was further probed in an interesting article by Shahriari *et al.*,[74] which strongly suggested that the ability of an aqueous salt in the phase-separation of an IL could be rationalized in terms of the Hofmeister series. Finally, a Finnish patent application by King *et al.*[75] was recently filed which described the use of phosphonium ILs, such as tetraoctylmethylphosphonium acetate ([P_{8881}][OAc]) that can dissolve cellulose and effectively solvate wood upon addition of polar aprotic solvents (e.g., DMSO). In the absence of co-solvent, the IL can be tuned to extract non-cellulosic components and with co-solvent the cellulosic component can be swollen and dissolved, allowing for recalcitrance reduction. Upon addition of alcohols, polymeric fractions can be precipitated and removed from the media. Upon addition of water, the hydrophobic IL phase-separates and can be further recycled, or purified if necessary. The cellulose solubility at the optimum DMSO composition for [P_{8881}][OAc] can reach approximately 20 wt% making it a very effective solvent for cellulose. Phosphonium ILs are also among the most stable cations, potentially minimizing side-reactions, which will reduce the sustainability of potential processes. Some potential drawbacks are ecotoxicity, toxicity to micro-organisms and, as with [emim][OAc], a buildup of lignin in the hydrophobic IL.

One article by Brennan *et al.*[76] describes the extraction of polysaccharides from aqueous IL solutions, using a naphthalene boronic acid auxiliary. This is a novel and effective method of removing low MW saccharides from IL but at this point it is not apparent how effective the method would be in the removal of oligosaccharides and larger. This will be an interesting topic to follow in the future and in relation to removal of lignin oligomers.

All these systems are at an early stage of understanding and need further refinement to see if they are techno-economically viable. A major goal that remains is to maximise recovery of the costly and potentially ecotoxic ILs.

6. Challenges and Future Outlook

In short, many of the above challenges are related to the high cost of ILs and their potential ecotoxicities. The high cost means that virtually all of the IL should be recovered after every process cycle. This will ease as the cost goes down, but the IL should still not be allowed to enter the environment if losses are made. The buildup of contaminants in the IL means that new recycling strategies must be considered, such as the development of distillable or phase-separable IL systems. It is not economically viable to dispose of the contaminated media at these costs. The inhibition of enzymes and micro-organisms is a major issue for downstream processing, in particular for bioethanol production. New enzymes and strains of micro-organism that can tolerate ILs are needed and are currently being developed and engineered. IL viscosities are rather high although co-solvents are very useful for reducing viscosity, increasing dissolution kinetics, reducing overall cost and in some cases stabilizing the ILs towards side reactions. Importantly, an academic understanding of lignocellulose dissolution and fractionation mechanisms, to yield proccessible pulps, should not be overlooked. This requires a good knowledge of lignocellulose chemical structure and the possible chemical interactions between the ILs, and the lignocellulosic functionalities. These interactions may or may not affect process sustainability. In the case of potential LCC cleavage, this may also affect the ability to separate lignin from polysaccharide during the pretreatment. This is of course very important for enzyme accessibility of the polysaccharide fraction and also processibility of the lignin-rich side-streams.

Ideally for biofuel production, from wood or grass-based materials, the ILs should dissolve cellulose under mild conditions and be non-reactive with lignocellulosic functionalities. They should preferably be cheap, low viscosity and low toxicity. Co-solvents may be used in the pretreatment step but this should not complicate the recovery of materials or the recycling of an IL. An ideal procedure will allow for the clean separation of lignin from polysaccharide. Recovery of lignin with native degrees of condensation and MW may provide for an additional necessary side stream, to mitigate the increased costs of the lower value biofuel. Reduction of cellulose I content, either to cellulose II or amorphous cellulose is preferable although significant reduction in MWs should be avoided in order to prevent their buildup in the recycled IL. ILs should be easily recycled by distillation or phase-separation, although minimizing the buildup of low MW materials will reduce the frequency at which this type of IL purification is employed after process cycles. So far all publications describe batch pretreatment processes. This will require large reactor volumes, considering product cost. Ultimately continuous processes should be developed which will reduce overhead costs drastically.

Aside from the previous and yet to be discovered challenges, the outlook is good, simply because ILs are superb solvators for lignocellulosic biomass. Research organizations such as the Joint Bioenergy Institute (JBEI) have made significant and rapid developments in this area, using ionic liquid pretreatments, for the production of bioethanol in particular. [emim][OAc] remains as one of the best solvents for cellulose, a major recalcitrant component of woody biomass. New and more recyclable systems are becoming available and will require the same scrutiny as what [emim][OAc] is receiving now, in terms of chemical reactivity, ecotoxicity, recyclability and recovery after use.

References

1. S. Gabriel and J. Weiner, Ueber einige Abkömmlinge des Propylamins. *Ber.*, **21(2)**, 2669–2679 (1888).
2. P. Walden, Molecular weight and electrical conductivity of several fused salts. *Bull. Acad. Imper. Sci.*, (St. Petersburg), **1800**, 405–422 (1914).
3. C. Graenacher, Cellulose solution and cellulose derivative and process of making same (*sic*). *US Patent*, 1924238 (1933).
4. R. P. Swatloski, S. K. Spear, J. D. Holbrey and R. D. Rogers, Dissolution of cellose (*sic*) with ionic liquids. *J. Am. Chem. Soc.*, **124**, 4974–4975 (2002).
5. R. P. Swatloski, R. D. Rogers and J. D. Holbrey, Dissolution and processing of cellulose using ionic liquids. *US Patent*, 6824599 (2004).
6. J. Wu, J. Zhang, H. Zhang, J. He, Q. Ren, M. Guo. Homogenous acetylation of cellulose in a new ionic liquid. *Biomacromol.*, **5**, 266–268 (2004).
7. H. Xie, A. W. T. King, I. Kilpeläinen, M. Granström and D. S. Argyropoulos, Thorough Chemical Modification of Wood-Based Lignocellulosic Materials in Ionic Liquids. *Biomacromol.,* **8**, 3740–3748 (2007).
8. A. W. T. King, I. Kilpeläinen, S. Heikkinen, P. Järvi and D. S. Argyropoulos, Hydrophobic Interactions Determining Functionalized Lignocellulose Solubility in Dialkylimidazolium Chlorides, as Probed by ^{31}P NMR; *Biomacromol.*, **10**, 458–463 (2009).
9. A. W. T. King, L. Zoia, I. Filpponen, A. Olszewska, H. Xie, I. Kilpeläinen, D. S. Argyropoulos. *In Situ* Determination of Lignin Phenolics and Wood Solubility in Imidazolium Chlorides Using ^{31}P NMR, *J. Agric. Food Chem.*, **57**, 8236–8243 (2009).
10. L. Zoia, A. W. T. King, D. S. Argyropoulos, Molecular Weight Distributions and Linkages in Lignocellulosic Materials Derivatized from Ionic Liquid Media. *J. Agric. Food Chem.*, **59**, 829–838 (2011).
11. A. Salanti, L. Zoia, E-L. Tolppa, M. Orlandi, Chromatographic Detection of Lignin-Carbohydrate Complexes in Annual Plants by Derivatization in Ionic Liquid. *Biomacromol.*, **13(2)**, 445–454 (2012).

12. Y. Fukaya, A. Sugimoto, H. Ohno, Superior Solubility of Polysacharides in Low Viscostiy, Polar, and Halogen-Free 1,3-Dialkylimidazolium Formates. *Biomacromol.*, **7(12)**, 3295–3297 (2006).
13. Y. Fukaya, K. Hayashi, M. Wada, H. Ohno, Cellulose dissolution with polar ionic liquids under mild conditions: required factors for anions. *Green Chem.*, **10**, 44–46 (2008).
14. Wilkes, J. S., Zaworotko, M. J., Air and Water Stable 1-Ethyl-3-methylimidazolium Based Ionic Liquids, *Chem. Commun.*, **13**, 965 (1992).
15. S. Spange, A. Reuter, E. Vilsmeier, T. Heinze, D. Keutel, W. Linert, Determination of empirical polarity parameters of the cellulose solvent N,N-dimethylacetamide/LiCl by means of the solvatochromic technique. *Polym. Chem.*, **36(11)**, 1945–1955 (1998).
16. M. J. Kamlet, J. L. M. Abboud, M. H. Abraham, R. W. Taft, Linear solvation energy relationships. 23. A comprehensive collection of the solvatochromic parameters,. pi.*, .alpha., and .beta., and some methods for simplifying the generalized solvatochromic equation. *J. Org. Chem.*, **48(17)**, 2877–2887 (1983).
17. L. K. J. Hauru, M. Hummel, A. W. T. King, I. Kilpeläinen, H. Sixta, Role of Solvent Parameters in the Regeneration of Cellulose from Ionic Liquid Solutions. *Biomacromol.*, **13(9)**, 2896–2905 (2012).
18. I. Kilpeläinen, H. Xie, A. King, M. Granström, S. Heikkinen, D. S. Argyropoulos, Dissolution of wood in ionic liquids. *J. Agric. Food Chem.*, **55(22)**, 9142–9148 (2007).
19. H. Xie, W. Liu, I. Beadham, N. Gathergood. Chapter 3, Biorefinery with Ionic Liquids, Edited by H. Xie, N. Gathergood, *The Role of Green Chemistry for the Biomass Processing and Conversion*, ISBN 978-0-470-64410-2 — John Wiley & Sons, (2012) p. 75–133.
20. A. Corma, O. de la Torre, M. Renz, Production of high quality diesel from cellulose and hemicellulose by the Sylvan process: catalysts and process variables, *Energy Environ. Sci.*, **5**, 6328–6344 (2012).
21. V. Garcia, J. Pakkila, H. Ojamo, E. Muurinen, R. L. Keiski, Challenges in biobutanol production: How to improve the efficiency? *Renew. Sustain. Energy Rev.*, **15**, 964–980 (2011).
22. Z. W. Gong, Q. Wang, H. W. Shen, C. M. Hu, G. J. Jin, Z. K. Zhao, Co-fermentation of cellobiose and xylose by *Lipomyces starkeyi* for lipid production. *Bioresource Technol.*, **117**, 20–24 (2012).
23. A. E. Atabani, A. S. Silitonga, I. A. Badruddin, T. M. I. Mahlia, H. H. Masjuki, S. Mekhilef, A comprehensive review on biodiesel as an alternative energy resource and its characteristics. *Renew. Sustain. Energy Rev.*, **16**, 2070–2093 (2012).
24. http://www.jbei.org.
25. V. Myllymäki, R. Aksela, Dissolution method for lignocellulosic materials, Patent, WO 2005/017001.
26. A. W. T. King, M. Lawoko, T. Leskinen, D. S. Argyropoulos, I. Kilpeläinen, Cellulose extraction from spruce sawdust with 1-allyl-3-methylimidazolium chloride (O5). 12th*European Workshop on Lignocellulosics and Pulp (Extended Abstract)*, Helsinki, Finland, 28–31 (2012).

27. T. Leskinen, A. W. T. King, I. Kilpeläinen, D. S. Argyropoulos, Fractionation of lignocellulosic materials with ionic liquids: Part 1. Effect of mechanical treatment. *Ind. Eng. Chem. Res.*, **50**, 12349–12357 (2011).
28. M. Lawoko, G. Henriksson, G. Gellerstedt, Characterisation of lignin-carbohydrate complexes (LCCs) of spruce wood (*Picea abies L.*) isolated with two methods. *Holzforschung,* **60**, 156–161 (2006).
29. L. K. J. Hauru, Y. Ma, M. Hummel, H. Sixta, In In Effect of pretreatment and cosolvent on the fractionation of birch with 1-ethyl-3-methylimidazolium acetate; *Books of Abstracts*, 243rd ACS National Meeting, San Diego, CA, March 25–29, 2012; American Chemical Society: Washington, DC, 2012. CELL-93 (2012).
30. H. Sixta. In *Prehydrolysis;* Sixta, H., Ed.; Handbook of Pulp, Vol. 1; WILEY-VCH: Weinheim (2006), p. 325–245.
31. N. Sun, M. Rahman, Y. Qin, M. L. Maxim, H. Rodríguez, R. D. Rogers, Complete dissolution and partial delignification of wood in the ionic liquid 1-ethyl-3-methylimidazolium acetate. *Green Chem.*, **11**, 646–655 (2009).
32. R. Rinaldi, R. Palkovits, F. Schüth, Depolymerisation of Cellulose Using Solid Catalysts in Ionic Liquids, *Angew. Chemie Int. Ed.*, **120**, 8167–8170 (2008).
33. L. M. Mihichuk, G. W. Driver, K. E. Johnson, Brønsted Acidity and the Medium: Fundamentals with a Focus on Ionic Liquids. *ChemPhysChem.*, **12(9)**, 1622–1632 (2011).
34. T. Liebert, T. Heinze, Interaction of ionic liquids with polysaccharides 5. Solvents and reaction media for the modification of cellulose. *Bioresources,* **3**, 576–601 (2008).
35. G. Ebner, S. Schiehser, A. Potthast, T. Rosenau, Side reaction of cellulose with common 1-alkyl-3-methylimidazolium-based ionic liquids. *Tetrahedron Lett.*, **49**, 7322–7324 (2008).
36. Ö. P. Çetinkol, D. C. Dibble, G. Cheng, M. S. Kent, B. Knierim, M. D. Auer, E. Wemmer, J. G. Pelton, Y. B. Melnichenko, J. Ralph, B. A. Simmons, B. M. Holmes, Understanding the impact of ionic liquid pretreatment on eucalyptus. *Biofuels*, **1**, 33–46 (2010).
37. L. K. J. Hauru, M. Hummel, H. Sixta, Fractionation of birch, spruce and pine in 1-ethyl-3-methylimidazolium acetate*; Books of Abstracts*, 241st ACS National Meeting, Anaheim, CA, March 27–31, 2011; American Chemical Society: Washington, DC, 2011; CELL-297 (2011).
38. G. A. Grasa, R. M. Kissling, S. P. Nolan, N-Heterocyclic Carbenes as Versatile Nucleophilic Catalysts for Transesterification/Acylation Reactions. *Org. Lett.*, **4(21)**, 3583–3586 (2002).
39. R. Rodríguez, G. Gurau, J. D. Holbrey, R. D. Rogers, Reaction of elemental calcogens with imidazolium acetates to yield 2-chalcogenones: direct evidence for ionic liquids as proto-carbenes. *Chem. Commun.*, **47**, 3222–3224 (2011).
40. G. Gurau, R. Rodríguez, J. D. Kelley, P. Janiczek, R. S. Kalb, R. D. Rogers, Demonstration of chemisorption of carbon dioxide in 1,3-dialkylimidazolium acetate. *Angew. Chemie Int. Ed.*, 47, 3222–3224 (2011).

41. A. W. T. King, A. Parviainen, P. Karhunen, J. Matikainen, L. K. J. Hauru, H. Sixta, I. Kilpeläinen, Relative and inherent reactivities of imidazolium-based ionic liquids: the implications for lignocellulose processing applications. *RSC Adv.*, **2**, 8020–8026 (2012).
42. K. M. Torr, K. T. Love, Ö. P. Çetinkol, L. A. Donaldson, A. George, B. M. Holmes, B. A. Simmons, The impact of ionic liquid pretreatment on the chemistry and enzymatic digestibility of *Pinus radiata* compression wood. *Green Chem.*, **14**, 778–787 (2012).
43. R. Rinaldi, Instantaneous dissolution of cellulose in organic electrolyte solutions. *Chem. Commun.*, **47**, 511–513 (2011).
44. H. B. Xie, H. W. Shen, Z. W. Gong, Q. Wang, Z. B. Zhao, F. W. Bai, Enzymatic hydrolysates of corn stover pretreated by a N-methylpyrrolidone-ionic liquid solution for microbial lipid production. *Green Chem.*, **14**, 1202–1210 (2012).
45. Z. L. Wu, H. B. Xie, X. Yu, E. H. Liu, The dissolution of wood in organic electrolytes solution and components fractionation, *Under review* (2012).
46. S. Kumar, R. Gupta, Y. Y. Lee, R. B. Gupta, Cellulose pretreatment in subcritical water: Effect of temperature on molecular structure and enzymatic reactivity. *Bioresource Technol.*, **101**, 1337 (2010).
47. M. Wada, M. Ike, K. Tokuyasu, Enzymatic hydrolysis of cellulose I is greatly accelerated via its conversion to the cellulose II hydrate form. *Polym. Degrad. Stabil.*, **95**, 543 (2010).
48. R. Kumar, C. E. Wyman, Access of cellulase to cellulose and lignin for poplar solids produced by leading pretreatment technologies. *Biotechnol. Progress*, **25(3)**, 807–819 (2009).
49. M. Mora-Pale, L. Meli, T. V. Doherty, R. J. Linhardt, J. S. Dordick, Room temperature ionic liquids as emerging solvents for the pretreatment of lignocellulosic biomass. *Biotechnol. Bioeng.*, **108(6)**, 1229–1245 (2011).
50. H. B. Xie, W. Liu, Z. Zhao, Chapter 3, Lignocellulose Pretreatment by Ionic Liquids: A Promising Start Point for Bio-energy Production in *Biomass Conversion: The Interface of Biotechnology, Chemistry and Materials Science*. ISBN: 978-3-642-28417-5 — Springer-Verlag, Germany (2012), p. 123–144.
51. Q. Li, Y. C. He, M. Xian, G. Jun, X. Xu, J. M. Yang, L. Z. Li, Improving enzymatic hydrolysis of wheat straw using ionic liquid 1-ethyl-3-methyl imidazolium diethyl phosphate pretreatment, *Bioresource Technol.,* **100**, 3570–3575 (2009).
52. T. A. D. Nguyen, K. R. Kim, S. J. Han, H. Y. Cho, J. W. Kim, S. M. Park, J. C. Park, S. J. Sim, Pretreatment of rice straw with ammonia and ionic liquid for lignocellulose conversion to fermentable sugars, *Bioresource Technol.,* **101**, 7432–7438 (2010).
53. A. Brandt, M. J. Ray, T. Q. To, D. J. Leak, R. J. Murphy, T. Welton, Ionic liquid pretreatment of lignocellulosic biomass with ionic liquid-water mixtures. *Green Chem.*, **13**, 2489–2499 (2011).
54. D. Fu, G. Mazza, Aqueous ionic liquid pretreatment of straw. *Bioresource Technol.*, **102(13)**, 7008–7011 (2011).

55. D. Fu, G. Mazza, Optimization of processing conditions for the pretreatment of wheat straw using aqueous ionic liquid. *Bioresource Technol.*, **102(17)**, 8003–8010 (2011).
56. Z. Zhang, I. M. O'Hara, W. O. S. Doherty, Pretreatment of sugarcane bagasse by acid-catalysed process in aqueous ionic liquid solutions. *Bioresource Technol.*, **120**, 149–156 (2012).
57. K. Shill, S. Padmanabhan, Q. Xin, J. M. Prausnitz, D. S. Clark, H. W. Blanch, Ionic liquid pretreatment of cellulosic biomass: Enzymatic hydrolysis and ionic liquid recycle. *Biotechnol. Bioeng.*, **108**, 511–520 (2011).
58. http://www.cimv.fr, (2012).
59. Y. Zhang, H. Xie, E. Liu, Q. Wang, Pretreatment of Corn Stover with a Carboxylic Acid-Functionalized Ionic Liquid Aqueous Solution to Selectively Remove Hemicellulose and Enhance Enzymatic Hydrolysis. *Under review* (2012).
60. H. Xie, Z. Kent Zhao, Y. Zhang. A method for lignocellulose pretreatment by a carboxylic acid-functionalized ionic liquids solution. *Patent Application*, CN 2012/10512392.0.
61. S. Datta, B. Holmes, J. I. Park, Z. Chen, D. C. Dibble, M. Hadi, H. W. Blanch, B. A. Simmons, R. Sapra, Ionic liquid tolerant hyperthermophilic cellulases for biomass pretreatment and hydrolysis. *Green Chem.*, **12**, 338–345 (2010).
62. Y. Venkata Nancharaiah, A. J. Francis, Alkyl-methylimidazolium ionic liquids affect the growth and fermentative metabolism of *Clostridium sp. Bioresource Technol.*, **102**, 6573–6578 (2011).
63. S. Zhu, P. Yu, Y. Tong, R. Chen, Y. Lv, R. Zhang, M. Lei, J. Ji, Q. Chen, Y. Wu, Effects of the Ionic Liquid 1-Butyl-3-methylimidazolium Chloride on the Growth and Ethanol Fermentation of *Saccharomyces cerevisiae* AY92022. *Chem. Biochem. Eng. Quarterly*, **26**, 105–109 (2012).
64. N. Ilmberger, D. Meske, J. Juergensen, M. Schulte, P. Barthen, U. Rabausch, A. Angelov, M. Mientus, W. Liebl, R. A. Schmitz, W. R. Streit, Metagenomic cellulases highly tolerant towards the presence of ionic liquids-linking thermostability and halotolerance. *Applied Microbiol. Biotechnol.*, **95(1)**, 135–146 (2012).
65. S. W. Singer, A. P. Reddy, J. M. Gladden, H. Guo, T. C. Hazen, B. A. Simmons, J. S. VanderGheynst, Enrichment, isolation and characterization of fungi tolerant to 1-ethyl-3-methylimidazolium acetate. *J. App. Microbiol.*, **110(4)**, 1023–1031 (2011).
66. Q. Huang, Q. Wang, Z. Gong, G. Jin, H. Shen, S. Xiao, H. Xie, S. Ye, J. Wang, Z. K. Zhao, *Bioresource Technol.*, 130, 339–344 (2013)
67. M. J. Dunlop, Z. Y. Dossani, H. L. Szmidt, H. C. Chu, T. S. Lee, J. D. Keasling, M. Z. Hadi, A. Mukhopadhyay. Engineering microbial biofuel tolerance and export using efflux pumps. Molecular Systems Biology 7, Article 487, (2011).
68. K. Massonne, M. Siemer, W. Mormann, W. Leng, Distillation of ionic liquids. WO 2009/027250.
69. A. W. T. King, J. Asikkala, I. Mutikainen, P. Järvi, I. Kilpeläinen, Distillable acid-base conjugate ionic liquids for cellulose dissolution and processing. *Angew. Chemie Int. Ed.*, **50**, 6301–6305 (2011).

70. A. Parviainen, A. W. T. King, I. Mutikainen, M. Hummel, C. Selg, L. K. J. Hauru, H.Sixta, I. Kilpeläinen, Predicting Cellulose Solvating Capabilities of Acid-Base Conjugate Ionic Liquids. *ChemSusChem*, DOI: 10.1002/CSSC.201300143 (2013).
71. K. E. Gutowski, G. A. Broker, H. D. Willauer, J. G. Huddleston, R. P. Swatloski, J. D. Holbrey, R. D. Rogers, Controlling the aqueous miscibility of ionic liquids: Aqueous biphasic systems of water-miscible ionic liquids and water-structuring salts for recycle, methathesis, and separations. *J. Am. Chem. Soc.*, **125**, 6632–6633 (2003).
72. K. Shill, S. Padmanabhan, Q. Xin, J. M. Prausnitz, D. S. Clark, H. W. Blanch, Ionic liquid pretreatment of cellulosic biomass: Enzymatic hydrolysis and ionic liquid recycle. *Biotechnol. Bioeng.*, **108**, 511–520 (2011).
73. B. A. Simmons, P. Varanasi, L. Sun, N. Sun, J. Shi, K.Tran, A. George, A. Cruz Gonzalez, S. Singh, Ionic liquids: Developing and realizing a scalable and cost-effective biomass pretreatment technology (K-11). *The 3rd Asian-Pacific Conference on Ionic Liquids and Green Processes* (APCIL'12), Sept 17–19, 2012, Beijing, China (2012).
74. S. Shahriari, C. M. S. S. Neves, M. G. Freire, J. A. P. Coutinho, Role of the Hofmeister series in the formation of ionic-liquid-based aqueous biphasic systems. *J. Phys. Chem. B.*, **115**, 7252–7258 (2012).
75. A. W. T. King, A. J. Holding, I. A. Kilpeläinen, Method of dissolving lignocellulosic materials, *Patent Application*, FI 2012/6078.
76. T. C. Brennan, S. Datta, H. W. Blanch, B. A. Simmons, B. M. Holmes, Recovery of Sugars from Ionic Liquid Biomass Liquor by Solvent Extraction, *Bioenerg. Res.*, 3, 123–133(2010).

CHAPTER 6

ENZYMATIC DECONSTRUCTION OF LIGNOCELLULOSE TO FERMENTABLE SUGARS

QINING SUN

Institute of Paper Science and Technology, School of chemistry and biochemistry, Georgia Institute of Technology, Atlanta, GA 30332, USA
qsun32@gatech.edu, sunqn515658@gmail.com

Lignocellulose is an intriguing potential feedstock for the production of fermentable sugars and generation of biofuels. Enzymatic hydrolysis of lignocellulose into fermentable sugars involves high quantity of a variety of enzymes. This section describes the fundamentals of enzymatic hydrolysis, multiple varieties of enzymes and their respective functions, activities, and mechanisms on carbohydrate-based substrates. Furthermore, novel methods and advances have been summarized and described in this section that increase the target sugar monomer yield for following stepwise fermentation. Hence, the objective of the work is to establish comprehensive understanding of enzyme-substrate interactions in the enzymatic hydrolysis of pretreated biomass for biofuel production.

1. Introduction

Fossil fuels are one of the limited primary energy resources stored in the Earth's crust, which have been used extensively to mediate nearly all of modern technology. The availability of this limited primary energy resource is expected to be reduced to such a degree that a global reliance on fossil fuels will no longer be sustainable in the future. In order to reduce the heavy reliance on fossil fuels and resolve the serious challenges in energy resources, a diversification in energy sources is important to ensure stable supply of domestic energy as well as for maintaining environmental habitation.[1] Lignocellulosic biomass, which incorporates woody crops, agricultural residues and cellulosic waste composed of up to 75% polysaccharide sugars, will play an important role as a suitable feedstock that can be utilized in the synthesis of second generation biofuels; lignocellulosic biorefinery effort will assist in the long term managing of greenhouse gas and energy security.[2–5]

Biofuel production can be categorized into a variety of novel conversion and refinery development technologies, including: direct combustion, thermochemical conversion, electrochemical conversion, indirect liquefaction, physical extraction and biochemical processes.[6] Among the three mature thermochemical processes, pyrolysis, gasification and liquefaction, are the most prominent due to a substantial potential for direct industrial applicability. Pyrolysis involves the thermal decomposition reaction of organic material via elevated temperatures (300–700°C) in the absence of oxygen.[7] Gasification undergoes the process to convert biomass into syngas, such as carbon dioxide, carbon monoxide, methane, and hydrogen gas at high temperatures (>700°C), without combustion and under a controlled amount of oxygen and/or steam. Liquefaction is a process in which biomass is liquefied and converted into small molecules via a complex sequence of physical structure and chemical changes under different catalysts, it is accomplished by natural, direct and indirect thermal, extraction and fermentation methods. These three processes display unique technical characteristics which render lignocellulosic biomass to a solid, liquid, gaseous or gaseous/liquid fuel.[6]

The initiating biochemical process involved in bioconversion from lignocellulosic biomass into biofuel incorporates a pretreatment step that reduces the crystalline integrity of the of the polymer chains, which leaves the biomass prone to enzymatic attack. Enzymatic hydrolysis cleaves hollocellulose polymers into respective monomer units made up of sugars. Bacteria and fungi can then convert these sugar fractions into ethanol via aerobic or anaerobic metabolic pathways in a process known as fermentation with this process outlined in Fig. 1.[6,8] Enzymatic hydrolysis is the preliminary challenging step towards converting lignocelluloses into fermentable sugar monomers. Here, the aspects of recent biomass enzymatic hydrolysis systems, their detailed mechanisms, and efforts made to optimize these technologies in order to pave the way for boosting more monomeric sugar yields will be identified.

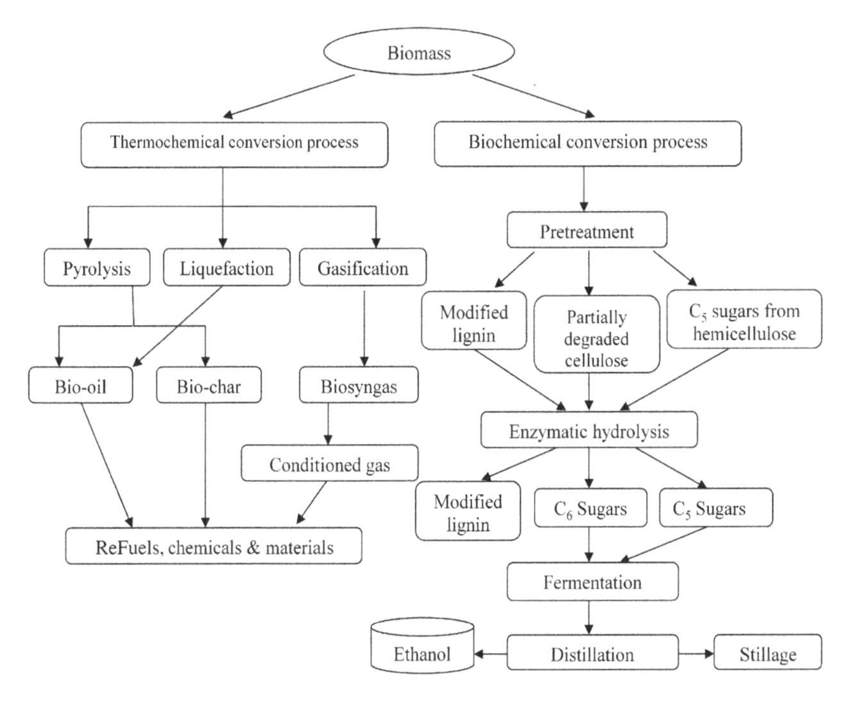

Fig. 1. A schematic illustration of biorefinery conversion of biomass to biofuels, materials and chemicals.

2. Enzymatic System

Enzymatic hydrolysis is a multi-step heterogenous reaction, in which lignocellulosic substrate is firstly broken down at the solid-liquid interface and followed by further liquid interface hydrolysis of soluble intermediates with specific catalytic proteins. For example, cellulose substrate is initally decomposed by the synergistic action of endoglucanases and/or exoglucanases/cellbiohydrolases, which is accompanied by further hydrolysis of soluble short celluloligosaccharies and cellobiose to produce glucose. Compared to more traditional hydrolysis methods such as acid-based methodologies, enzymatic hydrolysis has a substantial series of advantages. These include milder parameters, (pH 4.5–5.0 and temperature 40–50°C), low corrosion problems, low utility consumption, low toxicity, less formation of inhibitory by-products, and high sugar yield.[9,10]

There are many microorganisms in nature, mostly bacteria and fungi, producing biomass-degrading enzymes with different specificities, such as biomass composts, the rumen of cows, the hindgut of termite, and the extreme environmental niches. Bacteria involved in the hydrolytic degradation of cellulose (cellulolytic) include aerobic *Pseudomonas fluorescens subsp. cellulosa, Streptomyces lividans* and *Cellulomonas fimi.* Anaerobic variants include *Butyrivibrio fibrisolvens, Fibrobacter succinogenes, Ruminococcus flavefaciens, Clostridium cellulovorans, C. cellulolyticumy C. thermocellum.* All those aerobic and anaerobic bacteria utilize cellulases and hemicellulases to degrade lignocellulosic polymers.[11–17]

Furthermore, fungi play a significant role in generating different sets of hydrolytic enzymes, including cellulases, hemicellulases, phenolic acid esterases and ligninases for lignin degradation and modification.[18] Some are derived from aerobic fungi groups basidiomycetes and ascomycetes, such as *Trichoderma reesei, Trichoderma koningii, Trichoderma viride, Aspergillus niger, Aspergillus acculeatus, Sclerotium rolfsii, Phanerochaete chrysosporium, Fomitopsis palustris.* Of them *Trichoderma reesei* and *Phanerochaete chrysosporium* have a wide application in industry.[19–23] Alternatively, hydrolytic enzymes are also derived from anaerobic fungi belonging to anaeromyces, caecomyces, neocallimastix, orpinomyces and piromyces, such as *Anaeromyces mucrunatus, Caecomyces comunis, Neocalimastix frontalis, Piromyces sp., Orpinomyces sp., Anaeromyces elegans, Neocallimastic hurleyensis, Neocallimastic patriciarum.*[24–27]

All of the cellulolytic enzymes secreted by microbes are classes of glycoside hydrolases. This group consists of around 15 protein families and some subfamilies.[28–30] The combinations of enzyme and microbe can vary based on different biomass degrading ecosystems.

Enzymatic hydrolysis consists of multiphase reactions involving the soluble enzymes working on insoluble substrates in the beginning and further liquid-phase hydrolysis of soluble intermediates. Therefore, it is necessary to understand the effect and mechanism of varying cellulase, hemicellulase and lignin modifying enzyme systems on substrates, the influence of structural features of substrates and the mode of enzyme action on enzymatic hydrolysis.

2.1. *Cellulase enzyme system*

Cellulases are O-glucoside hydrolases hydrolyzing the β-1, 4 glucosidic linkages of cellulose. As described previously, there are three main types of enzymes involved in hydrolyzing cellulose microfibrils: endoglucoanase, exoglucanase, and β-glucosidase.

Endogluconases, e.g., 1, 4-β-D-glucan-4-glucanhydrolase (EC 3.2.1.4) cleave at random internal sites in the amorphous cellulose polysaccharide chain in a non-processive or processive manner, producing oligosaccharides of various sizes and new chain active ends that can be attacked by following exoglucanases. During hydrolysis rapid DP decrease of cellulose and cellooligomers dissolution are observed. In general, endoglucanases have a molecular weight that ranges from 22 to 45 kDa, optimum conditions are pH typically 4 to 5, temperature ranging from 50°C to 70°C. Endoglucanases have been isolated white and brown rot basidiomycetes, from the plant pathogen *S. rolfsii*, the yeast *Rhodotorula glutinis* and the termite symbiont *Termitomyces sp.*[4,31–34]

Exoglucanases, including 1, 4-β-D-glucan-glucanhydrolase (known as cellodextrinases, EC 3.2.1.74) and 1, 4-β-D-glucan-cellobiohydrolase (known as cellobihydrolases, EC 3.2.1.91) in a processive manner further catalyze the successive hydrolysis of residues from the reducing and non-reducing ends of the cellulose polysaccharide generated by endogluconases hydrolysis, liberating either glucose or cellobiose as main product of the reaction. There are two main forms in exoglucanases: CBH I proceeding from reducing end of the chain, and CBH II proceeding from the non-reducing end. Exoglucanases are monomeric proteins with a molecular weight ranging from 50 kDa to 65 kDa with low levels of glycosylation. They are able to hydrolyze crystalline cellulose, their optimum conditions are pH 4 to 5, temperature ranging from 37°C to 60°C. They have been found in basidiomycetes causing brown rot, such as *Fomitopsis palustris.*[31,32,35,36]

β-glucosidases (EC 3.2.1.21) carry out the exhaustive degradative hydrolysis of cellobiose into glucose, and also excecute the cission of glucose units from cellooligosachharides. β-glucosidases exist as homo-oligomers with molecular weights of approximately 100 kDa. The majority of these enzymes are glycosylated and takes place at moderate conditions including pH ranging from 3.5 to 5.5, and 45 to 75°C. They are synthesized by white and brown rot fungi, plant pathogens and yeast.[31,32,37]

2.2. Hemicellulase enzyme system

Hemicellulases are important components in degrading plant biomass, which are either glycoside hydrolases or carbohydrate esterases. There are multiple substituents linked to the hemicelluloses main backbone especially in xylan and mannan, such as arabinose, acetyl, galactose and glucose that require a large number of enzymes to be removed from the backbone and give access to the core enzymes thereby degrade the backbone into monosaccharides and acetic acid. There are two groups in enzymes degrading hemicelluloses: depolymerizing enzymes that cleave the backbone and the enzymes that remove substituents causing steric hindrances to the depolymerizing catalytic proteins, and include endo-xylanase, acetyl xylan esterase, β-xylosidase, endomannanase, β-mannosidase, α-L-arabinofuranosidase, α-glucuronidase, ferulic acid esterase, α-galactosidase, p-coumaric acid esterase.[38]

Xylanases are the main enzymes participating in the degradation of xylans, including the endoxylanases (EC 3.2.1.8; endo-1, 4-β-D-xylanases) that hydrolyze the xylan backbone linkages and release oligosaccharides, and β-xylosidases (EC 3.2.1.37; xylan 1, 4-β-xylosidase), which cleave xylo-oligosaccharides into xylose. Like xylanases, there are some similar enzymes for degradation of mannan, endo-mannanases and β-mannosidases.[31,38–40]

In addition, there are some other key hemicellases having specific functions in digesting carbohydrates, such as β-mannanases (EC 3.2.1.78) hydrolyzing mannan-based hemicelluloses and release short β-1,4-manno-oligomers, which can be further hydrolyzed to mannose by β-mannosidases (EC 3.2.1.25); α-L-Arabinofuranosidases (EC 3.2.1.55) and α-L-arabinanases (EC 3.2.1.99) hydrolyzing arabinofuranosyl-containing hemicelluloses; α-D-Glucuronidases cleaving the α-1,2-glycosidic bond of the 4-O-methyl-D-glucuronic acid sidechain of xylans; β-xylosidases (EC 3.2.1.37), exo-type glycosidases hydrolyzing short xylooligomers into single xylose units; and hemicellulolytic esterases including acetyl xylan esterases (EC 3.1.1.72) that hydrolyze the acetyl substitutions on xylose moieties, and feruloyl esterases (EC 3.1.1.73), which hydrolyze the ester bond between the arabinose substitutions and ferulic acid.[41–48]

2.3. *Lignin modifying enzymes*

Due to the fact that lignin complex polyphenolic macromolecular structure is composed mainly of β-O-4-aryl linkages, extracellular oxidative enzymes releasing highly unstable products that later undergo oxidation reactions are normally involved in lignin depolymerization, such as manganese peroxidase (MnP; EC 1.11.1.13), lignin peroxidase (LiP; EC 1.11.1.14), laccase (EC 1.10.3.1; p-diphenol dioxygen oxidoreductase). Most of them are secreted by white rot-causing fungi.

LiP is a glycoprotein with a heme group. As an oxidoreductase to boost hydrogen peroxide-dependent oxidative reaction, it can oxidize phenolic and non-phenolic lignin compounds, amines, aromatic ethers and aromatic polycyclic compounds.[39,40]

MnP, like LiP, is also a glycoprotein. Mn^{3+} is a strong oxidant that plays a key role in degrading phenolic lignin compounds.[49]

In addition, a study found basidiomycete *Pleurotus eryngii* yielding peroxidase combining the merits and activities of both LiP and MnP to better degrade lignin linkages.[50]

Laccase is widely distributed in higher plants, fungi and bacteria, which has been isolated from ascomycetes, deuteromycetes and basidiomycetes fungi.[51–53] Laccases are copper-containing glycosylated polyphenol oxidases that can oxidate the phenolic and non-phenolic compounds within mediators to produce active aromatic radicals, and thereby attack phenolic subunits of lignin, cleave the Cα-Cβ, aryl-alkyl and ether linkages, break aromatic rings and cause demethoxylation in lignin. The optimal conditions for laccase are highly dependent on the substrates, for example, pH of phenol ranges from 3 to 7 for fungal laccases and up to 9 for plant laccases. Temperature ranges from 10°C to 50°C, and a study found 50°C substantially increased the laccase activity. Laccases have been found in basidomycetes including *P. chrysosporium, Pleurotus ostreatus, T. versicolor* and *Pycnoporus sanguineus.*[31,54–57]

2.4. *Pectin degrading enzymes*

There are some enzymes degrading pectin, such as pectin methyl esterase, pectate lyase, polygalacturonase, rhamnogalacturonan lyase. Of them pectin methyl esterases remove methyl groups from pectin to open the access to depolymerising enzymes such as polygalacturonase, pectin lyase or pectate lyase. Rhamnogalacturonases and rhamnogalacturonan lyases mainly degrade the branched region in the pectin, while pectin acetylesterases remove acetyl groups from acetylated homogalacturonan and rhamnogalacturonan.[58-60]

3. Cellulose Enzymatic Saccharification

Over several decades there have been many studies investigating the effects and activities of carbohydrate degrading enzymes on depolymerization of lignocellulosic substrates into soluble fermentable sugars, although those understandings are still far from complete.[61–64] Reviewing the enzyme behaviors and mechanism on substrates paves the way for sugar yield enhancement.

3.1. *Enzyme behavior in hydrolysis*

Full enzymatic hydrolysis of cellulose to its glucose units is a comprehensive process carried out by the synergistic action of different glycohydrolases, in which enzymes play different roles at various stages.[63,64] Building on the studies of those enzymes on lignocelluloses, there are three processes that occur simultaneously during enzymatic on insoluble cellulosic substrates: chemical and physical changes in solid-phase cellulose, which are manifested as the DP and chain end concentration variation; initial hydrolysis in a slow reaction rate to release of soluble intermediates from the surface of cellulose mainly catalyzed by endoglucanases to increase chain ends and decrease DP significantly; and secondary hydrolysis relating to hydrolyze soluble intermediates to short oligosaccharides with various molecular weights and distributions in a fast rate with the main participation of exoglucanases, and finally degrade to glucose with the main participation of β-glucosidases.[65–68]

Enzymatic saccharification of cellulose is a heterogeneous process where cellulases react with macroscopic cellulose captaining crystalline region and amorphous regions. Detailed mechanism for cellulases deconstructing the cellulose is described as follows. Firstly, cellulases need to adsorb onto the surface of the insoluble cellulose substrates before hydrolysis (Fig. 2(a)). Secondly, under the solvent role and join action of enzymes on the inaccessible bulk of the substrate especially the crystalline cellulose regions, cellulose opens its matrix structure, which increase the molecular disorder of crystalline region in fibrous cellulosic network, and fragments its aggregation to short fibers, this initial stage is termed amorphogenesis Fig. 2(b).[69,70] Once getting the access into the cellulose network, joint action of endo-/exoglucanases enhances the fragmentation of accessible molecules to soluble short oligosaccharides, and which thereby are degraded into cellobiose and to glucose with the participation of β-glucosidases Figs. 2(c) and (d).[70]

3.2. *Cellulase adsorption and desorption*

Due to the multiphase reactions in the enzyme-substrate interaction, enzyme adsorption and desorption processes are highlighted. The factors involved have been subjected to considerable scrutiny, such as substrate features, purity, pretreatment influence, enzyme/substrate ratio, the affinity of enzyme complex system, and the topography of substrate changes, inactivation of the bound or free enzymes.

The classical models and kinetics assumption of uniformly mixed system cannot stand firm in the enzymatic cellulose deconstruction, as it is heterogeneous in nature where cellulose cleavage occurs at the interface between the solid cellulose surface and an aqueous enzyme solution, and it should take the spatial details of the cellulose substrate plus the specificity of binding sites into consideration. Recent studies show us a coarse-grained model for the multiple enzymes action on cellulose, and the cellulase adsorption and desorption reactivity kinetic models also have been set up to better understand the enzyme behavior during hydrolysis.[71–75]

Cellulose substrate model and model of cellulase with endo-/exo-activity are set in Figs. 3–5. Models set by Andrea Asztalos account for the mobility and action of a single cellulase enzyme as well as the synergy of multiple endo- and exo-cellulases on a cellulose surface, and take several reactive events into consideration, containing adsorption of cellulases on the solid cellulose substrate, inter-chain hydrogen bond breaking, hydrolysis of glycosidic bonds, and desorption of cellulases from cellulose.

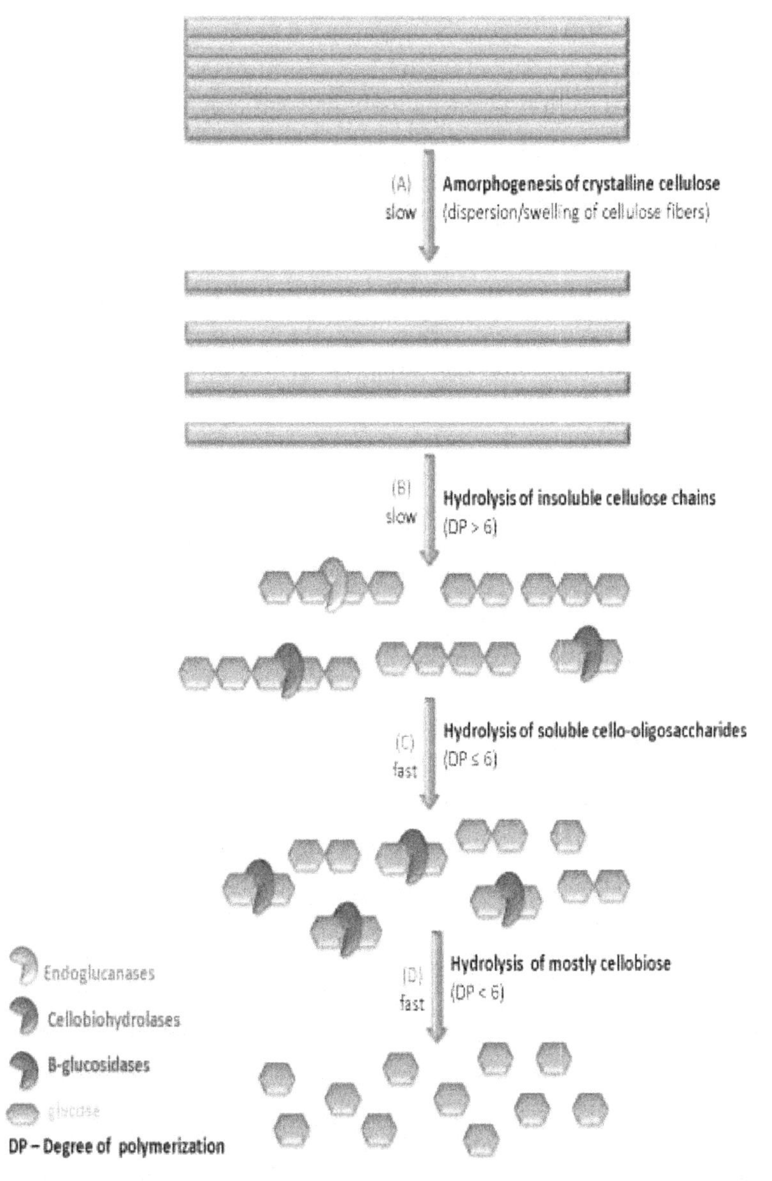

Fig. 2. Proposed mechanism for cellulose amorphogenesis/depolymerization by cellulases (Reproduced from Ref. 70 with permission).

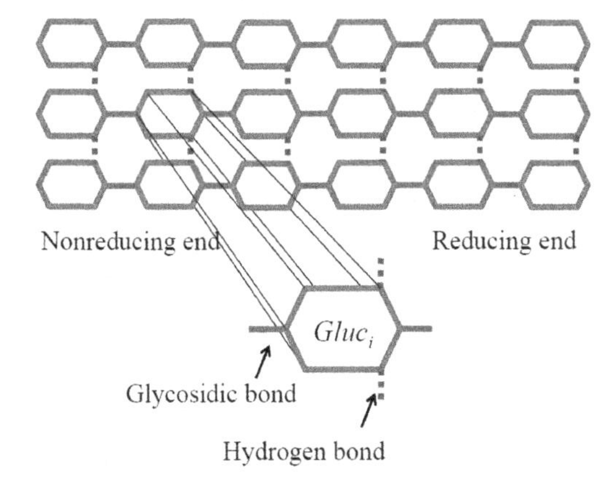

Fig. 3. Model for the cellulose surface composed of glucose (Reproduced from Ref. 73 with permission).

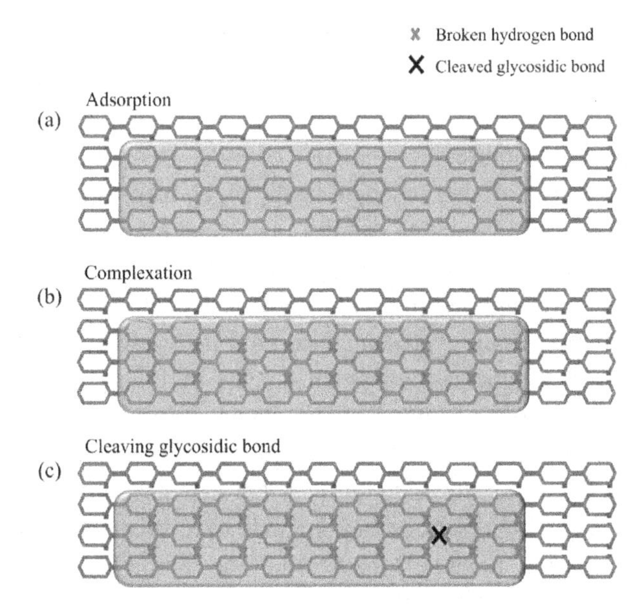

Fig. 4. Schematic representation of (a) an endo-cellulase adsorbed onto the cellulose surface, (b) hydrogen bonds breaking between the monomers covered by the enzyme and (c) hydrolysis of the glycosidic bond (Reproduced from Ref. 73 with permission).

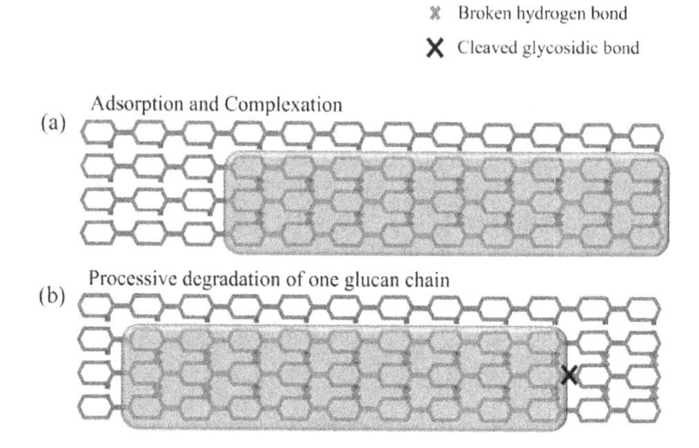

Fig. 5. Schematic representation of (a) an exo-R cellulase adsorbed to the cellulose surface followed immediately by the breaking of hydrogen bonds between the monomers covered by the cellulase and (b) the processivity of the glucan chain by an exo-R cellulase; it comprises the hydrolysis of the glycosidic bond and cellulase directed movement along the chain (Reproduced from Ref. 73 with permission).

Those models capture the interaction of endo- and exo-cellulases with crystalline cellulose at a mesoscopic level, which form a bridge between all-atom molecular dynamics studies and deterministic reaction-rate approaches and thereby improve our understanding of the molecular-level details of the enzymatic hydrolysis of crystalline cellulose.[76–79]

Furthermore, Maurer presented a modified Langmuir-Michaelis-Menten (LMM) kinetic model which consist of four processes in the interaction between the enzyme in the aqueous phase and a digestible solid substrate: (1) the reversible adsorption of enzyme in aqueous phase to and from the solid substrate surface; (2) enzyme-substrate complex formation; (3) surface reaction releasing dissolved sugars from solid substrate to liquid solution; and (4) the return of the complexed enzyme to the complexed enzyme to the relatively simple adsorbed state. Investigators utilize flow ellipsometry to quantify the adsorption and activity of cellulase on a flat, model cellulose surface, and for the first time, ascertain quantitative estimates of all kinetic parameters characterizing the reaction sequence.[74]

3.3. *Carbohydrate-bonding modules*

Cellulases are O-glucoside hydrolases attacking the β-1, 4 linkages of cellulose consisting of a catalytic domain and a cellulose binding module as show in Fig. 6.[70] Some hemicellulases such as xylanases also have the

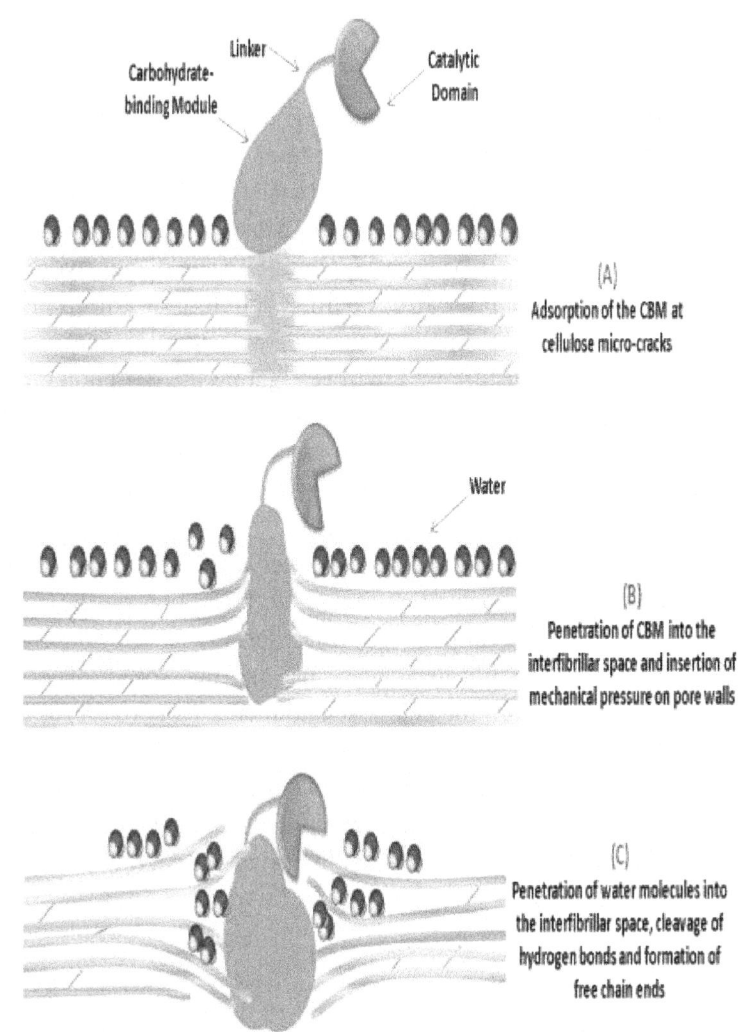

Fig. 6. Schematic representation of amorphogenesis of cellulose fibers mediated by the carbohydrate-binding module (CBM) of cellobiohydrolase I (CBHI) (Reproduced from Ref. 70 with permission).

same modular structures. The substrate-binding module or carbohydrate-binding module (CBM) with special targeting function is grouped into two categories: type A CBMs interact with crystalline cellulose, type B CBMs interact with amorphous cellulose. CBMs regulate docking of cellulases onto the substrates to form a complex held tight by specific, non-covalent, thermodynamically favorable bonds, and thereby catalytic module by determining specificity and reactivity towards the substrates establishes a high, local concentration of the enzyme on the substrates surface.[80–82]

CBM draws investigators' attention due to its special role in substrate targeting, selectivity in solid/liquid phases and disruption of non-hydrolytic crystalline substrates. CBM has 31 families, 18 of which have one solved three-dimensional structure. Most of them consists of a 'β-jelly-roll structure' of two β sheets, with either a planar hydrophobic surface that mediates binding to crystalline cellulose or a deep cleft that allows binding to a single polysaccharide molecule.[83,84] The disruptive function of CBM has been found by investigating the morphological and structural changes of cotton fibers plus the molecular dynamic simulations. CBM promotes non-hydrolytic disruption of crystalline cellulose by reducing the inter-fiber interaction, weakening and splitting the hydrogen bonds to increase the cellulose surface area.[85–87]

Mechanism has been proposed to better describe the dispersion of cellulose and cellulases behaviors including adsorption on the cellulose surface and penetration into interfibrillar spaces, as shown in Fig. 6. Larger CBM part inserts itself into the narrow space of cellulose crystalline matrix to swell the cellulose and increase the room for accommodating water among the microfibrils, which breaks the hydrogen bonds between cellulose chains and causes disassociation of individual microfibrils. Once anchoring the enzymes into the substrates via CBM, enzyme catalytic domain works fast on hydrolysis to degrade the substrate Fig. 6(c).[88–92]

3.4. Trichoderma reesei *system*

Most reported source of cellulases are derived from *Trichoderma reesei*, which drew much attention due to its high enzyme yield, availability in research, and high stability and efficiency of the cellulases it secreted. The optimal conditions for the maximum cellulase activity are normally set at 50°C to 55°C and pH ranging from 4.0 to 5.0. *T. reesei* consists of many active proteins with catalytic domain and CBM connected by a peptide linker, including two cellobiohydrolases (CBH I-II), five endoglucanases (EGI-V), two glucosidases (BGI-II), and some hemicellulases. Of them CBH I, CBH II and EG II are the principle components of the *T. reesei* cellulase system, taking up 60, 20 and 12%, respectively, of the total cellulase protein yielded by the fungus on the substrates.[93–100]

CBH I has four surface loops forming a tunnel of 50 Å adjacent to a β-sandwich structure, which works from the reducing end of the cellulose. CBH II, different structure with CBH I, has two well-ordered loops forming a 20 Å long tunnel adjacent to an α/β-barrel structure, which works from the cellulose non-reducing end. CBH I works processive, moving along a crystalline cellulose chain via CBM, chain and feeding it into the catalytic domain where β-1, 4-glycosidic linkages are degrading into cellobiose.[101–106] EG I results in a rapid decrease of DP by its active site groove allowing glucan chains to be cleaved to short oligosaccharides. In addition, for insoluble cellulosic substrates EG I displays relatively low rate of reducing sugar release on crystalline cellulose, and CBH II exhibits higher activity than CBH I. For soluble substrate, EG I shows significantly higher specific activities and reaction rate compared with CBH I and CBH II.[66,68,107,108]

4. Factors Influencing Lignocelluloses Enzymatic Hydrolysis

There are several constraint factors affecting the enzymatic hydrolysis, which not only result from experimental conditions, but also from the lignocellulosic substrates and enzymes features.[109] In order to clear the roadblock in lignocelluloses enzymatic hydrolysis for improving fermentable sugars, reviewing and summarizing important factors in both structural features of cellulose and the mode of enzyme action come to be very important.

4.1. *Experimental conditions involved factors*

It is apparent that various conditions within the lignocellulosic substrates and enzymes can limit the rate and degree of hydrolysis, such as substrate concentration, temperature, pH, reaction time and substrate/enzyme mixing ratio. Studies found some connections among the residence time and pH, and generally optimal conditions of difference enzymes were usually reported to be in the range of 35 to 55 °C for temperature, pH 4 to 5, and resident time around 150 h.[110,111] Substrate concentration in the slurry solution is a main factor influencing the yield and rate of enzymatic hydrolysis. Operating hydrolysis at high initial substrate concentration causes substrate inhibition for enzyme and thereby results in lower hydrolysis rate and problems in mixing and mass transfer.[112] Larger amounts of product at the initial hydrolysis stage including glucose, cellobiose and other oligosaccharides with different molecular weights and distribution inhibit the enzyme activity to different extents, including the glucose and cellobiose caused inhibition of β-glucosidases and cellobiohydrolases,[113–116] hemicelluloses and lignin association caused prevention for cellulases from accessing the cellulose surface, and some lignin-derived compounds such as vanillic acid, syringic acid and syringylaldehyde in addition to formic acid caused significant inhibition of xylanases.[117,118] Furthermore, enzyme loading affects the hydrolysis rate and yield to a large extent, it differs depending on the specific substrate and its composition. Higher load of enzymes, higher fermentable sugars yield. However, considering the process cost, some enzymes loading like cellulase are in the range of approximately 5 to 40 FPU per gram of substrate.[119–121] Enzyme recycling also appears to be an important factor affecting hydrolysis by reducing costs and saving enzymes. Typically up to 60~70% of total enzyme can be bound to lignin after complete hydrolysis.[122–124] Recovery of enzymes is influenced by lignin binding effect and inactivation of enzymes. Several methods work on reusing the enzymes, such as ultra-filtration to remove sugars that inhibit the enzyme action, immobilization to fixate the carbohydrolytic enzymes onto a solid matrix.[125–128]

4.2. *Substrate features involved factors*

Enzymatic hydrolysis efficiency also depends largely on the nature of original biomass feedstock, especially the physicochemical properties of lignocellulosic substrates.

With regard to chemical compositions, lignin acts like a physical barrier on restricting the access of enzymes to substrates and inhibiting the enzyme activity by adsorbing the enzymes unspecifically onto its particles or surfaces via hydrophobic interactions, electrostatic interactions and hydrogen bonding.[129–132] Hemicelluloses not only contribute to the xylan and glucan and mannan sugar yield, but also act as physical barriers to limit the accessibility of enzymes to attack cellulose by their association with lignin.[133–135] Acetyl groups also influence the hydrolysis to some extent by inhibiting the formation of productive binding between cellulose and catalytic domain of enzymes.[136–138]

Cellulose accessible surface area, specific surface area, pore size, crystallinity and degree of polymerization also play crucial roles influencing the enzymatic digestibility. Accessible surface area is one of the most important factor to affect the substrate accessibility to enzymes with tight relation with particle size, porosity and volume. Cellulose accessibility can be reflected by cellulases CBM attaching to the hydrophilic face of cellulose crystalline part and the anatomical structure of the cell wall, especially the pore amount, volume and size.[139–141] Specific surface area contains exterior surface area affected by substrate length and width and interior surface area influenced by lumen size and substrate pore and crack number.[131] Cellulose crystallinity and morphology are inherenthy connected with cellulase enzyme activity and therefore efficient enzymatic hydrolysis. Cellulose crystalline regions display rigid characteristic to outside change especially pretreatment and enzymatic hydrolysis. Cellulose crystallinity was reported to influence enzyme adsorption and effectiveness of adsorbed cellulase components, and thereby affect the hydrolysis rate and yield. Results showed decreasing crystallinity can increase the initial hydrolysis rate and reduce the hydrolysis time or the amount of enzyme required to get high digestibility.[82, 142–144] Degree of polymerization (DP) in enzymatic hydrolysis can reflect the extent to which cellulose depolymerization proceeded by the cellulase components. DP is not an independent factor, which is always accompanied by changes of crystallinity and porosity of the substrates. Results showed us that lower DP can provide more binding sites to cellulase enzymes such as CBH I, EG I and β-glucosidase therefore increase the hydrolysis rate.[131, 145–148]

4.3. *Enzyme related factors*

Hydrolysis inhabitation influenced by enzyme derives from the enzyme source, nature of enzyme, product inhibition, thermal inactivation, activity balance, enzyme processibility and compatibility.[82,149] Enzyme nature and its interaction with cellulosic substrates take a large and growing proportion. Typically the carbohydrate-binding module (CBM) contributes to enzyme efficiency by attaching to and disrupting the substrates. The overall adsorption binding efficiency of cellulase is markedly enhanced by presence of CBM. Furthermore, some enzyme structural variation in the hydrophobicity of the surface also affects substrate adsorption affinity.[150] In addition, the combined action of several enzymes results in a higher action rate than the sum of their individual action, which is termed synergistic cooperation. Study found both endoglucanases and cellobiohydrolases can hydrolyze amorphous cellulose, and cellobiohydrolases largely hydrolyze the crystalline cellulose, all of the above are highly influenced by synergism.[151]

5. Strategies to Enhance Enzymatic Hydrolysis

In order to overcome current obstacles and improve the enzymatic hydrolysis efficiency with a higher sugar yield and lower cost, there are several innovative and developing strategies that include synergetic cooperation studies and addition of new surfactants.

5.1. *Synergistic effects on enzymatic hydrolysis*

Synergistic interactions among various enzymes have been studied to design optimal combinations and ratios of enzymes for lignocellulosic substrates. Studies have been focused on the synergism of special enzymes in the cellulose and hemicelluloses related substrates hydrolysis and some related factors, such as enzyme compositions and concentration, enzyme selection, sequential or simultaneous synergy, inhibition factors, and enzymes with cross-specificity.[152–155]

Types of synergism in cellulose hydrolysis include endoglucanase and exoglucanase; exoglucanase and exoglucanase; endoglucanase and endoglucanase; exoglucanase or endoglucanase and β-glucosidase; cellulose-enzyme-microbe (CEM) synergism; cellulase complexes formation synergism.[82,156–162] Among them synergism between endoglucanases and exoglucanases draws the most attention and is very important for hydrolysis of crystalline cellulose.

Types of synergism in hemicellulose hydrolysis contains acetylxylan esterase and endoxylanase; feruroyl esterase and endoxylanase; xylanase, β-xylosidase, acetyl xylan esterase; ferulic acid esterase and Family 10 and 11 endoxylanase; exo- and endo-xylanase; endo-xylanase, β-xylosidase and α-L-arabinofuranosidase; α-L-arabinofuranosidase and endoxylanase; GH43 α-L-arabinofuranosidase and two GH51α-L-arabinofuranosidase; arabinofuranosidase and endo-arabinase; β-mannanase and β-mannosidase; α-galactosidase (GH27) and β-mannanase; xylanase, mannanase and α-galactosidase combination.[163–175] Typically it requires a large number of hemicellulases with synergic interactions hydrolyzing hemicellulose due to its complicated structure and the xylan, mannan and pectin caused inhibition on each other.

In addition, some other synergies on complex lignocellulosic substrates include endo-mannanase and endo-xylanase; arabinofuranosidase and endo-mannanase; arabinofuranosidase, endo-mannanase and endo-xylanase; endo-xylanase and endo-glucanase; endo-xylanase and cellulases; arabinofuranosidase, endo-xylanase, β-xylosidase and acetyl xylan esterase; cellobiohydrolase, endo-xylanase, acetyl xylan esterase and ferulic acid esterase; arabinofuranosidase, β-galactosidase and endo-xylanase; ferulic acid esterase and 2 family 10 and 11 endo-xylanases.[166,167,176–181] All types of synergism above are valuable and beneficial for accessory enzymes on lignocellulose degradation to enhance sugar yield.

5.2. *Additives and surfactants*

Addition of various additives including surfactants and non-catalytic proteins can modify substrate surface properties to aid the enzymatic hydrolysis and lower the enzyme loading. Surfactants are amphiphilic compounds containing a hydrophilic head and a hydrophobic tail. Fatty acid esters of sorbitan polyethoxylates (Tween® 20 & 80), polyethylene glycol are the most effective surfactants, which help to lower interface tension between substrate thereby to improve hydrolysis via reducing enzyme adsorption to the lignin.[182–184] In addition, those surfactants also hinder irreversible adsorption of enzymes to lignin, and/or improve enzyme thermal stability to avoid denaturation.[185–187]

6. Conclusions and Outlook

Lignocellulose enzymatic hydrolysis is a key step in biofuel production. Current research is investigating means to improve the performance cellulolytic enzymes for commercial processes via innovative strategies. However, some factors in both lignocellulosic substrate and enzyme can influence each other which therefore inhibit the enzymatic hydrolysis. Those bottlenecks in enzyme characteristic and its interaction with substrate inevitably require finely-tuned synergy among enzymes and advanced biotechnologies to break for achieving maximum sugar yield.

Acknowledgments

The author is grateful for the financial support from the Paper Science & Engineering (PSE) fellowship program at Institute of Paper Science & Technology (IPST) and the School of Chemistry and Biochemistry at Georgia Institute of Technology, Atlanta, GA, USA.

References

1. A. J. Ragauskas, C. K. Williams, B. H. Davison, G. Britovsek, J. Cairney, C. A. Eckert, W. J. Frederick, J. P. Hallett, D. J. Leak, C. L. Liotta, J. R. Mielenz, R. Murphy, R. Templer and T. Tschaplinski, *Science*, **311**, 484–489 (2006).
2. J. Lee, *J. Biotechnol.*, 56, 1–24 (1997).
3. E. A. Bayer, R. Lamed and M. E. Himmel, *Curr. Opin. Biotechnol.,* **18**, 237–245 (2007).
4. L. R. Lynd, J. H. Cushman, R. J. Nichols and C. E. Wyman, *Science*, **251**, 1318–1323 (1991).
5. L. D. Gomez, C. G. Steele-King and S. J. McQueen-Mason, *New Phytol.*, **178**, 473–485 (2008).
6. A. Demirbas, *Energy Convers. Manage.*, **50**, 2782–2801 (2009).
7. S. Katyal, *Energy Sources Part a-Recovery Utilization and Environmental Effects,* **29**, 1477–1485 (2007).
8. Y. H. P. Zhang, S. Y. Ding, J. R. Mielenz, J. B. Cui, R. T. Elander, M. Laser, M. E. Himmel, J. R. McMillan and L. R. Lynd, *Biotechnol. Bioeng.,* **97**, 214–223 (2007).

9. D. G. Shell, B. Duff, in *Handbook on Bioethanol: Production and Utilization*, Ed. C. E. Wyman, (Taylor & Francis, Washington, DC, 1996), p. 382.
10. Y. Y. Lee, P. Iyer, R. W. Torget, in *65 Advances in Biochemical Engineering/ Biotechnology. Recent Progress in Bioconversion of Lignocellulosics,* Eds. T. Scheper. G. T. Tsao. (Springer, Berlin, Heidelberg, New York, Barcelona, Hong Kong, London, Milan, Paris, Singapore, Tokyo, 1999), p. 93.
11. R. H. Doi in *Cellulases of mesophilic microorganisms — Cellulosome and noncellulosome producers*, Vol. 1125 Eds.: J. Wiegel, R. J. Maier and M. W. W. Adams, (Blackwell Publishing, Oxford, 2008), pp. 267–279.
12. S. Khanna and Gauri, *Enzyme Microb. Technol.*, 15, 990–995 (1993).
13. K. L. Braithwaite, G. W. Black, G. P. Hazlewood, B. R. S. Ali and H. J. Gilbert, *Biochem. J.*, **305**, 1005–1010 (1995).
14. N. Arcand, D. Kluepfel, F. W. Paradis, R. Morosoli and F. Shareck, *Biochem. J.*, **290**, 857–863 (1993).
15. M. V. S. Murty and T. S. Chandra, *Antonie. van. Leeuwenhoek.*, **61**, 35–41 (1992).
16. L. L. Lin and J. A. Thomson, *FEMS Microbiol. Lett.,* **84**, 197–204 (1991).
17. P. Tomme, R. A. J. Warren, N. R. Gilkes, *Adv. Microb. Physiol.*, **37**, 1–81 (1995).
18. M. E. Himmel, P. A. Karplus, J. Sakon, W. S. Adney, J. O. Baker and S. R. Thomas, *Appl. Biochem. Biotechnol.*, **63–5**, 315–325 (1997).
19. M. Dashtban, H. Schraft and W. S. Qin, *Int. J. Biol. Sci.*, **5**, 578–595 (2009).
20. R. E. Quiroz-Castaneda, E. Balcazar-Lopez, E. Dantan-Gonzalez, A. Martinez, J. Folch-Mallol and C. M. Anaya, *Electr. J. Biotechnol.*, 12 (2009).
21. V. A. Adisa and A. O. Fajola, *Mycopathologia.*, **82**, 23–27 (1983).
22. E. T. Reese, R. G. H. Siu and H. S. Levinson, *J. Bacteriol.*, **59**, 485–497 (1950).
23. D. F. Bateman, *Physiologial. Plant. Pathol.*, **2**, 175–184 (1972).
24. M. J. Nicholson, M. K. Theodorou and J. L. Brookman, *Microbiology-Sgm.*, **151**, 121–133 (2005).
25. Y. W. Ho, T. Bauchop, N.Abdullah, S. Jalaludin, Mycotaxon., 38, 397–405, (1990).
26. J. Webb and M. K. Theodorou, *Can. J. Bot.*, **69**, 1220–1224 (1991).
27. C. K. Pai, Z. Y. Wu, M. J. Chen, Y. F. Zeng, J. W. Chen, C. H. Duan, M. L. Li and J. R. Liu, *Appl. Microbiol. Biotechnol.*, **85**, 1451–1462 (2010).
28. M. L. Rabinovich, M. S. Melnik, and A. V. Bolobova, *Applied. Biochemistry and Microbiology.*, **38**, 305–321 (2002).
29. M. L. Rabinovich, M. S. Mel'nik, *Usp. Biol. Khim.*, **40**, 205–266 (2000).
30. B. Henrissat, A. Bairoch, *Biochem. J.*, **316**, 695–696 (1996).
31. N. Aro, T. Pakula and M. Penttila, *FEMS Microbiology Reviews.*, **29**, 719–739 (2005).
32. P. Baldrian and V. Valaskova, *FEMS Microbiology Reviews.*, **32**, 501–521 (2008).
33. S. J. Ding, W. Ge and J. A. Buswell, *Enzyme Microb. Technol.*, **31**, 621–626 (2002).
34. V. Valaskova and P. Baldrian, *Microbiology-Sgm.*, **152**, 3613–3622 (2006).
35. N. Hamada, K. Ishikawa, N. Fuse, R. Kodaira, M. Shimosaka, Y. Amano, T. Kanda and M. Okazaki, *J Biosci Bioeng.*, **87**, 442–451 (1999).

36. J. J. Yoon, Y. K. Kim, *Journal of Microbiology*, **43**, 487–49 (2005).
37. M. Dashtban, H. Schraft and W. S. Qin, *Int. J. Biol. Sci.*, **5**, 578–595 (2009).
38. A. S. Meyer, L. Rosgaard and H. R. Sorensen, *J. Cereal. Sci.*, **50**, 337–344 (2009).
39. C. Sanchez, *Biotechnol. Adv.*, **27**, 185–194 (2009).
40. J. Pérez, J. Muñoz-Dorado, T. de la Rubia, J. Martínez, *Int Microbiol.*, **5**, 53–63 (2002).
41. D. Shallom and Y. Shoham, *Curr Opin Microbiol.*, **6**, 219–228 (2003).
42. D. Hogg, E. J. Woo, D. N. Bolam, V. A. McKie, H. J. Gilbert and R. W. Pickersgill, *J. Biol. Chem.*, **276**, 31186–31192 (2001).
43. B. C. Saha, *Biotechnol. Adv.*, **18**, 403–423 (2000).
44. D. Nurizzo, T. Nagy, H. J. Gilbert and G. J. Davies, *Structure.*, **10**, 547–556 (2002).
45. V. Mai, J. Wiegel and W. W. Lorenz, *Gene.*, **247**, 137–143 (2000).
46. R. C. Lee, M. Hrmova, R. A. Burton, J. Lahnstein and G. B. Fincher, *J. Biol. Chem.*, **278**, 5377–5387 (2003).
47. J. A. M. Prates, N. Tarbouriech, S. J. Charnock, C. Fontes, L. M. A. Ferreira and G. J. Davies, *Structure.*, **9**, 1183–1190 (2001).
48. F. D. Schubot, I. A. Kataeva, D. L. Blum, A. K. Shah, L. G. Ljungdahl, J. P. Rose and B. C. Wang, *Biochemistry*, **40**, 12524–12532 (2001).
49. K. Ikehata, I. D. Buchanan and D. W. Smith, *J. Environ. Eng. Science.*, **3**, 1–19 (2004).
50. S. Camarero, S. Sarkar, F. J. Ruiz-Duenas, M. J. Martinez and A. T. Martinez, *J. Biol. Chem.*, **274**, 10324–10330 (1999).
51. G. Benfield, Sheila M. Bocks, K. Bromley, B. R. Brown, *Phytochemistry*, **3**, 79–88 (1964).
52. G. Diamantidis, A. Effosse, P. Potier and R. Bally, *Soil. Biol. Biochem.*, **32**, 919–927 (2000).
53. A. Assavanig, B. Amornkitticharoen, N. Ekpaisal, V. Meevootisom and T. W. Flegel, *Appl. Microbiol. Biotechnol.*, 1992, **38**, 198–202.
54. E. Dantan-Gonzalez, O. Vite-Vallejo, C. Martinez-Anaya, M. Mendez-Sanchez, M. C. Gonzalez, L. A. Palomares and J. Folch-Mallol, *Int. Microbiol.*, **11**, 163–169 (2008).
55. M. Heinzkill, L. Bech, T. Halkier, P. Schneider and T. Anke, *Appl Environ Microbiol.*, 1998, 64, 1601–1606.
56. E. M. Ko, Y. E. Leem and H. T. Choi, *Appl. Microbiol. Biotechnol.*, **57**, 98–102 (2001).
57. A. M. Farnet, S. Criquet, S. Tagger, G. Gil and J. Le Petit, *Can. J. Microbiol.*, **46**, 189–194 (2000).
58. R. S. Jayani, S. Saxena and R. Gupta, *Process Biochem.*, **40**, 2931–2944 (2005).
59. P. Michaud, A. Da Costa, B. Courtois and J. Courtois, *Crit. Rev. Biotechnol.*, **23**, 233–266 (2003).
60. E. Bonnin, K. Clavurier, S. Daniel, S. Kauppinen, J. D. M. Mikkelsen and J. F. Thibault, *Carbohydr. Polym.*, **74**, 411–418 (2008).
61. A. Berlin, V. Maximenko, N. Gilkes and J. Saddler, *Biotechnol. Bioeng.*, **97**, 287–296 (2007).
62. X. Zhang, W. J. Qin, M. G. Paice and J. N. Saddler, *Bioresour. Technol.*, **100**, 5890–5897 (2009).

63. L. P. Walker and D. B. Wilson, *Bioresour. Technol.,* **36**, 3–14 (1991).
64. M. P. Coughlan, *Bioresour. Technol.,* **39**, 107–115 (1992).
65. P. Valjamae, G. Pettersson and G. Johansson, *Eur. J. Biochem.*, **268**, 4520–4526 (2001).
66. K. M. Klemanleyer, N. R. Gilkes, R. C. Miller and T. K. Kirk, *Biochemical Journal* 1994, 302, 463–469.
67. K. M. KlemanLeyer, M. SiikaAho, T. T. Teeri and T. K. Kirk, *Applied and Environmental Microbiology*, 1996, 62, 2883–2887.
68. M. Srisodsuk, K. Kleman-Leyer, S. Keranen, T. K. Kirk and T. T. Teeri, *Eur. J. Biochem.*, **251**, 885–892 (1998).
69. M. P. Coughlan, *Biotechnol. Genet. Eng.*, **3**, 39–109 (1985).
70. V. Arantes and J. N. Saddler, *Biotechnol Biofuels*, 3, 4 (2010).
71. H. Ooshima, M. Sakata and Y. Harano, *Biotechnol. Bioeng.*, **25**, 3103–3114 (1983).
72. H. Ooshima, M. Kurakake, J. Kato and Y. Harano, *Appl. Biochem. Biotechnol.*, **31**, 253–266 (1991).
73. A. Asztalos, M. Daniels, A. Sethi, T. Shen, P. Langan, A. Redondo, S. Gnanakaran, *Biotechnol. Biofuels*, **5**, 55 (2012).
74. S. A. Maurer, C. N. Bedbrook, and C. J. Radke, *Ind. Eng. Chem. Res.*, (2012). DOI: 10.1021/ie3008538.
75. Q. Q. Wang, J. Y. Zhu, C. G. Hunt and H. Y. Zhan, *Biotechnol. Bioeng.*, **109**, 1965–1975 (2012).
76. G. T. Beckham, Y. J. Bomble, E. A. Bayer, M. E. Himmel and M. F. Crowley, *Curr. Opin. Biotechnol.*, **22**, 231–238 (2011).
77. L. T. Bu, G. T. Beckham, M. F. Crowley, C. H. Chang, J. F. Matthews, Y. J. Bomble, W. S. Adney, M. E. Himmel and M. R. Nimlos, *J. Phys. Chem. B.*, **113**, 10994–11002 (2009).
78. M. R. Nimlos, J. F. Matthews, M. F. Crowley, R. C. Walker, G. Chukkapalli, J. V. Brady, W. S. Adney, J. M. Clearyl, L. H. Zhong and M. E. Himmel, *Protein. Eng. Des. Sel.*, **20**, 179–187 (2007).
79. L. Zhong, J. F. Matthews, M. F. Crowley, T. Rignall, C. Talon, J. M. Cleary, R. C. Walker, G. Chukkapalli, C. McCabe, M. R. Nimlos, C. L. Brooks, M. E. Himmel and J. W. Brady, *Cellulose*, **15**, 261–273 (2008).
80. N. R. Gilkes, B. Henrissat, D. G. Kilburn, R. C. Miller and R. A. J. Warren, *Microbiol. Rev.*, **55**, 303–315 (1991).
81. A. B. Boraston, D. N. Bolam, H. J. Gilbert and G. J. Davies, *Biochem. J.*, **382**, 769–781 (2004).
82. L. R. Lynd, P. J. Weimer, W. H. van Zyl and I. S. Pretorius, *Microbiol. Mol. Biol. R.*, **66**, 506–577 (2002).
83. H. J. Gilbert, D. N. Bolam, L. Szabo, H. Xie, M. P. Williamson, P. J. Simpson, S. Jamal, A. B. Boraston, D. G. Kilburn, R. Anthony and J. Warren in *An update on carbohydrate binding modules*, Vol. Eds.: T. T. Teeri, B. Svensson, H. J. Gilbert and T. Feizi, (*Royal. Soc. Chemistry, Cambridge*, 2002) pp. 89–98.

84. P. Hagglund, T. Eriksson, A. Collen, W. Nerinckx, M. Claeyssens and H. Stalbrand, *J. Biotechnol.*, **101**, 37–48 (2003).
85. L. S. Wang, Y. Z. Zhang and P. J. Gao, *Sci. China. Ser. C-Life Sci.*, **51**, 620–629 (2008).
86. P. J. Gao, G. J. Chen, T. H. Wang, Y. S. Zhang and J. Liu, *Acta. Biochimica. Et. Biophysica. Sinica.*, **33**, 13–18 (2001).
87. R. Pinto, S. Moreira, M. Mota and M. Gama, *Langmuir.*, **20**, 1409–1413 (2004).
88. S. M. Southall, P. J. Simpson, H. J. Gilbert, G. Williamson and M. P. Williamson, *FEBS. Lett.*, **447**, 58–60 (1999).
89. K. Sorimachi, M. F. LeGalCoeffet, G. Williamson, D. B. Archer and M. P. Williamson, *Structure*, **5**, 647–661 (1997).
90. A. A. Klyosov, *Biochemistry*, **29**, 10577–10585 (1990).
91. J. F. Matthews, C. E. Skopec, P. E. Mason, P. Zuccato, R. W. Torget, J. Sugiyama, M. E. Himmel and J. W. Brady, *Carbohydr. Res.*, **341**, 138–152 (2006).
92. M. L. Rabinovich, V. V. Nguen, A. A. Klesov, *Biokhimiia*, **47**, 465–77 (1982).
93. H. Esterbauer, W. Steiner, I. Labudova, A. Hermann and M. Hayn, *Bioresour. Technol.*, **36**, 51–65 (1991).
94. R. A. Nieves, C. I. Ehrman, W. S. Adney, R. T. Elander and M. E. Himmel, *World J. Microbiol. Biotechnol.*, **14**, 301–304 (1998).
95. T. B. Vinzant, W. S. Adney, S. R. Decker, J. O. Baker, M. T. Kinter, N. E. Sherman, J. W. Fox and M. E. Himmel, *Appl. Biochem. Biotechnol.*, **91–3**, 99–107 (2001).
96. A. Goyal, B. Ghosh and D. Eveleigh, *Bioresour. Technol.*, **36**, 37–50 (1991).
97. B. Nidetzky and M. Claeyssens, *Biotechnol. Bioeng.*, **44**, 961–966 (1994).
98. J. O. Baker, C. I. Ehrman, W. S. Adney, S. R. Thomas and M. E. Himmel, *Appl. Biochem. Biotechnol.*, **70–2**, 395–403 (1998).
99. H. J. Lee and R. M. Brown, *J. Biotechnol.*, **57**, 127–136 (1997).
100. M. Linder and T. T. Teeri, *J. Biotechnol.*, **57**, 15–28 (1997).
101. J. Rouvinen, T. Bergfors, T. Teeri, J. K. C. Knowles and T. A. Jones, *Science*, **249**, 380–386 (1990).
102. C. Divne, I. Sinning, J. Stahlberg, G. Pettersson, M. Bailey, M. Siikaaho, E. Margollesclark, T. Teeri and T. A. Jones, *J. Mol. Biol.*, **234**, 905–907 (1993).
103. C. Divne, J. Stahlberg, T. Reinikainen, L. Ruohonen, G. Pettersson, J. K. C. Knowles, T. T. Teeri and T. A. Jones, *Science*, **265**, 524–528 (1994).
104. C. Divne, J. Stahlberg, T. T. Teeri and T. A. Jones, *J. Mol. Biol.*, **275**, 309–325 (1998).
105. M. Vrsanska and P. Biely, *Carbohydr. Res.*, **227**, 19–27 (1992).
106. B. K. Barr, Y. L. Hsieh, B. Ganem and D. B. Wilson, *Biochemistry*, **35**, 586–592 (1996).
107. K. Klemanleyer, E. Agosin, A. H. Conner and T. K. Kirk, *Appl. Environ. Microbiol.*, **58**, 1266–1270 (1992).
108. H. Henriksson, J. Stahlberg, R. Isaksson and G. Pettersson, *FEBS Lett.*, **390**, 339–344 (1996).
109. S. D. Mansfield, D. A. Swanson, N. Roberts, J. A. Olson and J. N. Saddler, *Tappi J.*, **82**, 152–158 (1999).

110. C. Tengborg, M. Galbe and G. Zacchi, *Biotechnol. Progr.,* **17**, 110–117 (2001).
111. L. Olsson and B. HahnHagerdal, *Enzyme Microb. Technol.*, **18**, 312–331 (1996).
112. Y. Sun and J. Y. Cheng, *Bioresour. Technol.*, **83**, 1–11 (2002).
113. S. I. Yun, C. S. Jeong, D. K. Chung and H. S. Choi, *Biosci., Biotechnol., Biochem.*, **65**, 2028–2032 (2001).
114. C. H. Decker, J. Visser and P. Schreier, *J. Agric. Food. Chem.,* **48**, 4929–4936 (2000).
115. M. Holtzapple, M. Cognata, Y. Shu and C. Hendrickson, *Biotechnol. Bioeng.*, **36**, 275–287 (1990).
116. J. S. Tolan, B. Foody, *Adv Biochem Eng Biotechnol.*, **65**, 41–67 (1999).
117. G. Panagiotou and L. Olsson, *Biotechnol. Bioeng.*, **96**, 250–258 (2007).
118. M. Cantarella, L. Cantarella, A. Gallifuoco, A. Spera and F. Alfani, *Biotechnol. Progr.*, **20**, 200–206 (2004).
119. J. B. Kristensen, J. Borjesson, M. H. Bruun, F. Tjerneld and H. Jorgensen, *Enzyme Microb. Technol.*, **40**, 888–895 (2007).
120. S. Kim and M. T. Holtzapple, *Bioresour. Technol.*, **96**, 1994–2006 (2005).
121. C. Tengborg, M. Galbe and G. Zacchi, *Biotechnol. Progr.*, **17**, 110–117 (2001).
122. H. Jorgensen and L. Olsson, *Enzyme Microb. Technol.,* **38**, 381–390 (2006).
123. Y. P. Lu, B. Yang, D. Gregg, J. N. Saddler and S. D. Mansfield, *Appl. Biochem. Biotechnol.,* **98**, 641–654 (2002).
124. A. Berlin, N. Gilkes, A. Kurabi, R. Bura, M. B. Tu, D. Kilburn and J. Saddler, *Appl. Biochem. Biotechnol.,* **121**, 163–170 (2005).
125. A. Singh, P. K. R. Kumar and K. Schugerl, *J. Biotechnol.*, **18**, 205–212 (1991).
126. A. H. C. Yu, D. Lee and J. N. Saddler, *Biotechnol. Appl. Biochem.,* **21**, 203–216 (1995).
127. F. Dourado, M. Bastos, M. Mota and F. M. Gama, *J. Biotechnol.*, **99**, 121–131 (2002).
128. R. Gaur, Lata and S. K. Khare, *World J. Microbiol. Biotechnol.*, **21**, 1123–1128 (2005).
129. C. A. Mooney, S. D. Mansfield, M. G. Touhy and J. N. Saddler, *Bioresour. Technol.,* **64**, 113–119 (1998).
130. L. Laureano-Perez, F. Teymouri, H. Alizadeh and B. E. Dale, *Appl. Biochem. Biotechnol.,* **121**, 1081–1099 (2005).
131. R. P. Chandra, R. Bura, W. E. Mabee, A. Berlin, X. Pan and J. N. Saddler in *Substrate pretreatment: The key to effective enzymatic hydrolysis of lignocellulosics?*, Vol. 108 (Ed. L. Olsson), (Springer-Verlag Berlin, Berlin, 2007), p. 67–93.
132. X. J. Pan, *J. Biobased Mater. Bioenergy.*, **2**, 25–32 (2008).
133. B. Yang and C. E. Wyman, *Biotechnol. Bioeng.*, **86**, 88–95 (2004).
134. S. I. Mussatto, M. Fernandes, A. M. F. Milagres and I. C. Roberto, *Enzyme Microb. Technol.*, **43**, 124–129 (2008).
135. S. D. Mansfield, E. deJong, R. S. Stephens and J. N. Saddler, *J. Biotechnol.*, **57**, 205–216 (1997).
136. L. Zhu, J. P. O'Dwyer, V. S. Chang, C. B. Granda and M. T. Holtzapple, *Bioresour. Technol.*, **99**, 3817–3828 (2008).

137. X. B. Zhao, L. Wang and D. H. Liu, *J. Chem. Technol. Biotechnol.*, **83**, 950–956 (2008).
138. X. J. Pan, N. Gilkes and J. N. Saddler, *Holzforschung*, **60**, 398–401 (2006).
139. S. Y. Ding, Q. Xu, M. K. Ali, J. O. Baker, E. A. Bayer, Y. Barak, R. Lamed, J. Sugiyama, G. Rumbles and M. E. Himmel, *Biotechniques*, **41**, 435 (2006).
140. Y. S. Liu, Y. N. Zeng, Y. H. Luo, Q. Xu, M. E. Himmel, S. J. Smith and S. Y. Ding, *Cellulose*, **16**, 587–597 (2009).
141. Q. Xu, M. P. Tucker, P. Arenkiel, X. Ai, G. Rumbles, J. Sugiyama, M. E. Himmel and S. Y. Ding, *Cellulose*, **16**, 19–26 (2009).
142. Y. H. P. Zhang and L. R. Lynd, *Biomacromolecules*, **6**, 1510–1515 (2005).
143. R. Pinto, J. Carvalho, M. Mota and M. Gama, *Cellulose*, **13**, 557–569 (2006).
144. T. Jeoh, C. I. Ishizawa, M. F. Davis, M. E. Himmel, W. S. Adney and D. K. Johnson, *Biotechnol. Bioeng.*, **98**, 112–122 (2007).
145. R. Gupta and Y. Y. Lee, *Biotechnol. Bioeng.*, 102, 1570–1581 (2009).
146. M. M. Nazhad, L. P. Ramos, L. Paszner and J. N. Saddler, *Enzyme Microb. Technol.*, **17**, 68–74 (1995).
147. B. Nidetzky, W. Zachariae, G. Gercken, M. Hayn and W. Steiner, *Enzyme Microb. Technol.,* **16**, 43–52 (1994).
148. C. A. Wilson, S. I. McCrae and T. M. Wood, *J. Biotechnol.,* **37**, 217–227 (1994).
149. J. G. Elkins, B. Raman and M. Keller, *Curr. Opin. Biotechnol.,* **21**, 657–662 (2010).
150. A. V. Gusakov, A. P. Sinitsyn, A. G. Berlin, A. V. Markov and N. V. Ankudimova, *Enzyme Microb. Technol.*, **27**, 664–671 (2000).
151. S. D. Mansfield, C. Mooney and J. N. Saddler, *Biotechnol. Progr.*, **15**, 804–816 (1999).
152. G. Banerjee, S. Car, J. S. Scott-Craig, M. S. Borrusch, M. Bongers and J. D. Walton, *Bioresour. Technol.*, **101**, 9097–9105 (2010).
153. G. Banerjee, S. Car, J. S. Scott-Craig, M. S. Borrusch and J. D. Walton, *Biotechnology for Biofuels*, **3**. 22 (2010).
154. A. Berlin, V. Maximenko, N. Gilkes and J. Saddler, *Biotechnol. Bioeng.*, **97**, 287–296 (2007).
155. S. P. S. Chundawat, V. Balan and B. E. Dale, *Biotechnol. Bioeng.*, **99**, 1281–1294 (2008).
156. P. Tomme, V. Heriban, M. Claeyssens. *Biotechnol. Lett.*, **12**, 525–530 (1990).
157. T. M. Wood, V. Garcia-Campayo, *Biodegradation*, **1**, 147–161 (1990).
158. S. D. Mansfield, J. N. Saddler, G. M. Gubitz, *Enzyme Microb Technol.*, **23**, 133–140 (1998).
159. L. P. Walker, D. B. Wilson, D. C. Irwin, *Biotechnol. Bioeng.*, **40**, 1019–1026 (1992).
160. H. P. Fierobe, A. Mechaly, C. Tardif, A. Belaich, R. Lamed, Y. Shoham, J. P. Belaich and E. A. Bayer, *J. Biol. Chem.*, **276**, 21257–21261 (2001).
161. H. P. Fierobe, E. A. Bayer, C. Tardif, M. Czjzek, A. Mechaly, A. Belaich, R. Lamed, Y. Shoham and J. P. Belaich, *J. Biol. Chem.,* **277**, 49621–49630 (2002).
162. V. Zverlov, S. Mahr, K. Riedel and K. Bronnenmeier, *Microbiology-Uk.*, **144**, 457–465 (1998).
163. D. H. Cybinski, I. Layton, J. B. Lowry and B. P. Dalrymple, *Appl. Microbiol. Biotechnol.,* **52**, 221–225 (1999).

164. C. B. Faulds, D. Zanichelli, V. F. Crepin, I. F. Connerton, N. Juge, M. K. Bhat and K. W. Waldron, *J. Cereal. Sci.,* **38**, 281–288 (2003).
165. M. Vardakou, P. Katapodis, E. Topakas, D. Kekos, B.J. Macris, P. Christakopoulos, *Innovative Food Sci Emerg Technol.*, **5**, 107–112 (2004).
166. P. Raweesri, P. Riangrungrojana and P. Pinphanichakarn, *Bioresour. Technol.*, **99**, 8981–8986 (2008).
167. S. D. McClendon, H. D. Shin and R. R. Chen, *Biotechnol. Lett.*, **33**, 47–54. (2011).
168. A. Gasparic, J. Martin, A. S. Daniel and H. J. Flint, *Appl. Environ. Microbiol.*, **61**, 2958–2964 (1995).
169. T. Hashimoto and Y. Nakata, *J. Biosci. Bioeng.*, **95**, 164–169 (2003).
170. P. Christakopoulos, P. Katapodis, D. G. Hatzinikolaou, D. Kekos and B. J. Macris, *Appl. Biochem. Biotechnol.*, **87**, 127–133 (2000).
171. H. R. Sorensen, S. Pedersen and A. S. Meyer, *Biotechnol. Progr.,* **22**, 505–513 (2006).
172. H. Yang, H. Ichinose, M. Nakajima, H. Kobayashi, S. Kaneko, *Food Sci. Technol. Res.*, **12**, 43–9 (2006).
173. G. M. Gubitz, M. Hayn, M. Sommerauer and W. Steiner, *Bioresour. Technol.,* **58**, 127–135 (1996).
174. H. Wang, H. Y. Luo, J. A. Li, Y. G. Bai, H. G. Huang, P. J. Shi, Y. L. Fan and B. Yao, *Bioresour. Technol.,* **101**, 8376–8382 (2010).
175. J. H. Clarke, K. Davidson, J. E. Rixon, J. R. Halstead, M. P. Fransen, H. J. Gilbert and G. P. Hazlewood, *Appl. Microbiol. Biotechnol.*, **53**, 661–667 (2000).
176. N. Beukes and B. I. Pletschke, *Bioresour. Technol.,* **102**, 5207–5213 (2011).
177. N. Beukes and B. I. Pletschke, *Bioresour. Technol.,* **101**, 4472–4478 (2010).
178. N. Beukes, H. Chan, R. H. Doi and B. I. Pletschke, *Enzyme Microb. Technol.,* **42**, 492–498 (2008).
179. K. Murashima, A. Kosugi and R. H. Doi, *J. Bacteriol.,* **185**, 1518–1524 (2003).
180. M. J. Selig, E. P. Knoshaug, W. S. Adney, M. E. Himmel and S. R. Decker, *Bioresource Technology.*, **99**, 4997–5005 (2008).
181. A. Kosugi, K. Murashima and R. H. Doi, *Appl. Environ. Microbiol.*, **68**, 6399–6402 (2002).
182. M. Alkasrawi, T. Eriksson, J. Borjesson, A. Wingren, M. Galbe, F. Tjerneld and G. Zacchi, *Enzyme Microb. Technol.*, **33**, 71–78 (2003).
183. J. Borjesson, R. Peterson and F. Tjerneld, *Enzyme Microb. Technol.,* **40**, 754–762 (2007).
184. S. B. Kim, H. J. Kim and C. J. Kim, *Appl. Biochem. Biotechnol.*, **130**, 486–495 (2006).
185. Y. Zheng, Z. L. Pan, R. H. Zhang, D. H. Wang and B. Jenkins, *Appl. Biochem. Biotechnol.*, **146**, 231–248 (2008).
186. B. Yang and C. E. Wyman, *Biotechnol. Bioeng.*, **94**, 611–617 (2006).
187. T. Eriksson, J. Borjesson and F. Tjerneld, *Enzyme Microb. Techno.*, **31**, 353–364 (2002).

CHAPTER 7

FERMENTATION TO BIOETHANOL/BIOBUTANOL

TAO MA*, MATYAS KOSA[†], QINING SUN*

**Institute of Paper Science and Technology,*
School of chemistry and biochemistry,
Georgia Institute of Technology, Atlanta, GA 30332, USA
*taoma@gatech.edu, taoma.anson@gmail.com, sunqn*515658@gmail.com
[†]*Renmatix Inc. 660 Allendale Road, King of Prussia, PA 19406*
matyas.kosa@renmatix.com

The conversion of biomass into biofuels has drawn considerable attention as a promising alternative for fossil fuels. In general sense, fermentation refers to the biochemical transformation from organic substances to simpler compounds by utilizing the activities of enzymes in anaerobic environment. While in industry, fermentation is a prevalent technology which converts simple sugars to alcohols (ethanol, butanol and also propanols, etc.) and other value-added products through the metabolisms of microorganisms such as yeasts and certain species of bacteria. As an important stage in biomass conversion, biofuel fermentation has been advanced significantly towards the technologies that enable commercialization. This chapter provides a comprehensive overview of the current status and prospects of biofuel fermentation (mainly bioethanol and biobutanol), including the basic concepts, microorganisms and technologies.

1. Introduction

Research on biofuels (bioethanol) production started in Germany as early as 1898 and was intensified in late 1980s when the oil price dramatically increased.[1] Biofuel production was first commercialized from sugar, starch and oil-seed based feedstocks and are named as first-generation biofuels.[2]

Continuous biofuel production from many of these feedstocks will generate serious competition and debate for food resources. Therefore, sustainable and non-food feedstock should be used for further biofuel production. The advent of second-generation biofuels expand the feedstocks to other resources like lignocellulosic biomass, including agri-residues (e.g., corn stover, wheat and straws), agro-industrial wastes, forestry and energy crops (e.g., poplar and switch grass), which are abundant, inexpensive and renewable nonfood feedstock.[2,3] Lignocellulosic biomass accumulates in huge quantities annually and cause significant environmental problems.[3] US has the potential of producing 1.3 billion dry tons of biomass every year, which could substitute one third of the transportation fuel demands.[4] Thus efficient utilization of these materials for production of biofuels and other valuable products will not only provide energy benefits but also helps to meet environmental concerning.

Multiple pathways are available for the conversion from biomass to fuels, especially from lignocellulosic biomass. Currently, biochemical and thermochemical pathways are the two predominant approaches employed for biofuel production from biomass. In biochemical conversions, biomass is pretreated by chemical or physical methods (e.g., acid, alkaline and steam explosion) so that plant polysaccharides are more accessible to enzyme hydrolysis in order to obtain fermentable sugars, which are then fermented to biofuels. This is also known as direct fermentation.[1,6] Thermochemical pathway starts with the gasification of biomass into synthesis gas or syngas in short, which is a gas mixture consisting of mainly CO, CO_2 and H_2. The syngas is then converted to liquid fuels through Ficher-Tropsch (FT) synthesis by using metal catalysts such as ferrous (Fe), copper (Cu), molybdenum (Mo), aluminum (Al), Zinc (Zn), nickel (Ni), rubidium (Ru) and ruthenium (Rh) or fermentation using microbial catalysts. The conversion from syngas to biofuels by using microorganisms is indirect fermentation.[1,2,4,7–9]

In many fermentation pathways, a variety of products released from hydrolysis or fermentation can act as inhibitors to microorganisms for biofuel production reactions. They can be categorized into three major groups: weak acids (acetic, formic, octanoic and levulinic acid), furan derivatives and phenolic compounds.[10,11] Removing these unwanted inhibitory compounds and preventing their accumulation in the fermentation broth are necessary to increase the biofuel yield and production rate. To improve the fermentation yield and productivity, the generation and identification of various inhibitors, their inhibiting mechanisms and advanced technologies are broadly investigated.

Current research is more focused on fermentation to ethanol due to its extensive applications in industry. Interest in biobutanol keeps growing for its similar characteristics to gasoline, which makes biobutanol a most suitable transport fuel alternatives.[12,13] Compared with ethanol, butanol offers greater advantages as biofuel. Butanol contains 22% oxygen and has higher energy content and low vapor pressure. It is more hydrophobic than ethanol and is able to be directly blended with gasoline in any proportion without specific requirements to engine modifications.[12–17] Documents demonstrated that ethanol can only be mixed with gasoline up to 10% in order engines and up 85% in E85 compatible engines that power flex-fuel vehicles. N-Butanol can be blended up to 100% in unmodified 4-cycle ignition engines, 30% in diesel compression engines or to 20% (80% kerosene) in a jet turbine engine.[13,18,19]

To make the best use of lignocellulosic feedstocks, both cellulose and hemicellulose need to be converted to fermentable sugars. In other words, pentose sugars such as xylose also have to be utilized in the biofuel production. This is due to the fact that is the second major carbohydrate component in most lignocellulosic biomass.[20]

Production of ethanol from lignocellulosic biomass is more complicated because the carbohydrates are more difficult to solubilize than starch in grain.[20] The process includes Separate Hydrolysis and Fermentation (SHF), in which enzyme hydrolysis and fermentation take place sequentially, and Simultaneously Saccharification and Fermentation (SSF), where the hydrolysis and fermentation occur in one step.[21] Furthermore, if the fermentation of hexose and pentose takes place simultaneously, it is a more consolidated approach known as simultaneous saccharification and co-fermentation (SSCF). The process which tries to combine the cellulase production, cellulase hydrolysis and fermentation into one single process is consolidated bioprocessing (CBP), which is the direction of biomass processing technologies.[22] The alternative fuel butanol is mainly produced via acetone-butanol-ethanol (ABE) fermentation from renewable resources using bateria strains like *Clostridium acetobutylicum* or *Clostridium beijerinckii* in anaerobic conditions.[12,13,16] In other studies, both ethanol and butanol can be produced via syngas fermentation.[4,21]

Fermentation pathways are advantageous because of highly specific reactions which can be directed to production of desired valuable chemicals. The major challenges involved in fermentation processes are the requirements to work with living microorganisms in commercial scale, which also demand tight temperature, pH and other operating conditions.

2. Biochemical Fermenting Microorganisms and Developments

The conversion of sugars to alcohols such as ethanol and butanol can be done in multiple fermentation schemes using various microorganisms. For industrial purpose, ethanol/butanol has to be the major end product of fermentation. Different groups of microorganisms have been investigated to achieve the target, including yeasts, mesophilic and thermophilic bacterias and filamentous fungi. All these microbial fermentation processes require the energy to feed the organisms, which comes from the biomass in the forms of sugars.[23] They ferment carbohydrates to ethanol/butanol under anaerobic conditions.

Yeasts are eukaryotic microorganisms most suitable for ethanol production because they can grow on a variety of sugars and have relative high substrate and ethanol tolerance.[24] One of the best-studied yeast for ethanol production is *Saccharomyces cerevisiae*, which are used in industry to convert hexoses to ethanol through Embern-Meyerhoff (glycolytic) pathway.[1,24–28] While many industrial yeast strains are not able to ferment pentose (mainly xylose), which are the major components in hemicelluloses, metabolic engineering is needed to obtain pentose fermentable yeast strains.[29]

Many bacteria systems such as *Escherichia coli* can ferment lignocellulosic-derived sugars to ethanol, butanol and acids.[29–32] One good thing about bacteria is that some bacterium family members can convert both hexose and pentose to bioethanol. Besides, biotransformation from cellulose to ethanol can also be implemented by a variety of thermophilic anaerobic bacteria such as *Clostridium thermocellum*[33,34] and filamentous fungi, which includes *Fusarium oxysporum,*[24] *Neurospora crassa,*[35] *Monilia sp.*[36] *Aspergillus sp.*[37] *Zygosaccharomyces rouxii*[38] and *Paecilomyces sp.*[39]

General criteria for ideal micro-organisms include broad substrate utilization, high efficiency to convert both hexose and pentose to biofuels, high biofuel yield and productivity, robust growth medium and high tolerance to inhibitors. However, there are no universal microbes satisfying all the features. Therefore, metabolic engineering of the well-studied micro-organisms is an efficient approach to improve the biofuel production.

2.1. *Yeast-Saccharomyces cerevisiae*

S. cerevisiae is the most common natural fermentative yeast for ethanol and has been almost exclusively employed in industrial for ethanolic production from sugars and starch, since it is much more robust than the other yeasts to perform well under large-scale conditions. It fulfils most of the required properties for industrial ethanologenic organism.[40,41] It produces ethanol with near stoichiometric yields from glucose and the process tolerates high substrate and ethanol level, a wide spectrum of inhibitors and osmotic pressure.[1,40,42] The stoichiometry of this anaerobic reaction is described in the simplest form as the following.[43,44]

$$C_6H_{12}O_6 \rightarrow 2C_2H_5OH + 2CO_2 + \text{energy}$$
$$C_6H_{12}O_6 + 2Pi + 2ADP \rightarrow 2C_2H_5OH + 2CO_2 + 2ATP + 2H_2O$$

About 90~95% of the theoretical ethanol yield can be achieved while the rest of the glucose were consumed by cell mass synthesis, cell maintenance and side reactions which produce lactic acids, acetic acids and glycerols.[45]

In *S. cerevisiae*, pyruvate is produced through glycolytic pathway (also known as Embden-Meyerhoff pathway or EMP) and converted to ethanol via pyruvate decarboxylase/alcohol dehydrogenase enzymes.[1,44] The EMP-pyruvate to ethanol conversion fermentation can be divided into three phases. It starts with the conversion of glucose to fructose-1,6-diphosphate, which is known as *activation phase*. Two reactions with ATP (energy) consumption are involved to add phosphate groups to sugar molecules so that they can be activated or destabilized for further processing. The following conversion from fructose-1,6-diphosphate to pyruvate is *ATP generating phase*. In this phase, NAD^+ is converted to NADH by triose phosphate dehydrogenase and ATP is generated from two reactions catalyzed by 3-phospoglycerate kinase (PGK) and pyruvate kinase (PK). The amount of ATP generated doubles that consumed in the activation phase. Therefore, the energy released from this phase not only compensate the energy used previously, but also produce energy benefits. The last phase converts pyruvate to ethanol and CO_2. Another significant feature in this phase is the regeneration of NAD^+ from NADH produced in ATP generation phase and thus enable the pathway circulation.[46]

Sugar transportation to yeast cells has been determined to be the rate-limiting step of glycolysis.[47] *S. cerevisiae* has a complex yet efficient sugar transportation system with 20 genes to transport hexose on the genome. The transportation rates between genes differ and can be classified to low-, moderate- and high-affinity transporters. All the transporters facilitate glucose transport and the transportation needs to be activated by a concentration gradient across the plasma membrane.[44,48] The other major sugar, sucrose, is hydrolyzed to glucose and fructose. Fructose enters the EMP pathway through the enzyme catalytic reactions.[44] *S. cerevisiae* always metabolizes glucose and sucrose first in industrial applications as they are the two most preferred substrates for metabolism of *S. cerevisiae.*[44,49]

The major drawback of *S. cerevisiae* for lignocellulosic fermentation is its inability to metabolize pentose sugars such as xylose and arabinose. Genetic engineering strategies of *S. cerevisiae* for pentose fermentation are well developed and the performance is evaluated in depth in relation to the industrial applications.

2.2. *Bacteria-Zymomonas mobilis*

Z. mobilis is a gram-negative, rod-like bacterium and notable for its bioethanol production capabilities. Althrough *S. cerevisiae* are widely used for bioethanol production, *Z. mobilis* has its own advantages. The ethanol production rate from *Z. mobilis* is much faster than that of *S. cerevisiae* and thus more tolerant to high concentration of substrates and it tolerates up to 120 g/L ethanol.[44,50–52] *Z. mobilis* metabolize glucose, sucrose and fructose to pyruvate through the Entner-Doudoroff pathway.[1,44] The pyruvate is then fermented to ethanol and carbon dioxide as the end products (analogous to yeasts).

Unlike glycolytic pathway which theoretically produces two ATP from one glucose molecule, the ED pathway yields only one ATP per glucose moluecule.[45] This explains the high catabolic rate in *Z. mobilis*.[52] Previous studies of ED pathway employed by *Z. mobilis* demonstrate that the stages catalyzed by glucose-6-phosphate dehydrogenase (GDPH) and phosphoglycerate mutase (PGM) are the rate-limiting steps.[53] The most distinct difference between the ED pathway used by *Z. mobilis* and EMP pathway employed by *S. cerevisiae* is the absence of the step catalyzed by the enzyme phosphofructokinase (PFK), which is a key enzyme and tightly regulated in yeasts. This allows the ethanol production and ATP generation to be uncoupled in *Z. mobilis*.[44,54] Therefore, cell growth is not mandatory for ethanol production and ethanol yield close to theoretical value can be achieved because it is not sensitive for cell concentrations.[55] In summary, the high ethanol tolerance, high sugar uptake, high ethanol yield and production rates, no requirements for aeration and its ease for genetic manipulation makes *Z. mobilis* a competitive microorganism to *S. cerevisiae*.[44,56,57]

Despite of so many attractive benefits, *Z. mobilis* is still not widely used in industry due to several inherent deficiencies which make it undesirable. The most important reason is that *Z. mobilis* is limited to glucose, fructose and sucrose. Wide-type *Z. mobilis* can't ferment pentose sugars like xylose and arabinose which are important components in hemicelluloses.[44] Secondly, sucrose also has to be converted to glucose and fructose by the catalysis of levansucrase, which also facilitate fructose polymerization to levan and possibly other polymer products.[58] The formations of levan and other polymers not only compete with ethanol production, but also cause serious problems in the reactor such as fouling of the distillation columns and other equipments.[44] In addition, *Z. mobilis* can't tolerant inhibitors which present in lignocellulosic substrates such as acetic acids or phenolic compounds. The fermentation media for *Z. mobilis* tends to stabilize under pH~4.5. However, most potential contaminating bacteria is not efficiently inhibited at pH~4.5. This results in costly requirements for sterilization of the system to ensure high ethanol yield.[57]

Nevertheless, abundant research efforts have been made to overcome its disadvantages. Recombinant *Z. mobilis* strains have been invented for ethanol production to extend the range of substrate utilization spectrum.[56]

Acetic acid resistance is also improved for some engineered strains of *Z. mobilis* by mutation techniques.[59,60] All the efforts make *Z. mobilis* of high interest to be a promising microorganism for ethanol fermentation.

S. cerevisiae and *Z. mobilis* are model-organisms yeast and bacteria microorganisms for bioethanol fermentation in biochemical pathways, respectively. In both microorganisms, sugar is transported to the cell; pyruvate is then produced by glycolytic pathway in *S. cerevisiae* or Entner-Douoroff pathway in *Z. mobilis* followed by its fermention to ethanol via pyruvate decarboxylase and alcohol dehydrogenase enzymes. However, both of these native microorganisms can work only on hexose substrates from cellulosic hydrolysates. *Z. mobilis* is even not materialized in industry due to the absence of its pathways for the metabolism of mannose and galactose. Hemicellulose, the precursor of pentose sugars like xylose and arabinose, is also a major biomass energy resource present in lignocellulosic plants (Table 1).[20] Fermentation of hemicellulose is usually much more complicated than the fermentation of cellulose due to the presence of pentose sugars. Although multiple microorganisms are found to be naturally capable to metabolize xylose, including anaerobic fungi, a certain groups of mesophilic and thermophilic bateria (e.g., *Thermoanaerobacterials*) and the yeast *Pichia stapis*, there are not many natural organisms suited for industrial fermentation of pentose sugars.[1,61] Due to the demand of microorganisms capable of metabolizing hexose and pentose sugars simultaneously, research efforts directing the use of this precursor of pentose sugars and the expansion of more fractions of lignocellulosic raw materials are never lacking. Genetic engineering aiming at industrial hexose- and pentose-fermentative strains is well developed and will be discussed.

2.3. *Genetically engineered microorganisms*

As mentioned earlier, many industrial yeast strains and other microorganisms lack pentose-fermenting pathways. The status of metabolic engineering of yeast/bacteria/fungi strains for mixed sugar utilization is reviewed by Hahn-Hägerdahl *et al.*,[29] A. K. Chandel *et al.*[24] and B. S. Dien *et al.*[52] Basically, two approaches have been developed to integrate hexose and pentose fermentation. The first approach is the construction of recombinant

Table 1. Percent dry weight compositions of various biomass. (Wayman, 1996, adapted from Ref. 20).

Feedstock	Glu	Man	Gal	Xyl	Ara	Lignin	Ext.	Ash
Tree Species								
Hybrid Poplar	48.6	0.5	0.3	14.6	0.3	21.8		0.7
Poplar	49.9	4.7	1.2	17.4	1.8	18.1		0.5
White Oak	43.6	2.9	0.4	18.0	2.4	23.2		0.6
Red Oak	43.4	2.7		18.9	1.9	25.8		0.4
Walnut	46.2	2.6		16.5	1.8	21.9		1.0
Maple	44.9	2.9		17.3	2.8	20.7		0.6
Paper Office Paper	68.6	7.8		12.4		11.3		
Herbaceous Species								
Corn stalkes + Cobs	36.4	0.6	1.0	18.0	3.0	16.6	7.3	9.7
Corn Stover	40.9		1.0	21.5	1.8	16.7		6.3
Bagasse	40.2	0.3	0.5	21.1	1.9	25.2	4.4	4.0
Wheat Straw	38.2	0.3	0.7	21.2	2.5	23.4	13.0	10.3
Rice Straw	34.2			24.5		11.9	17.9	16.1
Switch grass	31.0	0.3	0.9	20.4	2.8	17.6	17.0	5.8

Blank indicates no data available.

strains by inserting genes coding pentose transportation and metabolism, while the second one is aiming at improving biofuel yield through genetic engineering on micro-organisms which already have capabilities to co-ferment hexose and pentose sugars.[52,62,63] Directed by the initiation and objectives of these two approaches, the development strategies for engineering microorganisms usually include the following steps:[1,44,64–66]

- Selection of a native host, which either has good biofuel production capability but limited sugar utilization range or is a poor biofuel producer but capable to metabolize a wide spectrum of biomass-derived sugars.
- Inserting plasmids with genes which have desired capabilities and any possible additional requirements (improving the biofuel production,

expand the utilization of biomass-derived sugars and improving the genetic stability) to the selected host.
- In-laboratory experimental confirmation of the performance of recombinant strains.
- Further modification to the biosynthesic pathways to improve the biofuel production and suppress the yield of by-products.

Another advanced approach is the development of organisms which can directly metabolize cellulose and hemicellulose, with no need of enzymatic hydrolysis treatments. This is also recognized as the most promising way for biofuel production, which will be discussed later.[44]

The genetically modified strains for industrial application must have, but not limited to the following properties:[29,63]

- Capability to utilize a wide spectrum of sugars.
- High tolerance for substrates, inhibitors and alcohols.
- High ethanol yield and production rate.
- High specificity for ethanol (or other alcohols) production.

2.3.1. *Pentose metabolism in yeast, bacteria and fungi*

The common initial metabolic pathways for D-xylose and L-arabinose in bacteria, yeast and filamentous fungi are summarized as follows.[24,29,67] In metabolism, xylose and arabinose are transported across the cell membrane and initially converted to xylose-5-phosphate (X-5-P), which then connects the pentose metabolism to the central metabolic pathways such as Embden-Meyerhof-Parnas (EMP) or Entner-Doudoroff (ED) pathways through pentose phosphate pathway (PPP). Within the PPP, X-5-P is converted to glycolytic intermediates such as glyceraldehyde-3-phosphate and fructose-6-phosphate, both of which are then incorporated into central metabolic pathways mentioned above to produce pyruvate and give origin to ethanol through decarboxylation and reduction.[20]

The initial conversion of D-xylose and L-arabinose in yeast and fungi takes place by a flow of reduction and oxidation reactions. D-xylose is directly reduced to xylitol by NAD(P)H-dependent D-xylose reductase (XR). While L-arabinose is first reduced to L-arabitol and then converted to xylitol through a series of oxidation/reduction steps. Xylitol is oxidized to D-xylulose by NAD^+ dependent xylitol dehydrogenase (XDH) and subsequently phosphorylated to D-xylulose-5-phosphate.[29,68–70]

In most bacteria, the metabolism of D-xylose and L-arabinose proceeds via isomerisation to D-xylulose and L-ribulose catalyzed by xylose isomerase (XI) and L-arabinose isomerase, respectively. D-xylulose and L-ribulose are then phosphorylated to D-xylulose-5-phosphate and L-ribulose-5-phosphate; the latter is then epimerized to D-xylulose-5-phosphate.[29,71,72]

2.3.2. *Metabolic engineering of yeast strains*

S. cerevisiae is used universally for commercial production of ethanol from biomass feedstocks. Although it owns genes for utilization of xylose, they don't support the growth of *S. cerevisiae* on xylose. However, it has been shown that it can ferment xylulose. Therefore, attempts have been made on introducing the genes for xylose isomerase (XI) from a xylose fermenting system to the yeasts, but the results were disappointing because it failed to express functionally as desired or the reaction rate was insufficient to achieve industrially competitive levels.[29,44,73–75]

An alternative approach was taken by the introduction of *XYL1* and *XYL2* genes encoding xylose reductase (XR) and xylitol dehydrogenase (XDH) from the xylose metabolizing yeast *Pichia stipitis* to *S. cerevisiae*, which resulted in its growth on xylose. Xylulose kinase that already existed in *S. cerevisiae* was also overexpressed to alleviate possible imbalance between XR and XDH, which can cause the accumulation of xylitol and thus reduce the ethanol yield.[20,64,76]

On research group at Purdue University has created recombinant yeast strains *Saccharomyces* 1400(pLNH33) and other pLNH transformants by integrating multiple copies of the *XYL* genes into the yeast chromosome. It was able to produce 50 g/L ethanol by co-fermenting a synthetic mixture of glucose and xylose at 50 g/L of each.[44,64] Another successful strain they have developed, *Saccharomyces* 424A(LNH-ST), can efficiently ferment glucose and xylose present in hydrolysates from different cellulosic biomass to ethanol at the yield of around 80% of theoretic values.[77]

Another group from Delft University developed strain *S. cerevisiae* RWB217 by integrating plasmids encoding the xylose isomerase (XI) from the anaerobic fungus *Piromyces* sp. E2 to *S. cerevisiae*. They also overexpressed all the enzymes required for conversion of xylulose to the intermediates of the glycolytic pathway and deleted the genes responsible for xylitol synthesis. This strain they developed demonstrated 82% ethanol yield of the theoretic value and the xylitol production was reduced but not eliminated.[75]

While most engineering strategies were established to improve the xylose fermentation, many other strategies were created aiming at arabinose utilization.[78–80] Delft University developed the strategy by transforming *S. cerevisiae* strains with genes for arabinose metabolism from *Lactobacillus plantarum* and overexpressing those for non-oxidative pentose phosphate pathway in selected host. The created strain *S. cerevisiae* IMS0002 was able to ferment arabinose and produce ethanol at 0.29 g/g dry cell weight-hour and at 84% theoretic value.[44,81]

Genetic engineering is not only limited to the involvement of xylose and arabinose utilization pathways, further improvements have been made toward other aspects, including random mutagenesis,[82] evolutionary engineering and breeding.[83] All strains are genetically modified to harbor the *P. stipitis* XR and XDH enzymes and overexpress the endogenous XK.[29,40,64,77,82–85]

2.3.3. *Engineering of Z. mobilis for xylose and arabinose metabolism*

Zymomonas mobilis receives considerable attention due to its effectiveness in sugar uptake and ethanol yield, high ethanol tolerance and amenability for genetic manipulation.[44,86] However, wide type strains of *Z. mobilis* only utilize glucose, fructose and sucrose. Genetic modification is needed to broaden its utilizable range of substrates.[86] Over the last decade, the first recombinant *Z. mobilis* strain capable of fermenting xylose and arabinose was developed by researchers at the National Renewable Energy Laboratory (NREL). The metabolism was achieved by the introduction and expression of four E. coli genes: xylose isomerase (*xylA*), xylulose kinase (*xylB*), transketolase (*tktA*), and transaldolase (*talB*) in a *Z. mobilis* host.[44,52,87,88] The proposed xylose metabolizing pathway consists of several steps as follows.[44] Xylose is converted to xylulose-5-P by xylose isomerase and xylulose kinase. In the following steps, transketolase and transaldolase convert xylulose-5-P to fructose-6-P and Glyceraldehyde-3-P, which are intermediates of the subsequent ED pathway.[52] The genes were expressed on a plasmid using enolase or glyceraldhyde-3-phosphate dehydrogenase, which are strong constitutive promotors from *Z. mobilis* host.[44,52] The resultant strain CP4 (pZB5) can simultaneously ferment glucose and xylose with 86% ethanol yield of theoretic value.[52]

The same strategy was applied to engineer *Z. mobilis* strain for arabinose fermentation. In this constructed strain, five genes isolated from *E. coli* were incorporated into the host using a plasmid. They are L-arabinose isomerase (*araA*), L-ribulose kinase (*araB*), L-ribulose-5-phosphate-4-epimerase (*araD*), transketolase (*tktA*) and transaldolase (*talB*).[44,52] The first three enzymes convert arabinose to xylulose-5-phosphate, while the last two are responsible for the transformation from xylulose-5-phosphate to ED pathway intermediates. The resulting strain, ATCC39676 (pZB206), was able to successfully ferment 25 g/L arabinose to ethanol with 98% ethanol yield of theoretical value. However, the rate of this arabinose fermentation was rather low compared with that of xylose fermenting strain.[52]

Originated from the initial achievements on engineering of *Z. mobilis* strains, Zhang and collaborators developed many other *Z. mobilis* strains. The strain ATCC39767 was transformed with plasmid carrying genes for xylose fermentation ability.[88] This strain possesses good capability to grow on hydrolysate from yellow poplar wood and is an efficient candidate for biofuel production. The problem with 39767 strain is its low tolerance to inhibitors, especially acetic acids.[89,90] Later ATCC39767 strain was adapted by Lawford *et al.* to increase its tolerance to acetic acid and other inhibitors by around five-month culturing of the strain in continuous high levels of hydrolysates from dilute acid pretreated (DAP) Yellow Poplar wood.[90] Zhang and collaborators also made considerable improvements on their *Z. mobilis* strains. The relatively new strain AX101 (parent strain ATCC 39676) can ferment both xylose and arabinose by carrying seven recombinant genes in its chromosomal DNA.[52] Unfortunately, AX101 strain still shows very low tolerance to acetic acid. Improving the tolerance of acetic acid seems to be a challenging part for *Z. mobilis* strains engineering.[52]

2.3.4. *Engineering of* Escherichia coli *for ethanol/butanol production*

Compared with *S. cerevisiae* and *Z. mobilis* strains, *E. coli* is naturally capable to metabolize a wide spectrum of sugar substrates to ethanol.[52,91] Under anaerobic conditions, the conversion is completed via a heterofermentative process: pyruvate is converted to acetyl-CoA and formate by the catalysis of pyruvate formate lyase (*Pfl*). Acetyl-CoA is then reduced to ethanol by a bifunctional ethanol dehydrogenase (*AdhE*).[44] The major problem of this native pathway is that the ethanol yield with this organism is pretty low.[44,52,91] Ethanol production is always accompanied by the formation of formate and acetate as unwanted products. In addition, this pathway is unbalanced because two NADH for each pyruvate are consumed to convert pyruvate to ethanol, while only one NADH is produced from the sugar. *E. coli* balances it by itself by producing acetic and succinic acids.[52,91,92] Thus, an engineered *E. coli* strain was developed by the introduction of genes encoding pyruvate decarboxylase (*Pdc*) and alcohol dehydrogenase II (*AdhII*) from *Z. mobilis*, both of which are expressed under the regulation of the native *lac* promoter.[91–93] The resulting strain was capable to generate ethanol as its primary fermentative products. The construct built in this approach was named the PET (production of ethanol) operon. To further improve the ethanol production, the endogenous fumarate reductase (*frd*) genes which are responsible for succinate formation were also disrupted. The final mutant *E. coli* KO11 strain has been tested in both synthetic sugar solutions and actual hydrolysates from biomass. The strain has been found to be very stable in batch culture and can produce ethanol up to 45–50 g/L, which is nearly 90% of the theoretical yield.[44,91] The native and engineered fermentation pathways are shown in Fig. 1.[44,52]

Another distinct feature of *E. coli* strain engineering is its potential for butanol fermentation.[91] The recombinant strain for *n*-butanol production was developed by introducing genes for butanol fermentation from *C. acetobutylicum* to the host, including *Thl*, *Hbd*, *Crt*, *Bcd* and *AdhE2* genes.[91] To further increase the butanol yield, *Thl* gene was substituted by *E. coli atoB* gene, and the biobutanol productivity increased over three-fold.[94] Although the observed biobutanol productivity is still significantly lower than that in native *Clostridia* (1.2 g/L), which is a common organism towards butanol fermentation, it provides an opportunity for further engineering.[91,95]

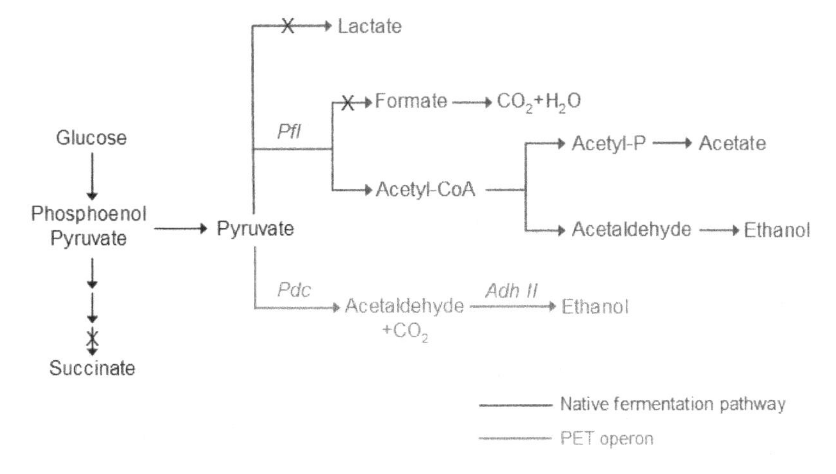

Fig. 1. Native heterofermentative pathway and engineered PET operon for ethanol production in *E. coli*. (Adapted from Refs. 44 and 52).

2.3.5. *Engineering* K. oxytoca *for ethanol production*

Klebsiella oxytoca is a gram-negative, rod-shaped bacterium which can grow on a variety of wood sources and show promise in industrial ethanol production.[52] This microorganism is of particular interest because the PET operon can also be introduced to *K. oxytoca* so it is able to metabolize monomeric sugars (hexoses and pentoses), as well as dimeric (cellobiose) and trimeric (cellotriose) sugars.[44,52] Therefore, this characteristic makes the strain appealing due to the economic advantages for the solubilization of cellulose, because it requires much fewer enzymes for the cellulose pretreatments. Another advantage of *K. oxytoca* is that it can grow at pH ~5.0 and 35°C, which is optimized for cellulose commercialization.[52] Ethanol is naturally synthesized under *Pfl* pathways like in *E. coli*, as shown in Fig. 1.[44] The ethanol fermentation engineering was implemented on the *K. oxytoca* strain M5A1 by transforming plasmids with PET operon into the strain chromosome. The final integrated strain was able to primarily produce ethanol at nearly 90% theoretical yield.[44,96] The best developed strain *K. oxytoca* P2 was able to consume 100 g/L cellubiose and produce 45 g/L ethanol at 88% theoretical yield without β-glucosidase.[44] It has been tested successfully with synthetic mixed sugars and actual hydrolysates from a variety of feedstocks, including mixed office paper, corn fiber, sugar beet pulp and sugarcane bagasse.[52] In addition, efforts were also made on utilizing cellulose by *K. oxytoca* P2 strain by inserting genes encoding endoglucanase from *E. chrysanthemi* into the chromosome of *K. oxytoca* P2 with the promoter from *Z. mobilis* for expression. The ethanol yield was improved by 22%.[97] All the results are encouraging for *K. oxytoca* as a promising candidate for commercialization.

We have discussed some of the leading microorganisms for ethanol production through biochemical fermentation, as well as their metabolic pathways and engineering strategies. There are plenty of other microbes which are considered promising for bioethanol fermentation. *Pichia stipitis* is the highest known yeast which is capable to ferment xylose and convert it to ethanol.[98] *Bacillus stearothermophilus* and *Thermoanaerobacter mathranii* are well known thermophilic strains for ethanol fermentation, and they provide several advantages such as low ethanol recovery cost, elimination of contamination and the potential to reduce ethanol inhibition.[44] Efforts were also made to improve their ethanol productivity.

Besides the microbes used in biochemical fermentation, many microorganisms, including mesophilic and thermophilic were also developed for thermochemical fermentation (biomass-derived syngas fermentation). The dominant organisms which are mostly studied are *Clostridium aceticum, Acetobacterium woodii, C. carboxydivorans* and *C. ljungdahlii*, etc.[2,4,99,100] In these mesophilic organisms, *C. ljungdahlii* is the most widely used homoacetogenic microorganism which can grow at a temperature around 37~40°C and pH 5.8~6.0. It is capable to produce ethanol as high as 48 g/L at pH 4.0~4.5 in a continuous stirred-tank reactor (CSTR).[4,101] *C. carboxydivorans* was found to be able to produce ethanol at 37°C and pH 5.75 in a bubble column reactor.[102] Thermophilic microorganisms for biomass-derived syngas fermentation include *Moorella thermoacetica, M. thermoautotrophica, Desulfotomaculum carboxydivorans* and *Carboxydocella sporoproducens* and the optimal operating temperature ranges 55~58°C.[2,4,7]

All the microorganisms reviewed above are focusing on bioethanol fermentation. Butanol is another excellent biofuel which can be produced from acetone-butanol-ethanol (ABE) fermentation. The development of microorganisms for biobutanol production also receives great attention. Among a variety of the microorganisms which can convert biomass feedstocks to butanol, the most commonly used strains are *Clostridium acetobutylicum* and *Clostridium beijerinckii*.[103,104] The advantage of these strains is that they can naturally utilize the hydrolysates from both hexose and pentose sugars.[13] Details about biomass-derived syngas ethanol fermentation and ABE fermentation for butanol production and their fermenting microorganisms will be discussed later in this chapter.

3. Direct Ethanol Fermentation Processing Strategies

The ethanol production from lignocellulosic biomass can be achieved by various coupling of the following major biological processes: (1) enzyme production; (2) pretreatments and enzymatic hydrolysis; (3) the fermentation of hexose and (4) the fermentation of pentose sugars.[105] Several factors have to be considered before the industrial process integration, including the feedstock availability/transportation cost, technical efficiency for every processing unit and their combination, economical feasibility and waste treatments and water cycle.[44] The processing configurations distinguished in research and industry are illustrated in Fig. 2.[106]

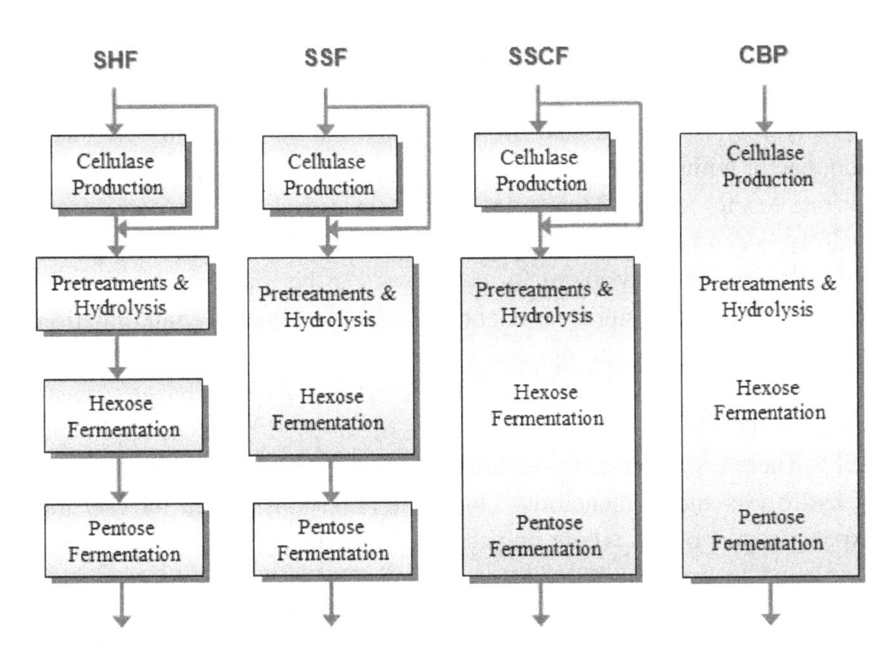

Fig. 2. Processing strategies in different configurations for ethanol production from lignocellulosic biomass. (Adapted from Ref. 106)

3.1. *Separate hydrolysis and fermentation (SHF)*

The simplest configuration is separate hydrolysis and fermentation (SHF). All the stages are physically separated. A small portion of the biomass feedstocks are used for enzyme production for subsequent processing. Then the enzymes enter the pretreatments and hydrolysis stage with the majority of the biomass. The stream from hydrolysis reactor then enters the fermenter where microbes are added to convert the hexose and pentose sugars to ethanol. The produced ethanol is then separated from the fermentation broth. The major advantage of SHF is the possibility to perform each step at its optimal conditions by adding different organisms according to the requirements.[105] Thus the overall efficiency may be higher than SSF (or SSCF). However, the sugars released from hydrolysis in the SHF inhibit enzymes significantly.[105,107]

3.2. *Simultaneous saccharification and fermentation (SSF)*

The simultaneous saccharification and fermentation (SSF) combines the pretreatments/hydrolysis and the hexose fermentation in one step. Compared with the separate hydrolysis and fermentation (SHF), the SSF process provides several advantages. In technical aspect, the substrates from hydrolysis are separated from the solids and transferred to fermentation reactor without washing or purification stages.[20] It eliminates the cost for one reactor vessel. More importantly, since the sugars released from hydrolysis can be consumed immediately by the fermenting microorganisms, the inhibition problem for cellulases can be minimized.[20,105,107] In addition, the low sugar concentration also reduces contamination problems.[105] Because the optimal temperatures for hydrolysis (40~50°C for common cellulases) and hexose fermentation (~30°C in baker's yeast) are different, a compromised temperature of 35~37°C has to be applied in SSF.[41] Therefore, more enzymes are needed to guarantee the completeness of hydrolysis and fermentation. This issue is responsible for most of the expensive part of the whole units.[105] Some thermo tolerant enzymes are developed to improve SSF performance.[41] The loss of the freedom in tuning the important parameters for pretreatments/hydrolysis and fermentation independently also requires the microorganism or enzymes to match to other operating conditions such as pH.[41]

3.3. *Simultaneous saccharification and co-fermentation (SSCF)*

Similar to the SSF process, a more consolidate process combines the hydrolysis and fermentation of hexoses and pentoses into one single step, which is known as simultaneous saccharification and co-fermentation (SSCF). This process further reduces the cost for equipment units and integration. Currently, this approach is not widely applicable in industry because there is a competition between hexose and pentose fermentation in the same vessel with shared organisms such as engineered *S. cerevisiae* and *Z. mobilis*, which always results in low ethanol yield, especially the xylose-derived ethanol production efficiency. Although various engineering strategies have been developed to improve the ethanol yield, it still has a long way to go for the purpose of industrialization.[105,108–110]

3.4. *Consolidated bioprocessing (CBP)*

In biofuel production industry, the approaches to reduce costs are always of great interests. Consolidated bioprocessing (CBP, also known as direct microbial conversion, DMC) attempts to integrate all the individual steps in conversion of plant materials to biofuels as shown in Fig. 2.[106] This is the most promising way for cost-effective biofuel production.[22] In addition, no substrates need to be consumed for enzyme production.[105]

CBP requires microbes or a combination of microbes which can directly utilize pretreated lignocellulosic biomass and produce ethanol at high rate and yield.[1,111] Aerobic fungi are capable to convert plant materials, but the major end product from aerobic microorganism is CO_2 rather than ethanol due to the preferred oxidative phosphorylation pathway.[1] Additionally, they are difficult to grow and maintain, and genetic manipulation of aerobic fungi for CBP is not commercialized neither.[1] The most promising candidate for CBP in native cellulolytic strategy is *C. thermocellum*, which is able to convert cellulose into ethanol directly, but can not utilize pentose sugars. Therefore, another thermophilic anaerobe capable to ferment pentose sugars is needed to be mixed with *C. thermocellum* in the CBP process. Studies have been extensively carried out with the cocultures belonging to the thermophilic gram-positive orders of *Clostridiales* (e.g., *C. thermocellum*) with *Thermoanaerobiales*. Shaw, Podkaminer *et al.* have developed a recombinant strain of

Thermoamaerobacterium saccharolyticum which can produce ethanol at high yield from lignocellulosic biomass. The strains were developed by eliminating the genes responsible for the formation of some types of inhibitors like organic acids.[105,112]

CBP is the direction of the development in biomass-to-biofuel industry, because the *integration* strategy has great potential to save cost. Previous studies have estimated that CBP can reduce the overall processing cost by twofold and the cost of biologically mediated processes by eightfold compared with SSF.[105] Lynd *et al.* have indicated that CBP can produce ethanol at costs of 1 US cent/L compared with 5 US cents/L from a typical SSCF configuration.[22,105,113]

4. Biomass-derived Syngas Fermentation to Biofuels

During biochemical fermentation pathways, the hydrolysates released from pretreated biomass is directly converted to ethanol. On the other hand, thermochemical fermentation involves the gasification of biomass to synthesis gas or syngas in short (usually a mixture of CO and H_2), and then the syngas is converted to biofuels by using either chemical catalysts (Fischer-Tropsch process) or microbial catalysts, the latter of which is known as syngas fermentation.

The syngas fermentation to bioethanol and other types of biofuels using different biological catalysts (e.g., *Clostridium ljungdahlii*, *Clostridum autoethanogenum*, *C. carboxydivorans*, *Butyribacterium methylotrophicum*, *Methanosarcina barkeri*, *Rhodospirillum rubrum* and *Peptostreptococcus productus*) is more attractive than biochemical approach and the FT synthesis process due to various inherent merits. These include (1) the elimination of the need for expensive metal catalysts and thus no issue for noble metal poisoning problems; (2) no requirements for complex pretreatment stages and expensive enzymes; (3) high specificity for biofuel production due to utilization of biological catalysts; (4) independence of H_2:CO ratio for bioconversion; (5) the possibility to operate bioreactors at ambient conditions.[2,4,99] The major drawbacks of syngas fermentation are the poor solubility of syngas in the liquid phase and low productivity, which prevent commercialization of the syngas fermentation process.[2,4]

4.1. *Biomass gasification*

Gasification of biomass feedstocks produces synthesis gas (syngas in short), which is a gas mixture predominantly CO and H_2. The composition and quality of syngas depend on multiple variables such as biomass feedstock, gasifier types and agents. Besides CO and H_2, biomass syngas may also contains carbon dioxide (CO_2), nitrogen (N_2), methane (CH_4), ethane (C_2H_6), ethylene (C_2H_4), acetylene (C_2H_2), tars, water vapor, sulfur containing compounds, dusts and higher hydrocarbons.[114]

There are two major types of gasifiers, namely fixed-bed and fluidized-bed gasifiers.[4] Fixed-bed gasifiers have a stationary reaction zone. Based on the direction of the biomass feeding and the oxidant utilized, fixed-bed gasifiers can be subcategorized to updraft and downdraft gasifiers.[4] Fluidized-bed gasifiers use hot solids (e.g., sand, ash or char) to increase the heat transfer and gasification efficiency.[4,44] They are further divided into bubbling fluidized-bed and circulating fluidized-bed gasifiers.[4] Several chemical reactions may take place in syngas production, and the overall gasification process is endothermic, thus heat is required as a driving force.[44] The operating conditions and evaluation of individual gasifiers are summarized by Ciferno and Marano.[44,115] Munasinghe *et al.* have summarized the composition of syngas produced from different gasifiers under different gasification conditions as illustrated in Table 2.[2,4]

4.2. *Metabolic pathways and biochemical reactions*

Both mesophilic and thermophilic microorganisms for syngas-fermentation have been discussed previously. Currently mesophilic microorganisms dominate.[2] Most of the syngas fermenting microorganisms such as *C. ljungdahlii* follow the acetyl-CoA pathway (also known as Wood-Ljungdahl pathway) for biofuel production and other byproducts such as acetate and butyrate.[4] The acetyl-CoA pathway which leads to ethanol, butanol, acetate and butyrate is illustrated in Fig. 3.[4] The major components of syngas, H_2 and CO, play important roles in providing the electrons required for the conversion by hydrogenase and CO dehydrogenase (CODH), respectively. CODH also produces carbonyl groups from CO_2, which is used to generate acetyle-CoA in subsequent steps. CO_2 is reduced to methyl groups through consecutive reductions in the presence of

Table 2. Gas composition of different gasification processes. (Adapted from Refs. 2 and 4)

Gasifier type	Fluidized bed	Updraft	Downdraft	Fluidized bed	Fluidized bed	Fluidized bed
Biomass type	n/a	n/a	n/a	Switch grass	Bark	Coal
Gasifying agent	Air	Air	Oxygen	Air	Air	Oxygen
N_2(%)	50.0	53.0	3.0	56.8	42.9	1
CO	14.0	24	48	14.7	19.6	67
CO_2 (%)	20	9	15	16.5	13.5	4
H_2 (%)	9	11	32	4.4	20.2	24
CH_4 (%)	7	3	2	4.2	3.8	0.02
H_2S(%)	n/a	n/a	N/a	n/a	Very low	1
Tars (g/m^3)	<10	>10	1	<1	<1	0
NH_3 (%)	n/a	n/a	n/a	n/a	n/a	0.04
H_2O(%)	n/a	n/a	n/a	n/a	Dry	a
Dust	High	Low	Low	n/a	n/a	n/a
C_2H_6, C_2H_4 and C_2H_2	n/a	n/a	n/a	3.2	n/a	n/a
H_2/CO	0.64	0.46	0.67	0.30	1.00	0.36
Ref	[116]	[116]	[116]	[114]	[9]	[9]

electrons provided by H_2/CO and adenosine triphosphate (ATP). The reducing equivalents carbonyl groups (–CO), methyl groups (–CH_3) and –CoA are then converted to acetyl-CoA by acetyl-CoA synthase complex (ASC). The produced acetyl-CoA is bifunctional, it plays the role of a precursor for cell macromolecules and serves as an energy source. Finally, acetyl-CoA is reduced to ethanol, butanol, acetate and butyrate. Acetate can be further reduced to ethanol.[2,4,7,99]

Munasinghe and Khanal have summarized the major chemical reactions that take place in syngas fermentation as the following.[2,4]

$$6CO + 3H_2O \rightarrow C_2H_5OH + 4CO_2, \qquad \Delta G^0 = -217.9\,\text{kJ/mol}$$

$$6H_2 + 2CO_2 \rightarrow C_2H_5OH + 3H_2O, \qquad \Delta G^0 = -97.3\,\text{kJ/mol}$$

$$4CO + 2H_2O \rightarrow CH_3COOH + 2CO_2, \qquad \Delta G^0 = -154.9\,\text{kJ/mo}$$

$$4H_2 + 2CO_2 \rightarrow CH_3COOH + 2H_2O, \qquad \Delta G^0 = -75.3\,\text{kJ/mol}$$

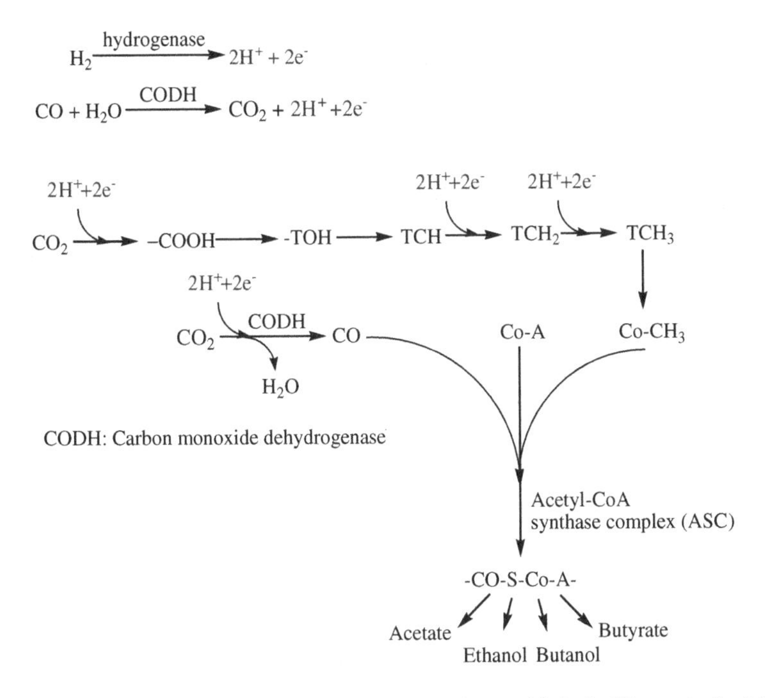

Fig. 3. Acetyl-CoA pathway for syngas fermentation to biofuel. (T: tetrahydrofolate; Co, corrinoid protein-methyl group carrier. Adapted from Ref. 4)

From the first two reactions, it can be observed that two thirds of CO can be converted to ethanol theoretically. The actual yield depends on syngas composition. Because a part of CO is used to supply electrons rather than biofuel production, it is important to maintain healthy concentrations of CO and hydrogen in syngas fermentation. H_2 plays an important role by supplying electrons required for the conversion by hydrogenase. If the enzyme is inhibited, the electrons will come from CO, which can significantly reduce the ethanol yield.[2,116]

4.3. *Reactor design for syngas fermentation*

Bioreactor configuration plays an important role for syngas fermentation. The key factors need to be considered for an efficient syngas fermentation bioreactor include high mass transfer rates, the ease for operation and scaling up, low operation and maintenance costs.[4] Some common reactor configurations are described in the following.

Continuous stirred-tank reactors (CSTR) are the most commonly employed reactor in syngas fermentation. The key feature of this type of reactors is the continuous gaseous injection into the liquid phase (nutrient flow) with agitation.[114] A higher level of agitator control helps to improve the gas-liquid mass transfer because it efficiently breaks gas bubbles into smaller ones, which make the gaseous substrates more accessible to the microbes in the liquid phase. However, it has to be noticed that higher agitation speed may significantly increase the power consumption of agitator and the operation cost.[2,4]

Bubble column reactors consist of large working columns and are designed for industrial applications.[4,114] The advantages of bubble column reactors are higher mass transfer rates and low operation and maintenance costs. The drawbacks are back-mixing and coalescence.[2,4]

Trickle-bed reactor is a continuous reactor with packed bed in which liquid culture flows downward through the packing media.[117] The syngas can flow either co-current or counter-current with respect to the liquid flow, depending on the applications. Since trickle-bed reactors don't require mechanical agitation and are operated under atmospheric pressure, the power consumption is much lower than CSTR.[4]

Microbubble dispersion stirred-tank reactor is a CSTR reactor equipped with a microbubble dispenser.[117] The decreased bubble size and increased interfacial surface area largely improve the gas-liquid mass transfer.[2,4]

Composite hollow fiber membranes (CHFM) can be used in membrane-based reactors to enhance the mass transfer in liquid culture.[118] In CHFM reactors, syngas diffuse through the walls of membranes without forming bubbles. The microbes grow as a film on the outer wall of the membranes and thus they can ferment H_2 and CO continuously. The merits of membrane-based reactor are higher mass transfer rate, reduced reactor volumes, higher microbial cell retention and higher tolerance to toxic compounds in syngas.[2,4]

4.4. *Important factors affecting syngas fermentation*

4.4.1. *Inhibitory compounds*

Syngas contains additional species including ethane (C_2H_6), ethylene (C_2H_4), acetylene (C_2H_2), tar, char particles, sulfur containing compounds, dusts and other higher hydrocarbons.[2,4,119] These impurities can affect the syngas fermentation efficiency by potential scaling in the pathways and inhibiting the microbial catalyst behavior, resulting in poor cell growth and product yield. Datar *et al.* have studied the cell dormancy, hydrogen uptake shutdowns and a shift in metabolic pathways from acidogenesis to solventogenesis. Tar, ash and other solid matter can be removed via filtration.[117] NO can inhibit hydrogenase enzyme activities and reduce the carbon availability for biofuel production.[4] The inhibitory effects of NO can be eliminated by using agents like sodium hydroxide, potassium permanganate or sodium hypochlorite.[2,4] The most common syngas fermenting organism *C. ljungdahlii* was found not significantly affected by H_2S concentrations as high as 5.2% (v/v)[2].

4.4.2. *Mass transfer*

Gas-liquid mass transfer is the rate-limiting step in syngas fermentation.[2] Several distinct stages are involved to transport gaseous substrates into the microbial cells, including the diffusion from bulk gas phase to the gas-liquid interface, traveling across the interface, its transport to the culture media around the microbes and the diffusion of the transported gaseous substrate into the microbial cells.[4] The gas-liquid mass transfer is the major resistance for during gaseous substrates transportation.[2,4,120] Poor diffusion of gaseous substrates into the liquid culture results in low level of substrate utilization and thus low productivity. Klasson *et al.* developed the mathematic model to estimate the mass transfer efficiency by calculating the coefficient k_L as the following:[2]

$$\frac{1}{V_L}\cdot\frac{dN_S^G}{dt}=\frac{k_La}{H}(P_S^G-P_S^L)$$

where N_S^G (mol) is the molar substrate transferred from the gas phase, V_L (L) is the volume of the reactor, P_S^G and P_S^L (atm) are the partial pressures of gaseous substrates in the gas and liquid phase, respectively. H (L . atm/mol) is the Henry's law constant and a (m²/L) is the gas-liquid interface surface area for unit volume. The driving force for mass transfer is the partial pressure difference of gaseous substrates. High pressure operation improves the mass transfer efficiency. However, high concentrations of

CO can inhibit anaerobic microorganisms. In order to evaluate the process kinetics, the gaseous substrate diffusion has to be correlated to its specific uptake rate (q_s/h^{-1}) by the following equation:[2]

$$q_s = \frac{q_m P_S^L}{K_p^{'} + P_S^L + (P_S^L)^2 / W^{'}}$$

where q_m (1/h), W' (atm) and K_p' (atm) are empirical constants. The substrate uptake rate Q_s (mg/L·h) can be calculated as: $Q_s = q_s \cdot X$, where X (mg/L) is the microbial cell concentration. By a further analysis of this mathematic relationship, it can be concluded that reactor operating pressure is inversely proportional to the cell concentration.

The gas solubility in the liquid culture media and gas-liquid mass transfer efficiency can also be enhanced by different bioreactor configurations which we have introduced previously, fabricated impeller design, modified fluid flow patterns and multiple mixing conditions (different speeds and times).[4,120]

4.4.3. *pH and temperature*

pH plays a crucial role in microbial activity in the liquid culture media. The pH suitable for biomass-derived syngas fermentation microbes ranges between 4.5 and 7.3, depending on the microorganisms and other species.[4] The optimum pH for *C. ljungdahlii* is 5.8~6.0.[2,4] Temperature affects the microbial growth kinetics and the solubility of syngas in aqueous culture media. For most of the mesophilic microorganisms, the favorable growth temperature is 37~40°C, while it is 55~80°C for most thermophilic microbes.[120]

4.4.4. *Types of microorganism and growth media*

Besides the most common strict anaerobe *C. ljungdahlii* mentioned previously, *C. aceticum*, *Acetobacterium woodii*, *C. autoethanogenum* and *C. carboxydeviron* are also widely used for syngas fermentation and engineering strategies have been developed to improve the production yields and efficiency.[2,4,120] The growth media provide all the culture required by the microbe propagation, including minerals, trace elements, vitamins and reducing agents. The selection of growth media and preparation protocols depends on the species and target products. Some media are provided by American Type Culture Collection (ATCC) while the rest are developed by individual researchers.

4.4.5. *Industrial-scale syngas fermentation and economics*

Due to the gas-liquid mass transfer problem, syngas fermentation technology is still not commercialized. The reactor design, process protocols optimization and microbial enzyme selection has to be considered together to increase the ethanol concentration. Currently the ethanol concentration from syngas fermentation is as low as 30 g/L, and this is responsible for the high cost of ethanol recovery. Piccolo and Bezzo have compared the techno-economics of both biochemical and thermochemical fermentation pathways. They conclude that biochemical fermentation is more economically feasible.

5. Biobutanol Fermentation

The use of butanol as biofuel offers several advantages in comparison with bioethanol, including higher energy content, higher hydrophobicity, lower flammability than ethanol and better blending capabilities without the need for mechanical modification to the engines using biobutonal. The comparison of properties for biobutanol and other fuels is illustrated in Table 3.[13]

Biobutanol can be produced from starch or lignocellulosic biomass by fermentation using bacterial strains from the *Clostridium* family in a typical ABE fermentation. We have discussed previously with regard to the

Table 3. Comparison of properties of different common fuels. (Reproduced from Ref. 13).

Fuel Properties	Gasoline	Methanol	Ethanol	Butanol
Energy density (MJ/L)	32.00	16.00	19.60	29.20
Air-Fuel ratio	14.60	6.50	9.00	11.20
Specific energy (MJ/kg air)	2.90	3.10	3.00	3.20
Heat of vaporization (MJ/kg)	0.36	1.20	0.92	0.43
Research Octane Number (RON)	91–99	136	129	96–105
Motor Octane Number (MON)	81–89	104	102	78–89

butanol fermentation with engineered *E. coli*, while clostridia produce solvent in biphasic fermentation, namely, acidogenic phase and solventogenic phase.[121] Acids formed in the pathways are activated in the acidogenic phase and reassimilated for the production of acetone, butanol and other alcohols. The solventogenic clostridia are capable to utilize a wide spectrum of carbohydrates and convert them to biofuels. The most commonly used strains are *C. acetobutylicum* and *C.beijerinckiid.*[4,121]

In order to improve the butanol yield, butanol tolerance and extending the substrate utilization, metabolic engineering strategies for the microorganisms of interest have been developed and employed. One of the successful engineering examples is the amplification of acetone formation pathway in *C. acetobutylicum.* The recombinant *C. acetobutylicum* carrying amplified genes of *adc* (encoding acetoacetate decarboxylase) and *ctfAB* (encoding CoA transferase) were activated earlier than normal, which led to earlier acetone production. Compared with the parent strains, this engineering strategy resulted in increased concentrations of acetone, butanol and ethanol by 95%, 37% and 90%, respectively.[121]

Besides the development of microorganisms, the fermentation approaches and technologies were also improved for butanol production. Productivity in batch reactors is normally very low because of butanol inhibition and long lag phase. Fed-batch and continuous fermentation coupled with novel product removal techniques is efficient to eliminate the low-productivity problem.[121]

Fed-batch fermentation is initiated by a batch mode with low substrate concentration (no inhibition), once the substrates are exhausted, it is followed by adding a concentrated substrates with slow rate so that keeping the substrate concentration below the toxic level.[121] This was employed for *C. beijerinckii* in a solution containing 500 g/L glucose; the product was recovered by gas stripping or pervaporation.[122]

Continuous culture technique is initiated keeping the cell growth until the exponential phase in a batch mode, and then the reactor is fed with continuous flow with medium and product stream. This technique was applied earlier on *C. beijerinckii* BA101 for butanol with a production rate of 1.74 g/L/h.[123] Later on, *C. beijerinckii* cells were immobilized on clay bricks particles by adsorption, which resulted in a reactor production rate of 15.8 g/L/h.[121,124] Due to the presence of high cell concentrations and thus high productivity, this immobilized cell continuous reactors demonstrate significant economic advantages.

In another reactor system, membrane cell recycling technique is employed. The reactor is initiated in a batch mode, and the fermentation broth is circulated through the membrane, which retains the cells and allows the aqueous solution pass by.[121] The cell concentration can be as high as 100 g/L and the reactor productivity can reach 6.5 g/L/h. However, membrane fouling is the major problem for this technique.[125]

The traditional product recovery process is utilizing distillation the fermentation broth with low butanol concentration. This is one of the factors responsible for high operation cost in biobutanol (and bioethanol) production.[121] To solve the problem, fermentation reactors can be coupled to an *in-situ* solvent recovery system. The well developed in-situ recovery technologies include gas stripping, liquid-liquid extraction and pervaporation.[121]

For gas stripping technology, either N_2 or fermentation gas (H_2 and CO_2) is bubbled into the fermentation broth, the bubbles are able to capture the butanol and other solvent molecules. Then the captured solvents condense in the condenser and are collected in the reservoir.[121] Gas stripping can be applied to batch reactor for solvents recovery from the fermentation broth of *C. beijerinckii* BA101.[121,126]

Liquid-liquid extraction employs extraction solvents to be mixed with the fermentation broth. The alcohol products (acetone, butanol and ethanol) are extracted by the external solvents and recovered by back extraction with a second extraction solvent or distillation. In order to avoid the extractant toxicity and emulsion formation problems, the extractant and fermentation broth are isolated by a membrane, which is specified for butanol exchange between the two phases.[121]

Pervaporation provides a membrane used for selective removal of volatile compounds from the fermentation broth. The volatile solvents diffuse through the membrane in vapor phase and recovered by condensation.[121] Both liquid (e.g., oleyl alcohol) and solid (e.g., silicon-silicate) membranes have been used.[121] Further purification like distillation is required since the membrane accepts acetone, butanol and ethanol as a whole. The solvent productivity in continuous fermentation with *C. acetobutylicum* reached up to 2.34 g/L/h in a pervaporation equipped with an ionic liquid polydimethylsiloxane (PDMS) ultrafiltration membrane.[127]

Butanol is a better renewable fuel than ethanol in many aspects. Various genetic modification strategies on clostridia family have been developed to improve the butanol productivity and yield. The developments of fermentation processes have also been explored extensively in order to reduce the cost and increase the fermentation efficiency. In the near future, much more investigation will be needed to make butanol fermentation process more economically competitive.

6. Summary and Outlook

The use of biofuels as the substitute for fossil fuels is the response to a number of issues people are facing, including the dramatic increase of petroleum price, depletion of fossil fuel resources and the environmental regulation requirements. Fermentation technology plays an important role in converting biomass to biofuels such as bioethanol and biobutanol. For fermentation in both biochemical and thermochemical pathways, the development and engineering of microorganisms always receive great attention in order to improve the biofuel productivity. Metabolic engineering on industrial applicable strains is an efficient way to achieve the utilization of a wide spectrum of biomass, improve the fuel production yield and rate, or minimize the inhibition effect. The development of different fermentation strategies and reactor configurations are needed for improving the fermentation efficiency and reducing the operation. A variety of integrated configurations and strategies have been applied in industry. However, in the near future, more research needs to be carried out on improving fermentation technologies. Major directions include optimizing the bioprocess and operation technologies, aiming at understanding the metabolic mechanisms involved in controlling the enzymes for degrading polysaccharides, inhibition elimination, and more reasonable and efficient reactor configuration design.

Acknowledgments

The author is grateful for the financial support from the Paper Science & Engineering (PSE) fellowship program at Institute of Paper Science & Technology (IPST) and the School of Chemistry and Biochemistry at Georgia Institute of Technology, Atlanta, GA, USA.

References

1. M. S. Elshahed, *J. Adv. Res.*, **1**, 103 (2010).
2. P. C. Munasinghe and S. K. Khanal, *Bioresour. Technol.*, **101**, 5013 (2010)
3. S. I. Mussatto and J. A. Teixeira, *Current Research, Technology and Educationopics in Applied Microbiology and Microbial Biotechnology*, **2**, 897 (2010).
4. P. C. Munasinghe and S. K. Khanal, Biomass-derived Syngas Fermentation into Biofuels. *Biofuels: Alternative Feedstocks and Conversion Processes.*, 79 (2011).
5. A project report sponsored by Department of Electrical Engineering, NTNU, Norwegian University of Science and Technology. 2005. Biofuels from lignocellulosic material, Trondheim. Available from: http://www.zero.no/transport/biodrivstoff/gasification/biofuels-from-lignocellulosic-material.pdf (accessed on: 09. 20. 2012).
6. D. Takara and S. K. Khanal, *Bioresour. Technol.*, **102**, 1582 (2011).
7. A. M. Henstra, J. Sipma, A. Rinzema and A. J. M. Stams, *Curr. Opin. Biotechnol.*, **18**, 200 (2007).
8. A. Demirbas, *Progress Energy Combust. Sci.*, **33**, 1–18 (2007).
9. V. Subramani and S. K. Gangwal, *Energy Fuels.*, **22**, 814 (2008).
10. E. Palmqvist and B. H. Hagerdal, *Bioresour. Technol.*, **74**, 17 (2000).
11. E. Palmqvist and B. H. Hagerdal, *Bioresour. Technol.*, **74**, 25 (2000).
12. S. A. Survase, E. Sklavounos, G. Jurgens, A. V. Heiningen and T. Granstrom, *Bioresour. Technol.*, **102**, 10996 (2011).
13. Y. Dahman, Sustainable Biobutanol and Working towards the Green Gasoline of the Future. *Ferment Technol.,* 1:e111. doi:10.4172/fmt.1000e111, 2012.
14. D. R. Aquino, *Chem. Eng. Prog.*, **3**, 8 (2007).
15. Anon*, Fuels & Lubes International Quarter.*, **3**, 19 (2006).
16. N. Qureshi and T. C. Ezeji, *Biofuels Bioprod. Bior.*, **2**, 319 (2008).
17. K. Kraemer, A. Harwardt, R. Bronneberg and W. Marquardt, 20th European Symposium on Computer Aided Process Engineering, ESCAPE20.
18. W. H. Schwarz and J. R. Gapes, *BioWorld.*, **16** (2006).
19. A. D. Zverlov and W. H. Schwarz, *Appl Microbiol Biotechnol.*, **77**, 23 (2007).
20. T. H. Kim, Dissertion: Bioconversion of Lignocellulosic material into Ethanol: Pretreatment, Enzymatic Hydrolysis, and Ethanol Fermentation. Auburn, Alabama (2004).
21. R. M. Worden, A. J. Grethlein, M. K. Jain and R. Datta, *Fuel*, **70**, 615 (1991).
22. L. R. Lynd, W. H. Van Zyl, *et al. Curr. Opin. Biotechnol.*, **16**, 577 (2005).
23. D. Antoni and V. V. Zverlov, *Appl Microbiol Biotechnol.*, **77**, 23 (2007).
24. A. K. Chandel, G. Chandrasekhar, K. Radhika, R. Ravinder, P. Ravindra, *Biotechnol. Mol. Biol. Rev.*, **6**, 8, (2011).
25. I. S. Pretorius, M. du Toit and P. van Rensburg, *Food Technol. Biotechnol.*, **41**, 3 (2003).
26. C. Vallet, R. said, C. Rabiller and M. L. Martin, *Bioorg Chem.*, **24**, 319 (1996).
27. A. G. Virginie, B. Bruno, D. Sylvie and S. Jean-Marie, *Biotechnol. Lett.*, **23**, 677 (2001).

28. Z. S. Yu and H. X. Zhang, *Bioresour Technol.*, **93**, 199 (2004).
29. B. H. Hagerdahl, K. Karhumaa, C. Fonseca, I. S. Martins and M. F. G. Grauslund, *Appl Microbiol Biotechnol.*, **74**, 937 (2007).
30. J. Wiegel and L. G. Ljungdahl, *Crit Rev Biotechnol.*, **3**, 39 (1986).
31. P. Sommer, T. Georgieva and B. K. Ahring, *Biochem Soc Trans.*, **32**, 283 (2004).
32. J. C. Fong, C. J. Svenson, K. Nakasugi, J. P. Bowman, B. Chen, D. R. Glenn, B. A. Neilan and P. L. Rogers, *Extremophiles,* **10**, 363 (2006).
33. Y. Lin and S. Tanaka, *Appl Microbiol Biotechnol.*, **69**, 627 (2006).
34. L. O. Ingram, T. Conway, D. P. Clark, G. W. Sewell and J. F. Preston, *Appl. Environ. Microbiol.*, **53**, 2420 (1987).
35. C. S. Gong, C. M. Maun and G. T. Tsao, *Biotechnol. Lett.*, **3**, 77 (1981).
36. J. N. Saddler and M. K. H. Chan, *Eur J. Appl. Microbiol. Biotechnol.*, **16**, 99 (1982).
37. E. Sugawara, S. Hashimono, Y. Sakurai and A. Kobayashi, *Biosci Biotechnol Biochem.*, **58**, 1134 (1994).
38. G. M. Pastore, Y. K. Park and D. B. Min, *Mycol. Res.*, **98**, 25 (1994).
39. P. Gervais and M. Sarrette, *J. Ferment Bioeng.*, **69**, 46 (1990).
40. B. H. Hagerdahl, K. Karhumaa, C. U. Larsson, M. G. Grauslund, J. Görgens and W. H. Zyl, *Microb Cell Fact*, B, 31 (2005).
41. Z. Barta, Doctroal Thesis: Experimental and Techno-Economic Approaches in Improvement of The Lignocellulosic-Ethanol Process. Budapest University of Technology and Economics. Department of Applied Biotechnology and Food Science (2011).
42. Z. L. Liu, B. C. Saha and P. J. Slininger, *Bioenergy.* Washington DC: ASM press (2008).
43. J. Slaa, M. Gnode and H. Else, *J. Org. Chem.: Chem. Dut. Aspects*, 134 (2009).
44. Caye M. Drapcho, Nghiem Phu Nhuan, Terry H. Walker, Biofuels Engineering Process Technology. McGraw Hill. DOI: 10.1036/0071487492 (2008).
45. Thomas, K. C., S. H. Hynes, and W. M. Ingledew. *Process Biochem.*, **31**, 321 (1996).
46. J. N. Pereira, M. Antonieta, G. Peixoto, C. Lídia, M. Maria, S. Anna, *Biomass of Lignocellulosic Composition for Fuel Ethanol Production within the Context of Biorefinery*, Rio de Janeiro (2008).
47. B. Adams, *J. Bacteriol.*, **111**, 308 (1972).
48. E. Reifenberger, E. Boles, M. Ciriacy, *Eur J Biochem.*, **245**, 324 (1997).
49. Verstrepen, K. J., *et al.*, *Trends Biotechnol.*, 22, 531 (2004).
50. P. S. Panesar, S. S. Marwaha and J. F. Kennedy, *J. Chem. Technol. Biotechnol.*, **81**, 623 (2006).
51. R. K. Scopes and G. Smith, *Biotechnol. Lett.*, **8**, 653 (1986).
52. B. S. Dien, M. A. Cotta and T. W. Jeffries, *Appl. Microbiol. Biotechnol.*, **63**, 258 (2003).
53. K. D. Barrow, J. G. Collins, D. A. Leigh, P. L. Rogers, R. G. Warr, *Appl. Microbiol. Biotechnol.*, **20**, 225 (1984).
54. K. Uldis, *Adv. Microb. Physiol.*, **51**, 73 (2006).
55. T. W. Jeffres, *Nat. Biotechnol.*, **23**, 40 (2005).
56. P. Gunasekaran, K. Chandra Raj, "Ethanol fermentation technology — Zymomonas mobilis (1999).

57. T. Karsch, U. Stahl and K. Esser, *Eur J. Appl. Microbiol. Biotechnol.*, **18**, 387 (1983).
58. H. W. Doelle, *et al. Crit Rev Biotechnol.*, **13**, 57 (1993).
59. E. L. Joachimsthal, P. L. Rogers, *Appl. Biochem. Biotechnol.*, **84**, 343 (2000).
60. M. Agrawal, Z. Mao, R. R. Chen, *Biotechnol. Bioeng.*, **108** (4), 777 (2011).
61. R. L. Uffen, J. Indust. *Microbiol. Biotechnol.*, 19(1), 1(1997).
62. T. W. Jeffries and Y. S. Jin, *Appl. Microbiol. Biotechnol.*, **63**, 495 (2004).
63. M. Galbe, G. Zacchi, *Appl. Microbiol. Mol. Biol. Rev.*, **62** (2), 334 (2002).
64. N. W. Y. Ho, Z. Chen, A. P. Brainard, *Appl. Environ. Microbiol.*, **64(5)**, 1852 (1998).
65. H. W. Wisselink, M. J. Toirkens, Q. Wu, J. T. Pronk, A. J. A. Van Maris, *Appl. Environ. Microbiol.*, **75(4)**, 907 (2009).
66. C. R. Shen, J. C. Liao, *Metab. Eng.*, **10(6)**, 312 (2008).
67. B. H. Hägerdal, N. Pammet, *Appl. Biochem. Biotech.*, **113**, 1207 (2004).
68. P. A. vanKuyk, M. J. Groot, G. J. Ruijter, R. P. Vries, *Eur. J. Biochem.*, **268(20)**, 5414 (2001).
69. R. Verho, J. Londesborough, M. Penttilä, P. Richard, *J. Biol. Chem.*, **279 (15)**, 14748 (2004).
70. P. Richard, M. Putkonen, R. Väänänen, J. Londesborough, M. Penttilä, *Biochem.*, **41(20)**, 6432 (2002).
71. S. Banerjee, A. Archana, T. Satyanarayana, *Curr. Micobiol.*, **29**, 349 (1994).
72. H. R. Harhangi, A. S. Akhmanova, R. Emmens, C. van der Drift, W. T. De Laat, J. P. Dijken, M. S. Jetten, J. T. Pronk, H. J. Op den Camp, *Arch. Microbiol.*, **180(2)**, 134 (2003).
73. T. Senac and B. H. Hägerdal, *Appl. Environ. Microbiol.*, **56(1)**, 120 (1990).
74. C. S. Gong, *et al. Appl. Environ. Microbiol.*, **41**, 430 (1981).
75. M. Kuyper, *et al. FEMS Yeast Res.*, **5**, 399–409 (2005).
76. M. H. Toivari, L. salusjärvi, L. Ruohonen, M. Penttilä, *Appl. Environ. Microbiol.*, **70(6)**, 3681 (2004).
77. M. Sedlak and N. W. Y. Ho, *Appl. Biochem. Biotech.*, **113**, 403 (2004).
78. K. Karhumaa, B. Wiedemann, E. Boles, B. H. Hägerdal and M. F. G. Grauslund, *Appl. Microbiol. Biotechnol.*, **73(5)**, 1039 (2007).
79. P. Richard, R. Verho, M. Putkonen, J. Londesborough and M. Penttilä, *FEMS Yeast Res.*, **3(2)**, 185 (2003).
80. J. Becker, E. Boles, *Appl. Environ. Microbiol.*, **69(7)**, 4144 (2003).
81. Wisselink, H. M, *et al.*, *Appl Environ. Microbiol.*, **73(15)**, 4881 (2007).
82. C. F. Wahlbom, W. H. Van Zyl, L. J. Jönsson, B. H. Hägerdal and R. R. Otero, *FEMS Yeast Res.*, **3(3)**, 319 (2003).
83. M. Sonderegger, M. Jeppsson, C. Larsson, M. F. G. Grauslund, E. Boles, L. Olsson, I. S. Martins, B. H. Hägerdal and U. Sauer, *Biotechnol. Bioeng.*, **87(1)**, 90 (2004b).
84. B. Johansson, Ph. D thesis: Metabolic engineering of the pentose phosphate pathway of xylose fermenting Saccharomyces cerevisiae. Department of Applied Microbiology, Lund University (2001).

85. J. Zaldivar, A. Borges, B. Johansson, H. P. Smits, S. G. V. Boas, J. Nielsen and L. Olsson, *Appl. Microbiol. Biotechnol.*, **59(4–5)**, 436 (2002).
86. Y. J. Woo, Enhanced bioethanol production by Zymomonas mobilis in response to the quorum sensing molecules AI-2, Durham theses, Durham University. Available at Durham E-Theses Online: http://etheses.dur.ac.uk/3231/ (2011).
87. M. Zhang, C. Eddy, K. Deanda, M. Finkestein and S. Picataggio, *Science*, **267**, 240 (1995a).
88. M. Zhang, M. A. Franden, M. Newman, J. Mcmillan, M. Finkelstein and S. Picataggio, *Appl. Biochem. Biotechnol.*, **51**, 527 (1995b).
89. H. G. Lawford and J. D. Rousseau, *Appl. Biochem. Biotechnol.*, **98**, 429 (2002).
90. H. G. Lawford, J. D. Rousseau, A. Mohagheghi and J. D. McMillan, *Appl. Biochem. Biotechnol.*, **77**, 191 (1999).
91. T. G. Liu and C. Khosla, *Annu. Rev. Genet.*, **44**, 53 (2010).
92. L. R. Jarboe, T. B. Grabar, L. P. Yomano, K. T. Shanmugan and L. O. Ingram, *Adv. Biochem. Eng. Biotechnol.*, **108**, 237 (2007).
93. L. O. Ingram, *et al.*, *Appl. Environ. Microbiol.*, **53**, 2420 (1987).
94. S. Atsumi, A. F. Cann, M. R. Connor, C. R. Shen and K. M. Smith, *et al.*, *Metab. Eng.*, **10**, 305 (2008).
95. M. Inui, M. Suda, S. Kimura, K. Yasuda and H. Suzuki, *et al.*, *Appl. Microbiol. Biotechnol.*, **77**, 1305 (2008).
96. B. E. Wood and L. O. Ingram, *Appl. Environ. Microbiol.*, **58**, 2103 (1992).
97. S. Zhou, F. C. Davis and L. O. Ingram, *Appl. Environ. Microbiol.*, **67**, 6 (2001).
98. T. W. Jeffries, *et al.*, *Nat. Biotechnol.*, **25 (3)**, 319 (2007).
99. H. Heiskanen, I. Virkajarvi and L. Viikari, *Enzyme Microb. Technol.*, **41**, 362 (2007).
100. H. Younesi, G. Najafpour and A. R. Mohameda, *Biochem. Eng. J.*, **27**, 110 (2005).
101. R. S. Tanner, L. M. Miller, and D. Yang, *Int. J. Syst. Bacteriol.*, **43**, 232 (1993).
102. S. Rajagopalan, R. P. Datar and R. S. Lewis, *Biomass Bioenergy*, **23**, 487 (2002).
103. N. Qureshi, B. C. Saha, B. Dien, R. E. Hector and M. A. Cotta, *Biomass and Bioenergy*, **34**, 559 (2010).
104. N. Qureshi, B. C. Saha, R. E. Hector, B. Dien, S. Hughes, S. Liu, L. Iten, M. J. Bowman, G. Sarath and M. A. Cotta, *Biomass and Bioenergy*, **34**, 566 (2010).
105. F. Goldschmidt, Term paper: From Cellulose to Ethanol: Engineering Microorganisms to Produce Biofuel (2008).
106. L. R. Lynd and P. J. Weimer, *et al.*, *Microbiol. Mol. Biol. R.*, **66(3)**, 506 (2002).
107. M. Galbe and G. Zacchi, *Appl. Microbiol. Biotechnol.*, **59(6)**, 618 (2002).
108. F. K. Kazi, J. Fortman, R. Anex, G. Kothandaraman, D. Hsu, A. Aden and A. Dutta, Techno-economic analysis of biochemical scenarios for production of cellulosic ethanol. Technical Report NREL/TP-6A2–46588, National Renewable Energy Laboratory, Golden, CO (2010).
109. C. A. Cardona and O. J. Sanchez, *Biores Technol.*, **98(12)**, 2415 (2007).
110. O. J. Sanchez and C. A. Cardona, *Bioresour. Technol.*, **99(13)**, 5270 (2008).

111. L. R. Lynd, D. Currie, N. Ciazza, C. Herring and N. Orem, Consolidated bioprocessing of cellulosic biomass to ethanol using thermophilic bacteria. In: J. D. Wall, C. S. Hardwood, A. L. Demain, editors. Bioenergy. Washington D.C. ASM Press (2008).
112. A. J. Shaw and K. K. Podkaminer, *et al.*, *Proc. Natl. Acad. Sci.*, **105(37)**, 13769 (2008).
113. W. H. van Zyl and L. R. Lynd, *et al.*, Biofuels, **108**, 205 (2007).
114. R. P. Datar, R. M. Shenkman, B. G. Cateni and R. L. Lewis, *Biotechnol. Bioeng.*, **86**, 587 (2004).
115. J. P. Ciferno and J. J. Marano. 2002. "Benchmarking biomass technologies for fuels, chemicals and hydrogen production." Report prepared for the National Renewable Energy Laboratory, U.S. Department of Energy (2002).
116. A. Ahmed and R. L. Lewis, *Biotechnol. Bioeng.*, **97(5)**, 1080 (2007).
117. M. D. Bredwell, P. Srivastava, R. M. Worden, *Biotechnol. Pro.*, **15**, 834 (1999).
118. R. Nerenberg, B. E. Rittmann, *Water Sci. Technol.*, **49**, 223 (2004).
119. A. Ahmed, R. L. Lewis, *Biotechnol. Bioeng.*, **97(5)**, 1080 (2007).
120. M. Mohammadi, G. D. Najafpourb, H. Younesic, P. Lahijanid, M. H. Uzira, A. R. Mohameda, *Renew. Sust. Energ. Rev.*, **15**, 4255 (2011).
121. C. Jin, M. F. Yao, H. F. Liu, C. F. Lee, J. Ji, *Renew. Sust. Energ. Rev.*, **15**, 4080 (2011).
122. T. C. Ezeji, N. Qureshi and H. P. Blaschek, *Appl. Microbiol. Biotechnol.*, **63**, 653 (2004).
123. J. Formanek, R. Mackie and H. P. Blaschek, *Appl. Environ. Microbiol.*, **63**, 2306 (1997).
124. N. Qureshi, J. Schiripsema, J. Lienharde and H. P. Blaschek, *World J. Microbiol. Biotechnol.*, **16**, 377 (2000).
125. T. C. Ezeji, N. Qureshi and H. P. Blaschek, *Chem Rec.*, **4**, 305 (2004).
126. S. Y. Lee, H. J. Park, L. K. Nielsen, J. Kim and K. S. Jung, *Biotechnol Bioeng.*, **101**, 209 (2008).
127. P. Izák, K. Schwarz, W. Ruth, H. Bahl and U. Kragl, *Appl. Microbiol. Biotechnol.*, **78**, 597 (2008).

CHAPTER 8

PYROLYSIS OF BIOMASS TO BIO-OILS*

HAOXI BEN

School of Chemistry and Biochemistry, Institute of Paper Science & Technology Georgia Institute of Technology 500 10th Street N.W, Atlanta, GA, 30332, United States
benhaoxi@gmail.com

Biomass pyrolysis oil has been reported as a potential renewable biofuel precursor. In this chapter, we will summarize the reported components in the pyrolysis oils produced from major biomass components — cellulose, hemicellulose, lignin and tannin. A comparison of analysis methods including GC-MS, FT-IR and NMR will also be presented. The challenging properties of pyrolysis oil including high molecular weight, acidity, oxygen content and viscosity, solid residue as well as aging problem will also be discussed to provide insight into the following upgrading process of pyrolysis oils.

1. Introduction

Over the last century, worldwide energy consumption has increased by 17 fold[1] and this demand is predicted to grow by more than 50% by 2025.[2] Known viable petroleum reserves are predicted to be consumed in less than fifty years at present rates of consumption. In addition, the carbon

*Approximately 80% of this book chapter was written as Haoxi Ben's PhD dissertation for Georgia Institute of Technology. It is entitled as "Thermal Conversion of Biomass and Biomass Components to Biofuels and Bio-Chemicals". The other 20% of this book chapter were accepted for publication in BioEnergy Research, 2013 (Ref. 99 of this book chapter). It is entitled as "Lignin Pyrolysis Components and Upgrading — Technology Review". The other authors are Arthur J. Ragauskas, Wei Mu and Yulin Deng. Reproduced with permission of Springer Science+Business Media.

dioxide emissions from the consumption of fossil fuels have been growing at an average rate of ~2 % per year and the rate continues to increase. Growing concerns about the effects of carbon dioxide emissions from fossil fuels call for sustainable energy sources, such as biomass. Biomass is a renewable resource for the sustainable production of fuels and chemicals that, to date, have been made primarily from fossil resources. Because of its carbon neutrality, relative abundance and non-food competition,[3] the use of biofuels and biochemicals could increase economics and provide environmental benefits. The U.S. Department of Agriculture and U.S. Department of Energy both strongly proposed to expand the usage of biomass as a resource of energy and chemicals, and they also established a vision to derive 25 % of chemicals and materials and 20 % of transportation fuels from biomass by 2030.[4] As the second most abundant biomass component and the primary renewable aromatic resource in nature, lignin, however, has received much less attention than plant polysaccharides as a resource for biofuels. Among the various conversion technologies being investigated, pyrolysis has been reported as one of the economic ways (i.e., low capital and operating costs) to utilize biomass for biofuels and biochemicals.[5] The pyrolysis of lignin yielding low-molecular weight compounds has been examined for the past fifty years. To understand the chemical components of pyrolysis products and seek an efficient characterization method will provide insight into the further upgrading process, which will produce biofuels and biochemicals. This review will summarize the reported pyrolysis products including gas and liquid phases and compare the various reported analysis methods.

The whole biomass pyrolysis products are complicated. To fully understand the whole biomass pyrolysis pathways call for fundamental analysis of pyrolysis of major biomass components — cellulose, hemicellulose, lignin and for barks — tannin. In addition, the complexity nature of pyrolysis oils brings a huge barrier for the traditional analysis methods such as GC-MS, GPC and FT-IR. GC-MS has always been used to analyze individual components in pyrolysis oil; however, only a small portion could be detected by GC due to the poor volatility. The spectroscopic technique such as FT-IR could give insights into the whole portions of pyrolysis oil. Unfortunately, the ability of such method to deal with complex mixture like pyrolysis oil is very limited. As a new way introduced into characterization of pyrolysis oil, NMR has some advantages compared to the traditional methods. NMR has the ability to analyze the whole portion of pyrolysis oil. In addition, it could characterize more than thirty different functional groups presented in the lignin pyrolysis oil and give quantitative results. Nevertheless, NMR could not provide exact structure of the components, which still make it difficult to accomplish the complete chemical characterization of pyrolysis oil.

Owing to the complicated components, characterization of the properties of pyrolysis oil will provide insight into the following upgrading process which is a challenging but crucial undertaking. The major drawback towards commercialization of bio-oil are several challenging properties including poor volatility, high oxygen content, acidity and viscosity, corrosiveness and cold flow problems.[6]

In this chapter, we will summarize the reported components in the pyrolysis oils produced from major biomass components, i.e., cellulose, hemicellulose, lignin and tannin. A comparison among different analysis methods including GC-MS, FT-IR and NMR will be presented. The challenging properties of pyrolysis oil such as high molecular weight, acidity, oxygen content and viscosity, solid residue will also be discussed.

2. Lignocellulose

Lignocellulosic biomass provides an abundant sustainable feedstock forbiofuel and it contains three major constituents: cellulose, hemicelluloses and lignin. Table 1 summarizes the distribution of the three major biopolymers in several hardwoods, softwoods and agricultural residue.[7,8]

2.1. *Cellulose*

Cellulose is the most abundant biopolymer and has an approximately100×10^9 metric tons annual biosynthesis rate.[2] Generally, biomasses contain around 35–50 % of cellulose, which is a linear polymer built up of β-D glucopyranose units covalently linked with 1→4 glycosidic bonds, with a degree of polymerization (DP) of ~300–15000. Table 2 summarizes the DP in different cellulose containing materials. Figure 1 shows the molecular structure of cellulose. In the native form of

Table 1. Major biomass components in different softwoods, hardwoods and residue species.[11,12]

Wood Species		Wood components[a]		
		Cellulose (%)	Lignin (%)	Hemicelluloses (%)
	Piceaglauca	41	27	31
	Abiesbalsamea	42	29	27
	Pinus strobes	41	29	27
	Tsugacanadensis	41	33	23
	Norway spruce	46	28	25
	Loblolly pine	39	31	25
Softwoods	*Thujaoccidentalis*	41	31	26
	Eucalyptus globulus	45	19	35
	Acer rubrum	45	24	29
	Ulmusamericana	51	24	23
	Populus tremuloides	48	21	27
	Betulapapyrifera	42	19	38
Hardwoods	*Fagusgrandifolia*	45	22	29
	Pine residues[b]	45	27	20
Residues	Pine bark	32	34	19

[a] all samples were analyzed extractives free.

[b] Pine residues are mixtures of stumps, limbs, tops, and dead trees.

Table 2. DP of native wood and non-woody celluloses after nitration using the viscometric method.[11–13]

Species	DP
Bagasse	925
Wheat straw	1045
E.regnans	1510
P. radiata	3063
Cotton linters	3170
Cotton stalks	1820
Aspen	4581
Nalita wood (12–30 months)	3181–3611
Jack pine	10300

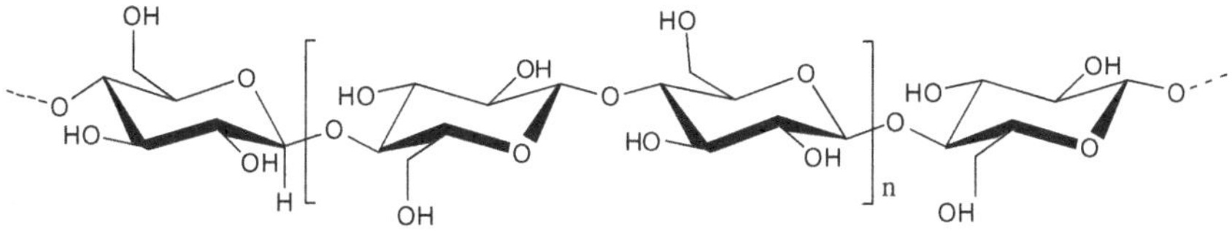

Fig. 1. Cellulose molecular structure.

Table 3. Cellulose crystallinity from several biomasses.[10,11,14]

Origin	Crystallinity (%)
Hybrid poplar	63[a]
Loblolly pine	63[a]
Alamo switchgrass	44[a]
Bagasse	37[b]
Wheat straw	35[b]
E.regnans	37[b]
P. radiata	34[b]
Cotton linters	80[b]
Norway spruce	67[b]

[a] Determined by CP/MAS ^{13}C-NMR spectroscopy.
[b] Determined by X-ray diffraction.

cellulose, the glucopyranose units in the cellulose chain are in the most thermodynamically stable conformation — chair conformation, which with the -CH_2OH and -OH groups in equatorial position.[9] Due to the highly organized structure and the inter- and intra-molecular hydrogen bonds, in the native form, cellulose has a relatively high-degree of crystallinity averaging 50–70%.[2,3,10] Table 3 summarizes the cellulose crystallinity from several biomasses.

2.2. *Hemicelluloses*

Hemicelluloses are low DP (~50–300) polysaccharides and contain arabinose, xylose, galactose, mannose, and glucose, along with side chains groups, such as acetyl, galacturonic acid, glucuronic acid, and

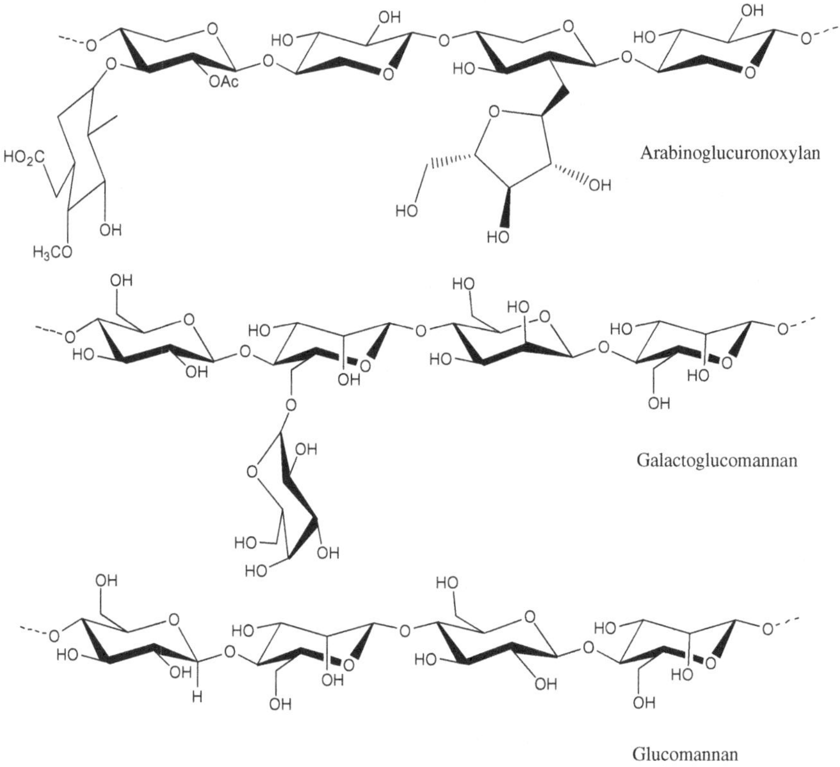

Arabinoglucuronoxylan

Galactoglucomannan

Glucomannan

Fig. 2. Major types of hemicelluloses in softwoods and hardwoods.

4-O-methylglucuronic acid. The major hemicelluloses in softwoods (SW) are galactoglucomannans and arabinoglucuronoxylan, and a small amount of arabinogalactan, xyloglucan, and other glucans. For hardwood, the predominant hemicellulose is glucuronoxylan and glucomannan with limited amount of galactans and glucans.[2] Figure 2 shows some of the major types of hemicelluloses in softwoods and hardwoods.

Willfor *et al.*[15,16] have examined the polysaccharides in 11 industrially important hardwood and 12 softwood species. They found that the most abundant sugar in the softwood hemicellulose is mannose; correspondingly, the most abundant sugar in hardwood hemicellulose is xylose. Representative data from this study is summarized in Table 4. The DP of some hemicelluloses in several hardwood and softwood species is summarized in Table 5.

Table 4. Amount and composition of sugar units (wt%) of different hardwood and softwood species.[15,16]

Wood Species	Ara	Xyl	Gal	Glc	Man	Rha	GlcA	GalA	4-*O*- MeGlcA
Hardwood									
Acacia crassicarpa	0.3	11.7	11.4	2.2	0.2	0.30	0.2	1.8	2.6
Betula pendula	0.5	23.6	1.3	1.8	0.9	0.5	0.3	2.2	3.7
Eucalyptus globulus	0.6	14.0	2.3	5.6	0.3	0.5	0.2	2.0	2.5
Populus deltoides	0.5	17.2	0.8	2.5	2.7	0.5	0.2	2.5	2.5
Populus tremula	0.5	19.6	1.1	4.2	1.3	0.5	0.1	2.4	2.5
Softwood									
Picea abies	1.4	5.8	2.0	3.3	9.0	0.2	0.3	1.6	9.8
Picea mariana	1.5	5.0	2.8	3.8	9.6	0.2	0.4	1.6	0.9
Abies lasiocarpa	1.6	5.0	4.0	4.0	10.0	0.3	0.4	1.5	0.9
Larix lariciana	2.7	5.1	8.6	4.0	10.5	0.2	0.6	1.4	1.1
Pinus resinosa	2.3	6.8	4.0	3.7	8.5	0.3	0.5	1.6	1.3

Table 5. DP of hemicelluloses in several biomasses.[17]

Origin	Hemicellulose	DP
Birch	Glucuronoxylan	~112
Aspen	Glucuronoxylan	~112
Spruce	Arabinoglucuronoxylan	~126
	Galactoglucomannan	~125
Pine	Arabinoglucuronoxylan	~126
	Galactoglucomannan	~125
Larch	Arabinoglucuronoxylan	~126
	Galactoglucomannan	~125

2.3. *Lignin*

As the second most abundant biomass component and the primary renewable aromatic resource in nature, lignin is distinctly different from cellulose and hemicelluloses. Lignin is one of the most complex natural polymers in regards of its chemical structure and composition. It is synthesized by enzymatic dehydrogenative polymerization of 4-hydroxyphenyl propanoid units. Figure 3 indicates the three phenyl propane precursors of lignin and

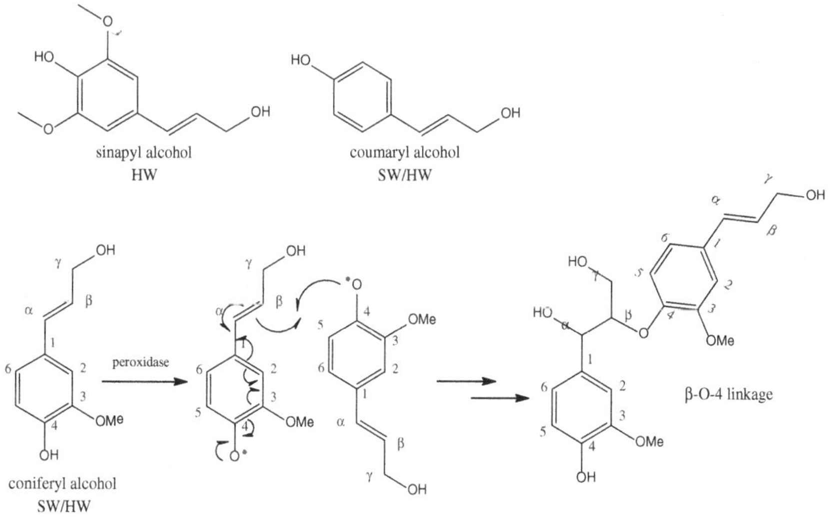

Fig. 3. Three phenyl propane precursors of lignin and the enzymatic synthesis pathway from coniferyl unit to one of the most abundant linkages in lignin structure — β-O-4 linkage.

the enzymatic synthesis pathway from coniferyl unit to one of the most abundant linkages in lignin structure — β-O-4 linkage. Some other major types of linkages and the reported abundance in softwood and hardwood lignins have been shown in Table 6.[18,19] Guerra *et al.* examined the molecular weights of various lignins isolated from different wood species and the results are summarized in Table 7.

3. Pyrolysis of Biomass Components

3.1 *Pyrolysis of lignin*

3.1.1. *Gas products of pyrolysis of lignin*

From the 1980's to the early 2000's, many lignin pyrolysis researches focused on the gas products.[20–26] Some major gas components reported in the literatures have been summarized in Table 8. Carbon monoxide and carbon dioxides are the top two most abundant components in the gas phase of pyrolysis of lignin. Normally, more than half percentages of gas products are these two components.[20–25] Methane has also been reported as another major gas component and the yield could be up to ~5 wt% of dry

Table 6. Reported abundance of major linkages in softwood and hardwood lignins.[18,19]

Linkage	β-O-4	4-O-5	Dibenzodioxocin
C-O linkage Abundance Per 100 C_9-units	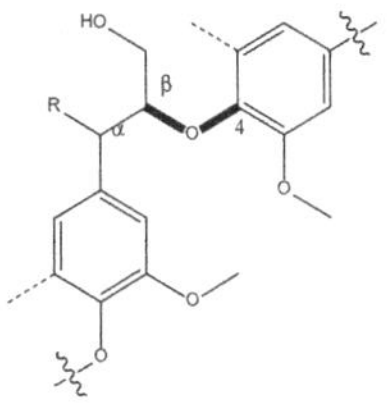	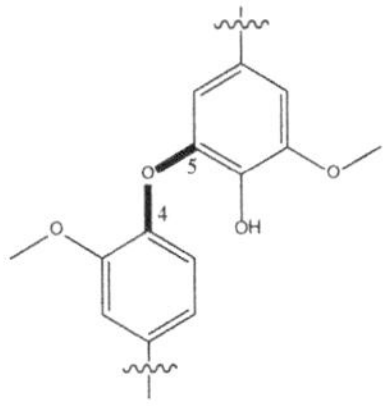	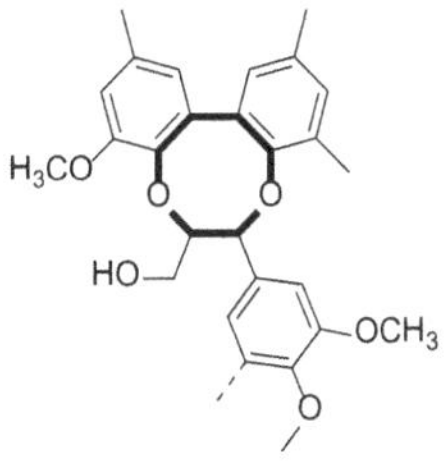
Softwood	45–50	4–7	5–7
Hardwood	60–62	7–9	0–2

Linkage	β -5	5–5	β -1	β-β
C-C linkage Abundance Per 100 C_9-units	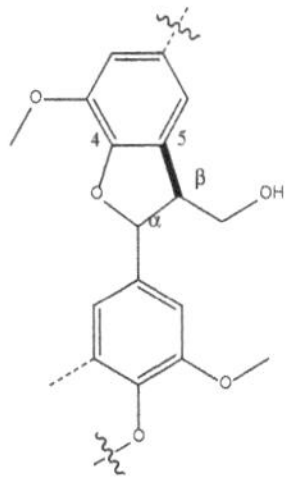	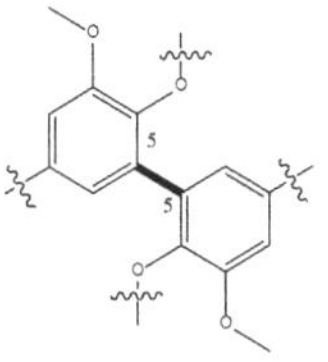	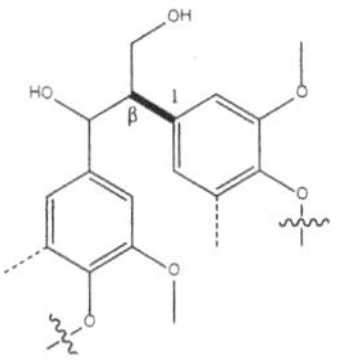	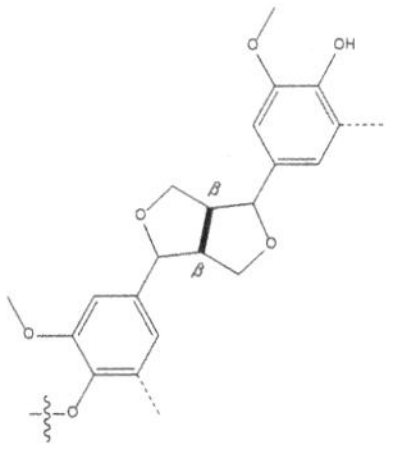
Softwood	9–12	19–22	7–9	2–4
Hardwood	3–11	3–9	1–7	3–12

Table 7. Weight-average molecular weight (M_w), number-average molecular weight (M_n), polydispersity (D) and approximate DP of various lignins isolated from different wood species.

Origin	M_w	M_n	D	DP[a]
Douglas Fir	38000	7600	5.0	42–211
White Fir	52000	6300	8.2	35–289
Redwood	30100	4700	6.4	26–167
E. globules	32000	8700	3.7	48–178
Southern Pine	57600	9700	5.9	54–320

[a]DP is calculated on the basis of phenyl propanoid units.

Table 8. Reported major gas products of pyrolysis from lignin.[99]

Major gas components of pyrolysis of lignin[a]			
CH_4 [20–26]	C_2H_4 [20–22,25,26]	C_2H_6 [20–22,25,26]	C_3H_6 [20–22,25,26]
CO[20–26]	CO_2 [20–25]	H_2 [23,24]	HCHO [20,22,26]
C_4H_8 [20–22,25,26]	CH_3CHO[20–22,26]		

[a]The pyrolysis temperatures are from 300–1000°C and lignins are isolated from both softwood and hardwood by kraft and ethanol-based pulping process.

lignin.[20, 21] Methane, carbon monoxide and carbon dioxides were found to increase in yield as the reactor temperature increased from 500 to 900 °C,[21, 22, 25, 26] nevertheless the content of methane decrease at a higher heating rate.[23] Most interestingly, Ferdous *et al.* [23, 24] indicated that pyrolysis of lignin also produces ~25 mol% of H_2 in the gas phase and the content significantly increased with increasing thermal conversion temperature. It is well known that H_2 and CO are the major components of syngas, which are used to produce synthetic petroleum, whereas the gas products of pyrolysis of lignin could also be used as syngas.

3.1.2. *Liquid products of pyrolysis of lignin*

Most of the pyrolysis works used GC-MS to analyze the liquid pyrolysis products.[27–43] By using pyrolysis (Py)-GC-MS, Jimenez *et al.*[43] indicated that softwood lignins yielded guaiacyl derivatives, coniferaldehyde and coniferyl alcohol were the major products and hardwood lignins produced guaiacyl and syringyl derivatives, syringaldehyde, coniferyl alcohol and sinapyl alcohol. Pyrolysis of bamboo lignin produced p-vinylphenol as the major compound. Similarly, Jiang *et al.* [31] also used Py-GC-MS to analyze pyrolysis products of lignin over a temperature range of 400–800 °C and indicated that the maximum yield of phenolic compounds was obtained at 600 °C. Most of the phenolic compounds had an individual yield of less than 1 wt% of lignin on a dry ash free basis. Greenwood *et al.*[38] pyrolyzed *Douglas fir* and *Q. nigra* water oak lignin in a laser micropyrolysis-GC-MS system. They found that guaiacol, 4-methyl-guaiacol, vinylguaiacol, eugenol, vanillin and coniferylaldehyde were the major components in the pyrolysis oil produced from *Douglas fir* lignin. For the *Q. nigra* water oak lignin pyrolysis oil, guaiacol, 4-methyl-guaiacol, vinylguaiacol, syringol, eugenol, 3, 5-dimethoxyacetophenone, 4-methyl 2, 5-dimethoxy benzaldehyde, 4-allyl-dimethoxylphenol, syringaldehyde, 2, 6-dimethoxyl-2-propylphenol and sinapaldehyde are found as the major components. Jegers *et al.*[40] also indicated that guaiacol, 4-methylguaiacol, 4-ethylguaiacol, catechol, 4-methylcatechol, 4-ethylcatechol, phenol, cresol and 4-ethylphenol were the major products of pyrolysis of lignin. As the most abundant products, the content of guaiacol and 4-methylguaiacol are ~5.3 wt% of dry lignin. Lou *et al.*[33] examined the effect of temperature on the composition of pyrolysis products and indicated that the contents of methoxyl contained components, such as guaiacol, 4-methylguaiacol, 4-vinylguaiacol and syringol decreased at higher pyrolysis temperature. In contrast, the contents of non-methoxyl contained compounds, such as cresols, ethyl-phenol, and 2, 6-dimethyl-phenol increased with increasing treatment temperature. A summary of lignin pyrolysis conditions and the yield of pyrolysis products are shown in Table 9.

To understand the possible decomposition pathways of lignin during the pyrolysis process and to find an effective upgrading method, many researchers choose to use pyrolysis oil model compounds to simplify

Table 9. Summary of lignin pyrolysis conditions and the yield of pyrolysis products.

Lignin	Temperature (°C)	Reactor	Tar (wt %)	Gas (wt %)	Char (wt %)
Kraft lignin (wheat straw and sarkanda grass)[41]	500	Fluidized bed	31	6	49
	410–560	Fluidized bed	31	12	34
	475–525	Fluidized bed	50	8	42
	700	Entrained flow	37	28	35
	480	Batch	22	30	48
Kraft lignin (pine)			23	39	41
Lignoboost™ (pine)	550	Fluidized bed[44]	22	49	29
EOL (*pinus radiate*)			16	21	63
Kraft lignin	800	Fixed bed[23]	3–5	49–52	43–48
EOL			14–21	41–44	35–44
Kraft lignin	650	Fix bed[24]	13	40	47
EOL			19	42	39
Klason lignin (almond shells)[22]	500	Micro pyroprobe reactor	53	7	34
	600		64	9	20
	700		55	17	17
	800		50	22	15
	900		43	29	14

the simulation model. Therefore, how to choose appropriate pyrolysis oil model compounds, which represent real pyrolysis oils, is challenging. To facilitate this part of work, a summary of GC-MS detected components in the lignin pyrolysis oils reported from seven literatures[31, 37–42] is shown in Table 10.

There are approximately one hundred compounds in Table 10 and almost all of them contain a phenol structure. Furthermore, phenol, cresols, guaiacol, 4-ethylphenol, syringaldehyde, acetosyringone,

Table 10. GC-MS detected components in the lignin pyrolysis oils.[a, 31, 37–42, 99]

Phenol[31, 37–42]	4-Allyl-dimethoxyphenol[38]	4′-Hydroxy-3′-methoxy acetophenone[31, 37–39, 41, 42] Acetovanillone Acetoguaiacone
2-Methylphenol[31, 37–42] o-Cresol	Dimethoxypropyl phenol[38]	5-Tert-butylpyrogallol[31]
4-Methylphenol[31, 37, 38, 40, 41] p-Cresol	Coniferylaldehyde[38, 42]	1-(4-Hydroxy-3-methoxyphenyl)-2-propanone[31, 41, 42] Guaiacyl acetone
2-Methoxyphenol[31, 37–42] Guaiacol	Sinapaldehyde[38, 42]	2-(3,4-Dimethoxyphenyl)-6-methyl-3, 4-chromanediol[31]
2,6-Dimethylphenol[31, 39, 41] 2,6-Xylenol	2,6-Dimethoxy-4-methyl-phenol[37, 42] 4-Methylsyringol	3,4-Dimethylbenzoic acid[31, 39]
4-Ethylphenol[31, 37, 40, 41] p-Ethylphenol	1-(4-Hydroxy-3-methoxyphenyl) propyne[37]	3-Methoxy-4-hydroxybenzoic acid[31]
3-Methylbenzaldehyde[31] m-Tolualdehyde	4-Ethyl-2,6-dimethoxy phenol[37, 42] 4-Ethylsyringol	4-Ethyl-1,2-dimethoxy-benzene[31]
2-Hydroxy-6-methylbenzaldehyde[31]	4-Vinyl-2,6-dimethoxy-phenol[37, 42] Vinylsyringol	4-Propenylsyringol[31, 37, 42] 4-Propenyl-2,6-dimethoxyphenol
2-Ethylphenol[31, 40, 41]	4-Propyl-2,6-dimethoxyphenol[37]	Ferulic acid[31] 3-Hydroxy-4-methoxycinnamic acid

(*Continued*)

Table 10. (*Continued*)

4-Methoxy-3-methylphenol[31]	Syringylacetone[37,42]	4-Hydroxy-3,5-dimethoxy-benzaldehyde[31,37,38,41,42] Syringaldehyde
2-Methoxy-4-methylphenol[31,37–42] 4-Methylguaiacol	m-Cresol [39,40]	Acetosyringone[31,37,38,41] 4′-Hydroxy-3′,5′-dimethoxyacetophenone
Catechol[31,38–41] 1,2-Benzenediol	p-Propylphenol[40]	1-(2,6-Dihydroxy-4-methoxyphenyl)-1-butanone[31] Desaspidinol
Benzofuran[31]	6-Ethylguaiacol[40]	Syringic acid[31] 4-Hydroxy-3,5-dimethoxybenzoic acid
p-Isopropylphenol[31] p-Cumenol	2-Methoxy-4-propyl-phenol[39,40,42] 4-Propylguaiacol	2,3,5-Trimethyl phenol[41]
2-Ethyl-4-methylphenol[31]	4-Methyl-1,2-benzenediol[39,40] 4-Methylcatechol	3-Ethyl phenol[39–41]
3-Methoxy-1,2-benzenediol[31,40,41] 3-Methoxycatechol	6-Ethylcatechol[40]	1,2,3-Trimethoxybenzene[41]
3-Methyl-1,2-benzenediol[31,38–41] 3-Methylpyrocatechol 3-Methylcatechol	3-Methylguaiacol[42]	Coniferyl alcohol[39–41]
2-Methoxy-4-ethylphenol[31,37–42]	3-Ethylguaiacol[42]	Methoxyeugenol[37,41] 4-Hydroxy-3,5-dimethoxyallylbenzene 4-Allyl-2,6-dimethoxyphenol
4-(2-Propenyl)phenol [31]	Propioguaiacone[42] 1-(4-Hydroxy-3-methoxy-phenyl)-propan-1-one	1-Methoxy-3-methylbenzene[38]
p-Isopropenylphenol[31]	6-Hydroxy-5,7-dimethoxy-indene[42]	Indene[38]
2-Methoxy-4-vinylphenol[31,37,38,41,42] 4-Vinylguaiacol	Dihydroconiferyl alcohol[42]	1,2,3-Trimethylbenzene[38]

(*Continued*)

Table 10. (*Continued*)

3-Methyl-5-methoxyphenol[31]	Propiosyringone[42]	1,2,4-Trimethylbenzene [38]
4-Ethyl-1,3-benzenediol[39] 4-Ethylresorcinol	Dihydrosinapyl alcohol[42]	Mesitylene[38]
2,6-Dimethoxyphenol[31,37,38,41,42] Syringol	Sinapyl alcohol[42]	4-Ethenylphenol[37,38] Vinylphenol
2,5-Dimethyl-1, 4-benzenediol[31]	2,3-dimethylphenol[39] 2,3-xylenol	m-Dimethoxybenzene[38]
2,4-Dimethoxyphenol[31]	Naphthalene[38,39]	Veratrole[38]
2′,4′-Dimethyl-acetophenone[31]	Benzene[41]	p-Dimethoxybenzene[38]
4-Ethyl-1,2-benzenediol[31,40] 4-Ethylpyrocatechol	Styrene[41]	Dimethyl catechol[38]
Eugenol[31,37–39,41,42]	p-Xylene[41]	Vinylcatechol[38]
3-Hydroxy-4-methoxy-benzaldehyde[31,38,41] Isovanillin	Ethylbenzene[41]	Vanillin [37–39,42]
2,5-Dimethoxy-benzylalcohol[31]	Toluene[41]	3′,5′-Dimethoxy-acetophenone[38]
2-Methoxy-4-(1-propenyl) phenol[31,37–39,41,42] Isoeugenol	Fluorene[38]	4-Methyl 2,5-dimethoxy benzaldehyde[38]
4′-Hydroxy-3′-methoxyacetophenone[31,37–39,41,42] Acetovanillone Acetoguaiacone		

[a] The pyrolysis temperatures are from 400–800 °C

catechol, 3-methylcatechol, 4-methylguaiacol, 4-vinylguaiacol, vanillin, syringol, eugenol, isoeugenol and acetovanillone have been reported in more than four references and many of them could also be found in other references.[29, 32–34, 36] Therefore, these components could be used as potential candidates for the model compounds study on pyrolysis of lignin. Figure 4 summarizes the reported possible decomposition pathways of lignin during the pyrolysis.

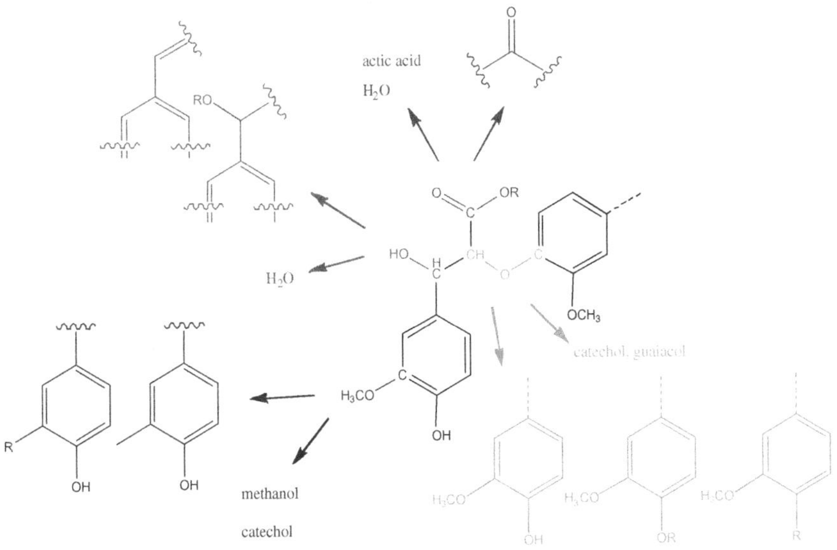

Fig. 4. The possible decomposition pathways of lignin during the pyrolysis.[45–51,99]

3.2. *Pyrolysis of cellulose*

Cellulose is the most abundant biopolymer. Generally, biomasses contain around 35–50% of cellulose, which is a polymer of β-(1, 4)-glucan with a degree of polymerization of ~300–15000. Figure 1 shows the model structure of cellulose. Other than the biochemical conversion of cellulose to ethanol — a second generation bio-ethanol, some researchers also reported to convert this biopolymer and its model compound such as glucose to liquid fuels and chemicals. For example, Huber *et al.*[52] examined the conversion of biomass-derived carbohydrates to the liquid alkanes by acid-catalyzed dehydration, which was followed by aldol condensation. The liquid alkanes have the molecular weight in the transportation fuel range and contain 90% of energy of the carbohydrate and H_2 feeds. Figure 5 shows this conversion process. One step conversion of cellobiose to C_6-alcohols with Pt, Pd, Rh or Ru nanocluster catalyst at 120°C, 4 MPa and different pHs has also be investigated in the literature and is illustrated in Fig. 6.[53] By using a Ru nanocluster catalyst at pH 2, cellobiose could be completely converted to sorbitol.

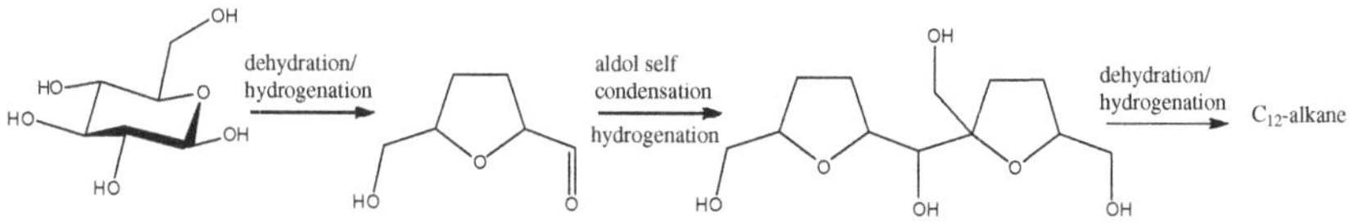

Fig. 5. Conversion of biomass-derived carbohydrates to the liquid alkanes by acid -catalyzed dehydration which was followed by aldol condensation.[52]

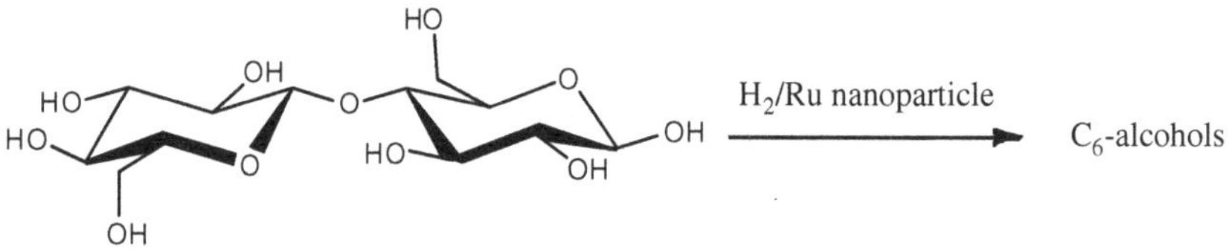

Fig. 6. One step conversion of cellobiose to C_6-alcohols with Pt, Pd, Rh or Ru nanocluster catalyst.[53]

As the major component in the biomass, cellulose plays an important role in understanding pyrolysis pathways of the whole biomass. Shen *et al.*[54] employed pyrolysis-GC-MS to investigate the mechanism of cellulose pyrolysis and the effects of temperature and residence time. The main products in the cellulose pyrolysis oil include levoglucosan, hydroxyacetaldehyde, hydroxyacetone, pyruvic aldehyde, glyceraldehyde, 5-hydroxymethyl-furfural and furfural. The contents of most of these main products increased when the temperature or residence time was elevated, nevertheless, the yield of pyrolysis oil decreased when temperature exceeded 570°C. The contents of the major components in cellulose pyrolysis oil — levoglucosan was more than 50% molar fraction of pyrolysis oil and decreased with the elevated temperature. In the meanwhile, the yield of char reached the minimum and remained constant when temperature exceeded 550°C. The authors also indicated that the formation of CO was increased with increased temperature and residence time, whereas slight change was observed for the yield of CO_2. The content of CO was up to 20.2 wt% of cellulose, while the yield of H_2 was ~0.9 wt% of cellulose. This result indicated that the gas products of pyrolysis of cellulose could also be used as the syngas to produce synthetic petroleum.

On the basis of the study on mechanism, most of hydroxyacetone was found to be produced from the direct conversion of cellulose molecules,

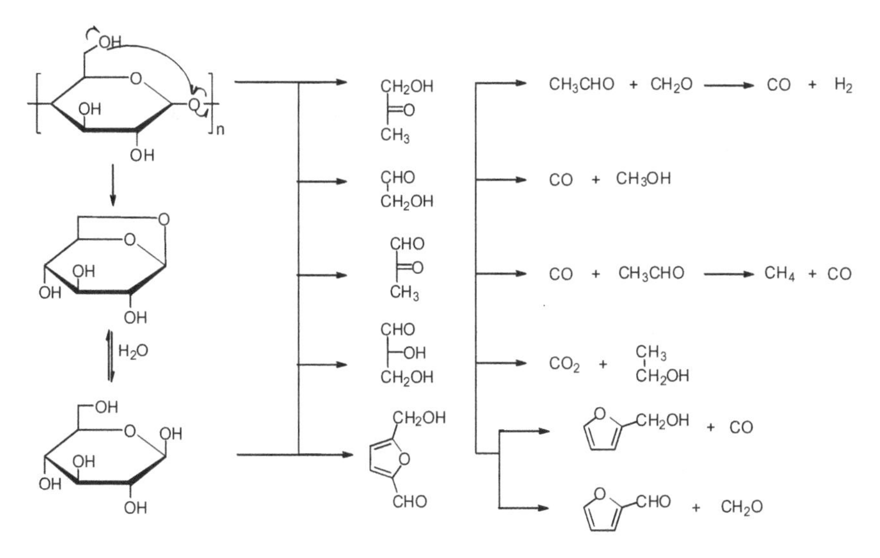

Fig. 7. Proposed primary decomposition of cellulose and secondary decomposition of the anhydrosugars such as levoglucosan.[54]

while the formation of pyruvic aldehyde is mainly from the secondary decomposition of levoglucosan. The authors[54] have proposed the primary decomposition of cellulose and secondary decomposition of the anhydrosugars such as levoglucosan, which have been summarized in Fig. 7.

Wang *et al.*[55] investigated the interaction of biomass components including cellulose, hemicellulose and lignin during the pyrolysis process. By the use of GC-MS, levoglucosan, 2, 5-dimethoxytetrahydrofuran, 1-hydroxy-2-propanone, altrose and 2-hydroxy-2-cyclopenten-1-one have been detected as major components in the cellulose pyrolysis oil. In addition, the authors indicated that the interaction of cellulose and hemicellulose significantly improved the formation of 2, 5-dimethoxytetrahydrofuran but inhibited the formation of altrose and levoglucosan. Similarly, Alen *et al.*[56] indicated that the cellulose pyrolysis oil could be summarized as several compound groups including the volatiles, which mainly contains carbon monoxide, carbon dioxide, methanol, acetaldehyde, acetic acid, hydroxyacetaldehyde (glycolaldehyde), 1-hydroxy-2-propanone (acetol), and some $<C_5$-hydrocarbons; the anhydroglucopyranose (levoglucosan); the anhydroglucofuranose (1,6-anhydro-β-D-glucofuranose); the dianhydroglucopyranose (1,4;3,6-dianhydro-α-D-gludopyranose) and the furans, which include (2H)-furan-3-one,

5-methyl-(3H)-furan-2-one (α-angelicalactone), 2-furaldehyde (furfural), 5-methyl-2-furaldehyde, and 5-hydroxymethyl-3-furaldehyde. They also concluded that at lower temperatures (400–600 °C), the main products of pyrolysis of cellulose were anhydrosugars; however, at higher pyrolysis temperatures (800–1000 °C), these anhydro sugars would be secondarily decomposed to volatiles.

To facilitate the analysis of pyrolysis of cellulose and further upgrading process, a summary of the GC-MS detected components in the cellulose pyrolysis oils reported from six references is shown in Table 11.[54,55,57–60] Levoglucosan has been reported as the most abundant component in the cellulose pyrolysis oil, the content is more than 50 wt% of pyrolysis oil. Furfural and 5-hydroxymethylfurfural are also the major components in the cellulose pyrolysis oil.

3.3. *Pyrolysis of hemicellulose*

Hemicellulose is a short-chain branched and substituted polymer of sugars with a degree of polymerization of ~50–300; typically biomass contains 25–30 wt% of hemicellulose. In contrast to cellulose, hemicellulose is a polymer of several different sugar monomers. It contains five-carbon sugars, i.e., xylose, arabinose and six-carbon sugars including galactose, glucose and mannose. All of these sugar monomers are substituted with acetate group. The most abundant sugar in the softwood hemicellulose is mannose; correspondingly, xylose is most abundant in the hardwood hemicellulose.[2]

Surprisingly, limited information on pyrolysis of hemicellulose was mentioned in literatures although hemicellulose is one of the major components in biomass. Patwardhan *et al.*[61] examined the pyrolytic behavior of switchgrass hemicellulose and a total of 16 different products were quantitatively indentified. The authors indicated that the pyrolysis products from pure hemicellulose could be classified into three categories: a) low-molecular-weight compounds including CO (2.8 wt%), CO_2 (18.8 wt%), formic acid (11.0 wt%), acetaldehyde (0.7 wt%), acetic acid (1.1 wt%), and acetol (3.0 wt%); b) furan/pyran ring derivatives including 2-methyl furan (1.5 wt%), 2-furaldehyde (2.2 wt%) and dianhydro xyloses (9.2 wt%) and c) anhydro sugars — anhydro xylopyranoses (3.4 wt%). In the meanwhile, there are 10.7 wt% of char, 15.1 wt% of water

Table 11. GC-MS detected components in the cellulose pyrolysis oils[a].[54,55,57–60]

Acetone	Resorcinol	2-butanone
Phenol	4-methyl-5h-furan-2-one	propenic acid acrylic acid
propanal	1,2,4-cyclopentantrione	ethylene glycol
2-propenal	2-hydroxy-3-methyl-2-cyclopenten-1-one	2,3-pentandione 2,3-pentanedione
Furan	1,4-benzenediol hydroquinone	acetyl methyl carbinol acetoin
formic acid	2-hydroxybenzaldehyde salicylaldehyde	2-cyclopentenone 2-cyclopenten-1-one
3-buten-2-one	4-methylphenol p-cresol	2-furanemethanol furfuryl alcohol
acetic acid	3-furancarboxylic acid methyl ester	2-furanone
2-butenal Crotonaldehyde	(2-methoxyethyl) cyclohexane	2-hydroxy-2-cyclopentene-1-one
1-hydroxy-2-propanone hydroxyacetone	methyl 2-furoate methyl-2-furancarboxylate	3-methyl-2,5-furanedione citraconic anhydride
2,5-dimethylfuran	3-methylphenol m-cresol	5-methylfurfural 5-methyl-2-furaldehyde
2-methylfuran	3,5-dihydroxytoluene 5-methylresorcinol	o-cresol
acetic acid methyl ester methyl acetate	1-(2-furanyl)-2-hydroxy-ethanone	2-hydroxy-3-methyl-4-pyranone
2-oxopropanoic acid methyl ester	4-hydroxy-2, 5-dimethyl-3(2H)-furanone	dihydro-4-hydroxy-2-furanone
3-furaldehyde	dihydro-3-methylene-2(3H)-furanone alpha-methylene-gamma- butyrolactone	benzoic acid
furfural 2-furaldehyde	maltol 3-hydroxy-2-methyl-4-pyrone	catechol 1,2-benzenediol
4-cyclopentene-1,3-dione cyclopenten-1,3-dione	3-ethyl-2-hydroxy-2-cyclopenten-1-one	1,2-dihydroxy-3-methylbenzene 3-methyl pyrocatechol 3-methylcatechol

(*Continued*)

Table 11. (*Continued*)

2-propylfuran	5,6-dihydro-6-methyl-2H-pyran-3(4H)-one	hexane
2(3H)-furanone,5-methyl-3-penten-4-olide	5-hydroxy-2-methyl-4H-pyran-4-one	hydroxyacetaldehyde
2-propanone,1-(acetyloxy)-acetonyl acetate	2,3-dihydro-3,5-dihydroxy-6-methyl-4H-pyran-4-one	2-hydroxy-2-cyclopenten-1-one
Furfuryl formate	3,5-dihydroxy-2-methyl-4H-pyran-4-one	1,3-butadiene-1-carboxylic acid
2-acetylfuran	1,2,3-benzentriol pyrogallol	ethylbenzene
2,3,5-trimethylfuran	Hydroxyquinol	5-methyl-2(5H)-furanone 2-penten-4-olide
2(3H)-furanone	5-hydroxymethyl-2-furancarboxaldehyde 5-(hydroxymethyl) furfural	2-methyl-pentanone
1,2-cyclohexanediol	2,3-dihydro-1h-inden-1-one 1-indanone	3-methyl-1,2-cyclopentadion
2H-pyran-2-one 2-pyrone	levoglucosan	2,4-dimethylphenol 2,4-xylenol
4H-pyran-4-one	2,3-butandione	stilbene

[a] The pyrolysis temperatures are from 350–730 °C.

and 4.9 wt % of xylose. The authors also investigated the influence of temperatures on products from hemicellulose pyrolysis. The results showed that when the pyrolysis temperatures increased from to 250°C 600 °C, the content of CO and CO_2 almost linearly increased from 0.5–4.7 wt% and from 2.7–24 wt%, respectively. Similarly, the yield of formic acid linearly increased from 5.4–15.6 wt% when the thermal treatment temperature increased from 350 °C to 600 °C. In contrast with the increasing gas yield at higher pyrolysis temperatures, the yield of char decreased from 24 wt% at 300 °C to 6.3 wt% at 600 °C. Based on the literature reports, the authors also proposed tentative thermal decomposition pathways of hemicellulose, which have been summarized in Fig. 8.

Lv *et al.*[62] investigated the effects of pyrolysis temperatures (400–900 °C) on the pyrolysis products from corn stalk hemicellulose. They indicated

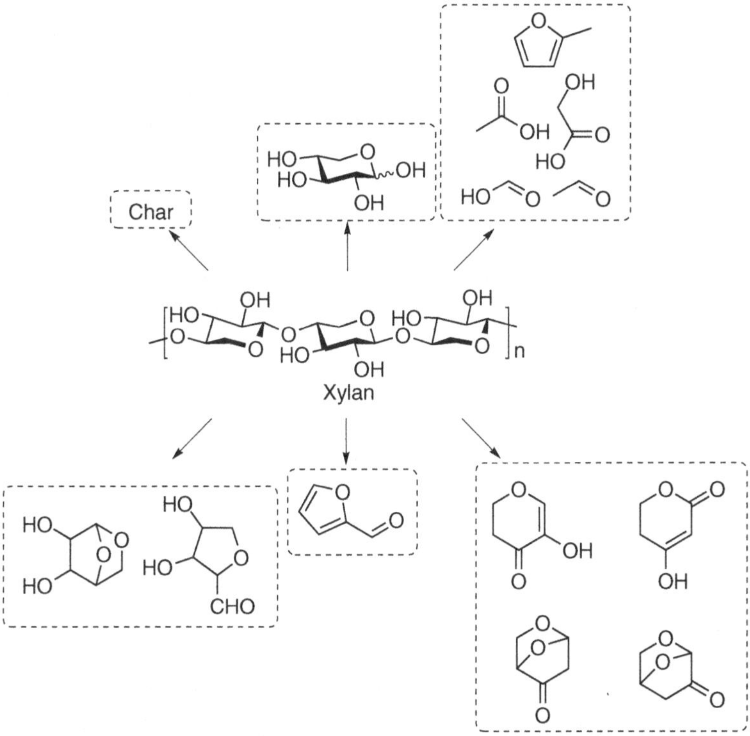

Fig. 8. Tentative thermal decomposition pathways of hemicellulose.[61]

that the yield of char linearly decreased but the yield of gas products increased when the pyrolysis temperature elevated. The maximum yield of liquid products (48.2 wt%) achieved at 450 °C. The pyrolysis oils were composed of ketones, furans, carboxylic acids, and alcohols. Ketones have been reported as the most abundant type of liquid products, which was up to 71 area% of pyrolysis oil detected by GC-MS. The 1-hydroxy-2-propanone (19 area% at 450 °C ~26 area% at 900 °C), 1-hydroxy-2-butanone (12 area% at 450 °C ~11 area% at 900 °C), 2-cyclopenten-1-one (5 area% at 450 °C ~7 area% at 900 °C), 2-methyl-2-cyclopenten-1-one (4 area% at 450 °C ~6 area% at 900 °C), 3-methyl-1,2-cyclopentanedione (4 area% at 450 °C ~7 area% at 900 °C) and 4-hydroxy-3-hexanone (3 area% at 450 °C ~0 area% at 900 °C) were the major components as the ketones type pyrolysis products. Furfural (15 area% at 450 °C ~14 area% at 900 °C), dihydro-2(3H)-furanone (~3 area% from 450 °C to 900 °C) and

2,5-dihydro-3,5-dimethyl-2-furanone (~2 area% from 450 °C to 900 °C) were the major types of furans in hemicellulose pyrolysis products. Acetic acid (16 area% at 450 °C ~11 area% at 900 °C) and propanoic acid (3 area% at 450 °C ~0 area% at 900 °C) were the main acid products from pyrolysis of hemicellulose. Cyclobutanol (2 area% at 450 °C ~0 area% at 900 °C) has been listed as the major alcohol type product. The top four gas products were CO_2, CO, H_2, and CH_4 (sorted by content), and there were also small amount of C_2H_4 and C_2H_6 produced from hemicellulose. Similar to Patwardhan,[61] Lv also indicated that the formation of CO_2 was due to the cleavage of acetyl and carboxyl groups in hemicellulose and a higher pyrolysis temperature favored such cracking, and thus increased CO_2 formation. The authors also indicated that the formations of CO, H_2, and CH_4 and other gas products such as C_2H_4 and C_2H_6 were related to the secondary decomposition of liquid products, which would also be improved at higher reactor temperature and lead to an increasing yield of such gas products. The authors examined the char by SEM and found that the surface of the higher temperature produced char was more smooth and porous.

Peng *et al.*[63] characterized pyrolysis of wheat straw hemicellulose by TG, DTG and pyrolysis-GC-MS. They indicated that the major weight loss of hemicellulose occurred at 190–315 °C and the main components in the hemicellulose pyrolysis oil were acetic acid, 2-furaldehyde, cyclopenten-1-one and small amounts of aromatic compounds including 2-ethyl-5-methyl-phenol and 2,5-dimethyl-phenol. In the meanwhile, Shen *et al.*[64] also investigated the pyrolytic behaviors of beech wood hemicellulose by TG-FTIR and pyrolysis-GC-FTIR. They found that methanol (0.81 wt% at 425 °C ~1.14 wt% at 690 °C), acetic acid (4.44 wt% at 425 °C ~2.45 wt% at 690 °C), acetone (0.69 wt% at 425 °C – 1.06 wt% at 690 °C), furfural (1.94 wt% at 425 °C ~3.16 wt% at 690 °C) and 1,4-Anhydro-D-xylopyranose (3.84 wt% at 425 °C ~0.69 wt% at 690 °C) were the major components in the pyrolysis oil produced from beech wood hemicellulose. In addition, the authors discussed the formation pathways of some major pyrolysis products. For instance, the authors indicated that the decomposition of xylan units produced furfural and 1,4-anhydro-D-xylopyranose, and the formation of acetic acid and CO_2 was due to the primary cleavage of O-acetylxylan unit. The xylan unit was also proposed as the main precursor for the formation of <C3 fragments and gas products including CO, H_2 and CH_4. The primary decomposition of 4-O-methyglucuronic acid unit produces methanol and the decomposition of the ring-opened intermediate products or secondary decomposition of the fragments form acetone and CO.

3.4. *Pyrolysis of tannin*

As we have discussed before, cellulose, hemicellulose and lignin are the major components in the biomass. However, in soft tissues including leaves, needles as well as barks etc., there are another type of biopolymer — tannins that could be more abundant than lignin occasionally.[8,65] There are two types of tannin, hydrolyzable tannins are derivatives of gallic acid, which are esterified to polyols such as glucose (Fig. 9), and condensed tannins (Fig. 10) that are polymers of flavonoids and are much more complicated than hydrolyzable tannins. Peter *et al.*[65] did a very detailed research about tannin signatures of 117 tissues from 77 different biomass species at the molecular level.

There are only very limited references investigated the pyrolysis of tannins. Ohara *et al.*[66] examined pyrolysis of various tannins from several

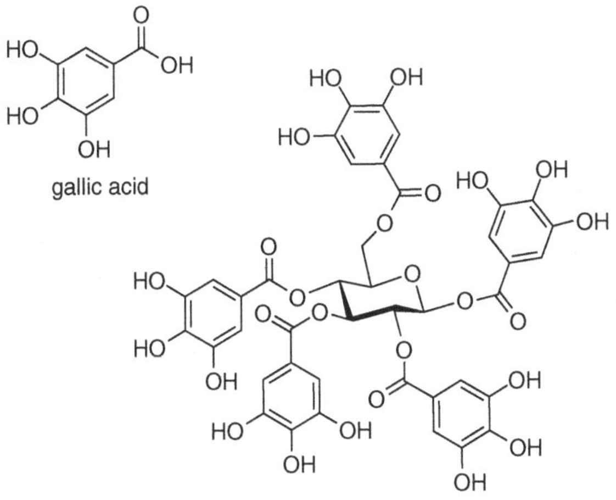

Fig. 9. Model structure of hydrolyzable tannin.

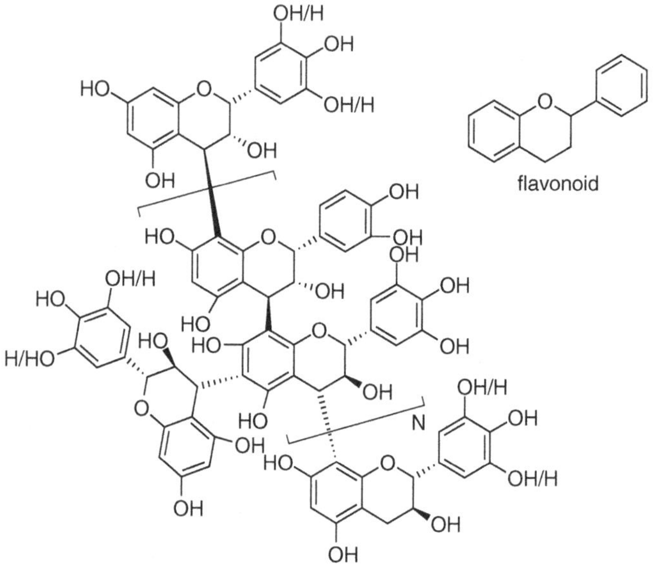

Fig. 10. Model structure of condensed tannin.

barks including *Acacia mearnsii De Wild, Salix rorida Lackschewitz, S. pet-susu Kimura, S. miyabeana Miq and S. sachalinensis Seem* as well as several tannin model compounds such as catechin and epicatechin. The authors indicated that catechol and 4-methylcatechol are the major products from pyrolysis of catechin and epicatechin and they also proposed the possible formation pathways of these two products. The formation of catechol is due to the cleavage of C2-C1′ bond of flavanols, and the fission of the pyran ring will produce 4-methylcatechol. For the pyrolysis of various bark tannins, phenol, catechol, resorcinol, 4-methylcatechol, pyrogallol and 5-methylpyrogallol were reported as the major components in the pyrolysis oils. Ben *et al.*[67] investigated the pyrolysis of pine bark tannin, pine bark and pine residue at 400 °C, 500 °C and 600 °C. The analysis of the yields of pyrolysis products indicated that pine bark tannin yielded large amounts of a light oil, which contained water and water soluble components. All the pyrolysis oils were analyzed by HSQC-NMR and the results indicated that catechol and methyl-catechol were the major components in the pine bark tannin pyrolysis oils. Most of the aromatic C-H and aliphatic C-H bonds in the pyrolysis oils produced from pine bark and pine residue were produced from the lignin and tannin components. Gaugler *et al.* [68] used TGA to analyze the thermal degradation of various condensed tannins and tannin model compounds. The highest weight loss of catechin occurred at 197 °C, sulfited tannin occurred at 159 °C, tannin acetate occurred at 189 °C and Quebracho tannin (commercial tannin extracted from Quebracho) occurred at 271 °C. Reported pyrolysis products from tannins have been summarized in Table 12.[66,67]

4. Characterization Methods of Pyrolysis Oil

4.1. *FT-IR analysis of lignin pyrolysis oil*

Most compounds in Tables 10, 11 and 12 have a molecular weight below 220 g/mol. However, some researchers have detected the average molecular weight of pyrolysis oil by GPC and reported that the molecular weights

Table 12. GC-MS detected components in the tannin pyrolysis oils.[66,67]

phenol	catechol 1,2-benzenediol	4-ethyl-1,3-benzenediol 1,3-benzenediol, 4-ethyl- 4-ethylresorcinol
4-methylphenol p-cresol	4-Methoxyphenol mequinol	1,4-benzenediol, 2-methoxy-
m-cresol phenol, 3-methyl-	4-methyl-1,2-benzenediol 1,2-benzenediol, 4-methyl- 4-methylcatechol	benzoic acid, 3, 4-dihydroxy-, methyl ester
2-methoxyphenol phenol, 2-methoxy- guaiacol	Hydroquinone benzene-1,4-diol	2-methoxy-4-methylphenol phenol, 2-methoxy- 4-methyl- 4-methylguaiacol
2,6-dimethylphenol phenol, 2,6-dimethyl- 2,6-xylenol	2-methoxy-4-ethylphenol phenol, 4-ethyl- 2-methoxy-	1,4-benzenediol, 2-methyl-
2,3-dimethylphenol phenol, 2,3-dimethyl- 2,3-xylenol	Benzene-1,3-diol resorcinol	
benzoic acid	1,3-benzenediol, 2-methyl-	

were from 210–1700 g/mol.[69–72] Due to the limitation of volatility of high molecular weight components in the pyrolysis oil, it has been suggested that only about 40 % of pyrolysis oil could be detected by GC.[73] Therefore, many researchers have also tried to find alternative characterization method, which could analyze the whole portion of pyrolysis oil, such as FT-IR.[42,74–79] Liu *et al.*[76] did a mechanistic study of hardwood and softwood lignin pyrolysis using a thermogravimetric analyzer coupled with a Fourier transform infrared spectrometry (TG-FTIR). They indicated that lignin underwent three consecutive sets of reactions during pyrolysis including the evaporation of water, the formation of primary volatiles and the release of small molecular gases. At first, the absorbed water was released by evaporation, and then at a higher temperature (above 100 °C) water was generated by the dehydration of lignin aliphatic hydroxyl groups. The authors also indicated that phenols, in addition to alcohols, aldehydes, acids and CO, CO_2 CH_4 were the major gaseous products. Scholze *et al.*[42] also characterized pyrolysis oil by FT-IR and indicated a correlation between carbonyl absorption bands and oxygen content as well as carbon content. Lievens *et al.*[80] demonstrated the deconvolution of the region of 1490–1850 cm^{-1} in the FT-IR spectra, which could provide detailed information about various carbonyl groups in various pyrolysis oils. To facilitate this promising fast analysis method of pyrolysis oils, a summary of the assignment ranges from seven references[42,74–76,78,79,81] are shown in Table 13.

4.2. *NMR analysis of pyrolysis oil*

It is well known that pyrolysis oil is a very complex mixture, whereas the ability of FT-IR to comprehend the details of pyrolysis oil is limited. Most recently, some research[44,71,82–89] have introduced NMR, including quantitative ^{1}H, ^{31}P and ^{13}C-NMR, and semi-quantitative HSQC-NMR to characterize pyrolysis oils. Mullen *et al.*[82] analyzed various pyrolysis oils produced from switchgrass, alfalfa stems, corn stover, guayule (whole plant and latex-extracted bagasse), and chicken litter by ^{1}H, ^{13}C, and ^{13}C-DEPT (Distortionless Enhancement Polarization Transfer)-NMR. They found that pyrolysis oil from chicken litter had the lowest overall amount of methyl groups and the highest ketone content of the pyrolysis

Table 13. FT-IR assignments of biomass pyrolysis oil.[99]

Assignments [42,74–76,78,79,81]	Wave number (cm^{-1})
O-H stretching vibration, H_2O	3429
Aromatic ring C-H stretching	3000–3040
Asymmetrical C-H stretching, vibration of aliphatic CH_3 and CH_2	2930–2980
Symmetrical C-H stretching, vibration of aliphatic CH_3 and CH_2	2850–2870
C=O stretch in unconjugated ketones, carbonyl and ester groups	1701–1734
C=O stretch in conjugated aryl ketones	1652–1666
Aromatic ring vibrations and C=O stretch	1593–1609
Aromatic ring vibrations	1504–1515
Asymmetric C-H bending (in CH_3 and $-CH_2$-)	1462–1464
Aromatic ring vibrations	1420–1424
Symmetric deformation of C-H in methyl groups	1365
Phenolic hydroxyl vibrations	1360
Vibrations of guaiacyl rings and stretching vibrations of C-O bonds	1270
C-C, C-O and C=O stretching	1214–1233
Vibrations of methoxyl group	1190
Deformation vibrations of C-H bonds in benzene rings	1160
Deformation vibration of C-H bonds in guaiacyl rings	1140
Vibrations of ester linkage	1115
Aromatic in-plane C-H bending	1114–1125
Deformation vibrations of C-O bonds in secondary alcohols and aliphatic ethers	1075–1090
Deformation vibrations of C-H bonds in aromatic rings	1030–1033
Aromatic out-of-plane C-H bending	914–919
Aromatic out-of-plane C-H bending in positions 2, 5 and 6 of guaiacyl units	852–859
Vibrations of C-H bonds in syringyl units	833
Bending out of the plane C-H	720–740
Bending out of the plane = CH	730–675

oils studied. The ^{13}C and DEPT-NMR analyses indicated that the pyrolysis oils from corn stover and switchgrass had the fewest aliphatic carbons. The large amount of methine (CH_1) groups in the corn stover pyrolysis oil suggested that its aliphatic groups were highly branched. However, there were almost the same amount of methyl ($-CH_3$) groups as its methine groups. While the percentage of $-CH_2-$ was low, it indicated that these branches were very short and could mostly be methyl groups. Conversely, pyrolysis oil from switchgrass appeared to have more straight-chain aliphatics. The authors also indicated that the aromatic region of these pyrolysis oils had CH_0:CH_1 ratios of >2:1, which represents highly complex substituted (at least four substituents) benzene rings. On the basis of 50 pyrolysis oil model compounds, Joseph *et al.*[84] have measured the T1 and proposed assignments range for both ^{13}C-NMR and ^{1}H-NMR.

Our previous work[90] used ^{31}P and ^{13}C-NMR to characterize pyrolysis oils produced from softwood (SW) kraft lignin at 400 °C, 500 °C, 600 °C and 700 °C. A ^{13}C-NMR database was created to provide a more accurate chemical shift assignment for analysis of pyrolysis oils. This analysis showed that the carbonyl group content was reduced after pyrolysis, and methoxyl group was significantly eliminated after pyrolysis, especially at higher pyrolysis temperatures. Nearly 70%–80% of the carbons from water insoluble portion of pyrolysis oil (heavy oil) are aromatic carbon. By using ^{31}P-NMR, the results indicated that the heavy oils contained less aliphatic hydroxyl group and carboxyl acid group than the lignin. The decreased concentration of aliphatic hydroxyl and acid groups was significant as it indicated that the lignin side chain hydroxyl groups were readily eliminated during the thermal treatment. In contrast, the content of guaiacyl, p-hydroxyphenyl and catechol type hydroxyl groups increased after pyrolysis. The ^{31}P-NMR results for the water soluble part of pyrolysis oil (light oil) showed that it contained nearly 80 w/w % water and another 10 w/w % was methanol, catechol and acetic acid.

To solve spectral overlapping problems when using ^{13}C-NMR to analyze the pyrolysis oils, our previous work[91] demonstrated that HSQC-NMR was uniquely well suited to analyze various C-H bonds presented in the pyrolysis oils. The fingerprint analysis of HSQC-NMR spectral data provided chemical shift assignment of twenty-seven (fourteen from lignin pyrolysis oil) different types of C-H bonds present in pyrolysis oils produced from cellulose, lignin and pine wood. The HSQC-NMR for the lignin pyrolysis oils showed that there were two different types of methoxyl group presented in the pyrolysis oils, which indicated that the native methoxyl group in the kraft lignin would rearrange to another type during the thermal treatment. As the main products of pyrolysis cellulose — levoglucosan[91,92] was shown to increase with the increasing pyrolysis temperature from 400 °C to 600°C, furfurals and phenols were also found to be the major components in the cellulose pyrolysis oils. The content of 5-hydroxymethyl-furfural (HMF) in the cellulose pyrolysis oil decreased at higher thermal treatment temperatures, which indicated that HMF could be further decomposed at higher temperatures. The content of aromatic C-H and aliphatic C-H bonds in the lignin pyrolysis oils increased with increasing pyrolysis temperature, which was attributed to the rearrangement and the cleavage of ether bonds or methoxyl groups in the lignin structure. The analysis of HSQC-NMR for the pine wood pyrolysis oils indicated that levoglucosan was also one of the major components presented in the pyrolysis oils and most of aromatic C-H and aliphatic C-H bonds in the pine wood pyrolysis oils produced from lignin component. Table 14 summarizs the functional groups presented in pyrolysis oils, which could be analyzed by NMR.

4.3. *Elemental analysis, viscosity, acidity, heating value and solid residue of pyrolysis oil*

The major drawback towards commercialization of pyrolysis oils are several challenging properties including poor volatility, high oxygen content, acidity and viscosity, corrosiveness, cold flow and aging problems.[6] The reported results of elemental analysis, viscosity, acidity, heating value and solid residue for various pyrolysis oils have been summarized in Tables 15 and 16. The elemental analysis results show that the oxygen contents in the whole biomass pyrolysis oils are much higher than the lignin pyrolysis oils, which may be due to the higher oxygen contents in the carbohydrates in the whole biomass. Limited amount of sulfur has been found in some whole biomass pyrolysis oils, especially in the barks pyrolysis oils. Due to the kraft pulping process, the kraft lignins always contain a small amount of sulfur, which explains the relatively higher sulfur contents in the kraft lignins pyrolysis oils. Table 15 exhibits the relatively high viscosity and acidity of various pyrolysis oils, which bring huge barriers to the usage as the precursor of biofuels. The relatively lower heating values for some biomass pyrolysis oils may be due to the production of water and mixed in the pyrolysis oils. The solid residue in pyrolysis oils could also be a problem during the storage and transportation.

Table 14. NMR detectable functional groups in pyrolysis oils.[44,71,82–93,99]

^{1}H-NMR[44,71,82,83,86]	^{13}C-NMR [a, 44,82,84–89]	^{31}P-NMR[85,87,88,90,93]	HSQC [a, 91]
-CHO, -COOH	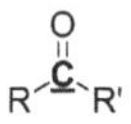	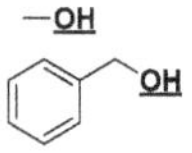	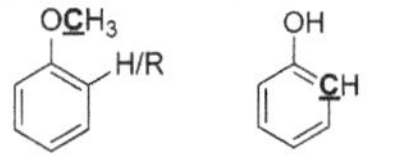
ArH, HC=C-	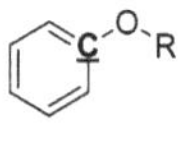	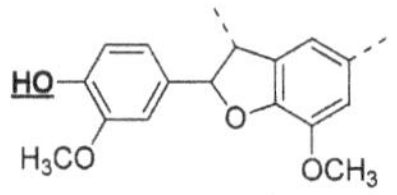	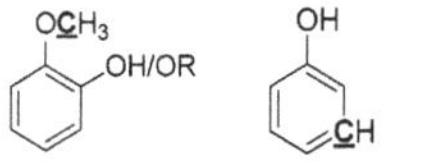
-CH$_n$-O-, CH$_n$-O-	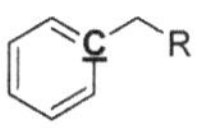	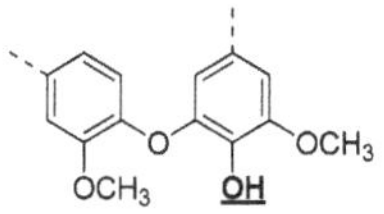	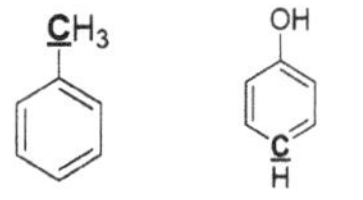
-CH$_3$, -CH$_n$-	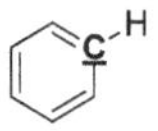	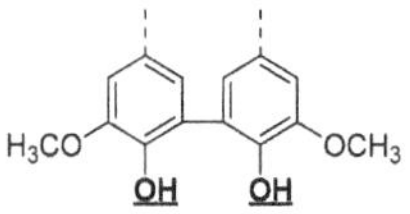	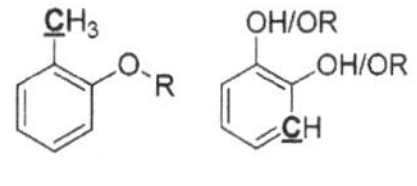
	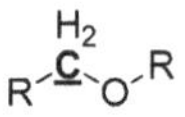	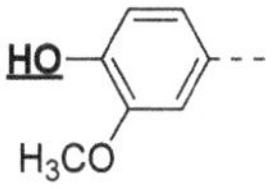	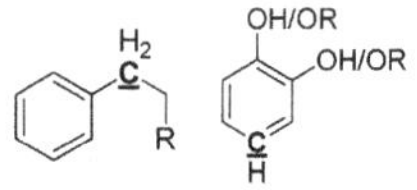
	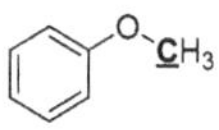	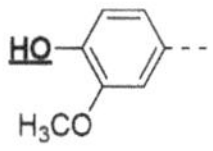	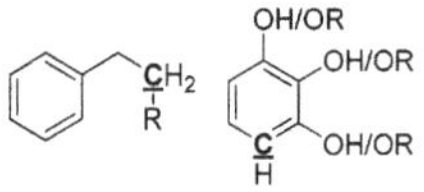
	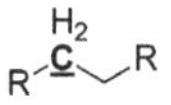	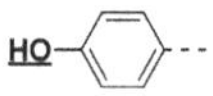	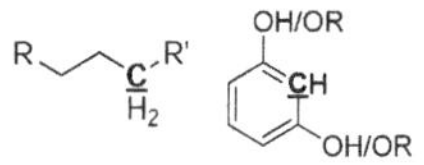
	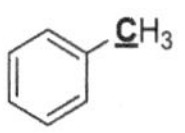	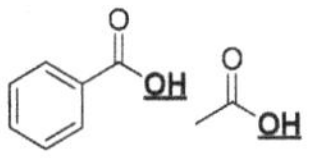	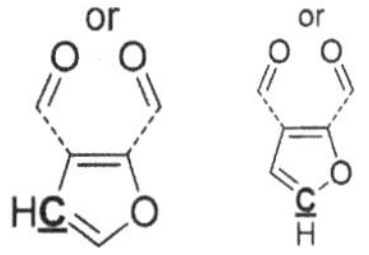
	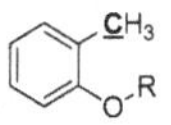		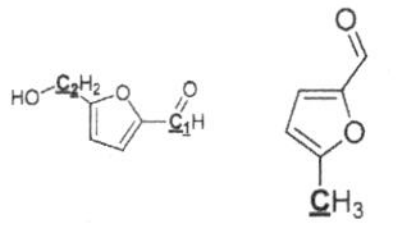
	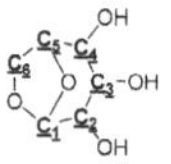		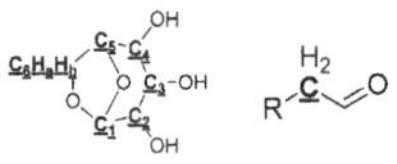

[a] The HSQC-NMR and ^{13}C-NMR detectable functional groups are based on our previous work which are the first reported efforts in the literature to fully characterize pyrolysis oil by NMR.

Table 15. Elemental analysis of pyrolysis oils.[a]

Biomass		C	H	O	S	N
Switchgrass [82]		52.97	6.43	39.13	—	0.38
Corn stover [82]		53.97	6.92	37.94	<0.05	1.18
Alfalfa stems [82]		57.00	7.89	31.30	0.07	3.75
Guayule (whole)[82]		69.93	8.54	19.31	0.20	2.92
Guayule bagasse[82]		69.97	7.96	21.38	0.07	0.82
Pine (pinusstrobes)[86]		59.2	6.6	34.2	—	<0.5
Pistachio shell[81]		67.35	8.36	23.70	—	<0.59
Indulin AT lignin[44]		69.79	3.39	15.02	2.17	0.57
Acetocell lignin [44]		71.42	4.38	21.43	<0.05	<0.5
Pine wood[39]		52.64	7.53	39.52	0.019	0.09
Oak wood[39]		47.19	4.51	47.97	0.022	0.12
Pine bark[39]		53.99	6.97	38.21	0.035	0.37
Oak bark[39]		45.57	6.05	47.75	0.28	0.32
Forestry residue[94]		38–44	7–8	48–53	—	0.1–0.3
Pine [94]		44–47	~7	46–49	—	<0.1
Pine bark (pinusbrutia Ten.)[95]		63.93	7.61	28.36	—	0.10
Pine[96]		40.6	7.6	51.7	0.01	<0.1
Forestry residue brown[96]		41.4	7.4	50.9	0.03	0.3
Forestry residue green[96]		41.2	8.0	50.5	0.02	0.3
Eucalyptus Crandis[96]		42.3	7.5	50.1	0.02	0.1
Barley straw[96]		26.5	9.0	62.7	—	0.9
Timothy hay[96]		32.1	8.5	58.7	—	0.6
Reed canary grass[96]		39.3	7.7	51.8	—	0.6
Hybrid poplar [97]		57.3	6.3	36.2	0.02	0.18
Willow[98]		43.17	7.15	49.49	0.10	0.10
Switch grass[98]		38.30	7.42	54.08	0.10	0.10
Reed canary grass[98]		38.42	7.89	53.49	0.10	0.10
Straw[98]		28.2	8.78	62.83	0.10	0.10
Dactylisglomerata[98]		36.75	8.82	52.46	0.10	1.88
Festucaarundinace[98]		32.05	9.76	56.69	0.10	1.41
Loliumperenne[98]		30.64	9.63	58.86	0.10	0.77
Soft wood kraft Lignin[71]		66.4	6.5	22.0	4.7	—
Lignoboost lignin[71]		69.0	6.6	22.8	1.2	—
Common fuels [b]	Gasoline	85–88	12–15	0	0	0
	Diesel	87	13	0	0	0

[a] The pyrolysis temperatures are from 400–800 °C

[b] Data is based on http://www.afdc.energy.gov/pdfs/fueltable.pdf

Table 16. Viscosity, acidity, higher heating value (HHV) and solid residue in the pyrolysis oils.[a]

Biomass	Viscosity (cSt)	Acidity (pH)	HHV (MJ/kg)	Solid residue (wt%)
Pine[96]	50°C 27.5	2.4	18.7	—
Forestry residue brown[96]	40°C 17	3.2	16.9	—
Forestry residue green[96]	40°C 24	-	16.7	—
Eucalyptus crandis[96]	40°C 23	2.2	17.3	—
Barley straw[96]	—	3.7	11.1	—
Timothy hay[96]	40°C 5	3.4	13.3	—
Reed canary grass[96]	—	3.6	16.0	—
Pine[86]	20°C, 22.60 Pa S	2.66	24.73	0.11
Pine wood[39]	50°C, 154 cP under 0.05 shear rate in s^{-1}	3.1	21.9	0.197
Oak wood[39]	50°C, 171 cP under 0.05 shear rate in s^{-1}	3.1	18.7	0.184
Pine bark[39]	50°C, 2529 cP under 0.05 shear rate in s^{-1}	3.2	18.3	0.428
Oak bark[39]	50°C, 5047 cP under 0.05 shear rate in s^{-1}	3.2	19.0	0.080
Forestry residue[94]	—	2.9–3.3	15.7–17.4	0.02–0.11
Pine[94]	—	2.4–2.6	17.8–19.0	0.03–0.29
Pine bark (pinusbrutia Ten.)[95]	—	—	31.03	—
Gasoline[b]	20°C, 0.5–0.6 mm^2/s	—	46.52	—
Diesel[b]	20°C, 2.8–5.0 mm^2/s	—	45.76	—

[a] The pyrolysis temperatures are from 400–800°C.

[b] Data is based on http://www.afdc.energy.gov/pdfs/fueltable.pdf

Acknowledgments

The authors are grateful for financial support from the Paper Science & Engineering (PSE) Fellowship program at Institute of Paper Science & Technology (IPST) and the School of Chemistry and Biochemistry at Georgia Institute of Technology.

References

1. A. Demirbas. Progress and recent trends in biofuels. *Progress in Energy and Combustion Science,* **33,** 1–18 (2007).
2. A. J. Ragauskas, C. K. Williams, B. H. Davison, *et al.* The path forward for biofuels and biomaterials. *Science,* **311,** 484–489 (2006)
3. K. David, A. J. Ragauskas. Switchgrass as an energy crop for biofuel production: A review of its ligno-cellulosic chemical properties. *Energy & Environmental Science,* **3**, 1182–1190. (2010).
4. R. D. Perlack, L. L. Wright, A. F. Turhollow, *et al.* Biomass as Feedstock for a bio-energy and bioproducts Industry: the technical feasibility of a billion-ton annual supply. *Biomass as Feedstock for a bioenergy and bioproducts Industry: the technical feasibility of a billion-ton annual supply* 2005.
5. R. P. Anex, A. Aden, F. K. Kazi, *et al.* Techno-economic comparison of biomass-to-transportation fuels via pyrolysis, gasification, and biochemical pathways. *Fuel* **89**, Supplement 1:S29-S35. (2010).
6. S. Czernik, A. V. Bridgwater. Overview of Applications of Biomass Fast Pyrolysis Oil. *Energy & Fuels*, **18** 590–598 (2004).
7. A. J. Ragauskas, M. Nagy, D. H. Kim , *et al.* From wood to fuels: Integrating biofuels and pulp production. *Industrial Biotechnology*, **2**, 55–65 (2006).
8. F. Huang, P. M. Singh, A. J. Ragauskas. Characterization of Milled Wood Lignin (MWL) in Loblolly Pine Stem Wood, Residue, and Bark. *Journal of Agricultural and Food Chemistry*, **59**, 12910–12916 (2011).
9. G. W. Huber, S. Iborra, A. Corma. Synthesis of transportation fuels from biomass: chemistry, catalysts, and engineering. *Chemical Reviews*, **106**, 4044–4098 (2006).
10. P. Sannigrahi, A. J. Ragauskas, G.A. Tuskan. Poplar as a feedstock for biofuels: A review of compositional characteristics. *Biofuels, Bioproducts anf Biofining,* **4**, 209–226 (2010).
11. V. P. Puri. Effect of crystallinity and degree of polymerization of cellulose on enzymatic saccharification. *Biotechnology and Bioengineering*, **26**, 1219–1222 (1984).
12. K. Kleman-leyer, E. Agosin, A. H. Conner, *et al.* Changes in Molecular Size Distribution of Cellulose during Attack by White Rot and Brown Rot Fungi. *Applied and Environmental Microbiology*, **58**, 1266–1270 (1992).

13. M. S. Jahan, S. P. Mun. Effect of tree age on the cellulose structure of Nalita wood (Trema orientalis). *Wood Science and Technology*, **39**, 367–373 (2005).
14. A. Thygesen, J. Oddershede, H. Lilholt, *et al.* On the determination of crystallinity and cellulose content in plant fibres. *Cellulose*, **12**, 563–576 (2005).
15. S. Willför, A. Sundberg, A. Pranovich, *et al.* Polysaccharides in some industrially important hardwood species. *Wood Science and Technology*, **39**, 601–617 (2005).
16. S. Willför, A. Sundberg, J. Hemming, *et al.* Polysaccharides in some industrially important softwood species. *Wood Science and Technology*, **39**, 245–257 (2005).
17. A. Jacobs, O. Dahlman. Characterization of the Molar Masses of Hemicelluloses from Wood and Pulps Employing Size Exclusion Chromatography and Matrix-Assisted Laser Desorption Ionization Time-of-Flight Mass Spectrometry. *Biomacromolecules*, **2**, 894–905 (2001).
18. J. Zakzeski, P. C. A. Bruijnincx, A. L. Jongerius, *et al.* Catalytic valorization of lignin for the production of renewable chemicals. *Chemical Rev*iews, **220**, 3552–3599 (2010).
19. F. S. Chakar, A. J. Ragauskas. Review of current and future softwood kraft lignin process chemistry. *Industrial Crops and Products*, **20**, 131–141 (2004).
20. J. A. Caballero, R. Font, A. Marcilla. Study of the primary pyrolysis of Kraft lignin at high heating rates: yields and kinetics. *Journal of Analytical and Applied Pyrolysis*, **36**, 159–178 (1996).
21. J. A. Caballero, R. Font, A. Marcilla. Pyrolysis of Kraft lignin: yields and correlations. *Journal of Analytical and Applied Pyrolysis*, **39**, 161–183 (1997).
22. J. A. Caballero, R. Font, A. Marcilla, *et al.* Flash pyrolysis of Klason lignin in a Pyroprobe 1000. *Journal of Analytical and Applied Pyrolysis*, **27**, 221–244 (1993).
23. D. Ferdous, A. K. Dalai, S. K. Bej, *et al.* Pyrolysis of Lignins: Experimental and Kinetics Studies. *Energy & Fuels*, **16**,1405–1412 (2002).
24. D. Ferdous, A. K. Dalai, S. K. Bej, *et al.* Production of H_2 and medium Btu gas via pyrolysis of lignins in a fixed-bed reactor. *Fuel Processing Technology*, **70**, 9–26 (2001).
25. B. Iatridis, G. R. Gavalas. Pyrolysis of a Precipitated Kraft Lignin. *Industrial & Engineering Chemistry Product Research and Development*, **18**, 127–130 (1979).
26. T. R. Nunn, J. B. Howard, J. P. Longwell, *et al.* Product compositions and kinetics in the rapid pyrolysis of milled wood lignin. Industrial & Engineering Chemistry Process Design and Development, **24**(3), 844–852 (1985).
27. M. Asmadi, H. Kawamoto, S. Saka. Gas- and solid/liquid-phase reactions during pyrolysis of softwood and hardwood lignins. *Journal of Analytical and Applied Pyrolysis*, **92**, 417–425 (2011).
28. H-W. Chen, Q-H. Song, B. Liao, *et al.* Further Separation, Characterization, and Upgrading for Upper and Bottom Layers from Phase Separation of Biomass Pyrolysis Oils. *Energy & Fuels*, **25**, 4655–4661 (2011).
29. T. Hosoya, H. Kawamoto, S. Saka. Solid/liquid- and vapor-phase interactions between cellulose- and lignin-derived pyrolysis products. *Journal of Analytical and Applied Pyrolysis*, **85**, 237–246 (2009).

30. M. Hyder, J. Å. Jönsson. Hollow-fiber liquid phase microextraction for lignin pyrolysis acids in aerosol samples and gas chromatography–mass spectrometry analysis. *Journal of Chromatography A.*, **1249**, 48–53 (2012).
31. G. Jiang, D. J. Nowakowski, A. V. Bridgwater. Effect of the Temperature on the Composition of Lignin Pyrolysis Products. *Energy & Fuels*, **24**, 4470–4475 (2010).
32. R. Lou, S-B. Wu, G-J. Lv. Effect of conditions on fast pyrolysis of bamboo lignin. *Journal of Analytical and Applied Pyrolysis*, **89**, 191–196 (2010).
33. R. Lou, S-B. Wu, G-J. Lv, *et al.* pyrolytic products from rice straw and enzymatic/mild acidolysis lignin. *Bio Resources*, **5**, 2184–2194 (2010).
34. C. A. Mullen, A. A. Boateng. Catalytic pyrolysis-GC/MS of lignin from several sources. *Fuel Processing Technology*, **91**, 1446–1458 (2010).
35. P. R. Patwardhan, R. C. Brown, B. H. Shanks. Understanding the Fast Pyrolysis of Lignin. *ChemSusChem*, **4**, 1629–1636 (2011).
36. Q. yang, S. Wu, R. Lou, *et al.* Analysis of wheat straw lignin by thermogravimetry and pyrolysis–gas chromatography/mass spectrometry. *Journal of Analytical and Applied Pyrolysis*, **87**, 65–69 (2010).
37. P. Bocchini, G. C. Galletti, S. Camarero, *et al.* Absolute quantitation of lignin pyrolysis products using an internal standard. *Journal of Chromatography A.*, **773**, 227–232 (1997).
38. P. F. Greenwood, J. D. H. van Heemst, E. A. Guthrie, *et al.* Laser micropyrolysis GC–MS of lignin. *Journal of Analytical and Applied Pyrolysis*, **62**, 365–373 (2002).
39. L. Ingram, D. Mohan, M. Bricka, *et al.* Pyrolysis of Wood and Bark in an Auger Reactor: Physical Properties and Chemical Analysis of the Produced Bio-oils. *Energy & Fuels*, **22**, 614–625 (2007).
40. H. E. Jegers, M. T. Klein. Primary and secondary lignin pyrolysis reaction pathways. *Industrial & Engineering Chemistry Process Design and Development*, **24**, 173–183 (1985).
41. D. J. Nowakowski, A. V. Bridgwater, D. C. Elliott, *et al.* Lignin fast pyrolysis: Results from an international collaboration. *Journal of Analytical and Applied Pyrolysis*, **88**, 53–72 (2010).
42. B. Scholze, D. Meier. Characterization of the water-insoluble fraction from pyrolysis oil (pyrolytic lignin). Part I. PY–GC/MS, FTIR, and functional groups. *Journal of Analytical and Applied Pyrolysis*, **60**, 41–54 (2001).
43. C. Saiz-Jimenez, J. W. De Leeuw. Lignin pyrolysis products: Their structures and their significance as biomarkers. *Organic Geochemistry*, **10**, 869–876 (1986).
44. S. H. Beis, S. Mukkamala, N. Hill, *et al.* Fast pyrolysis of lignins. *Bio Resources*, **5**, 1408–1424 (2010).
45. P. F. Britt, A. C. Buchanan, M. J. Cooney, *et al.* Flash vacuum pyrolysis of methoxy-substituted lignin model compounds. *The Journal of Organic Chemistry*, **65**, 1376–1389 (2000).
46. P. F. Britt, A. C. Buchanan, E. A. Malcolm. Impact of Restricted Mass Transport on Pyrolysis Pathways for Aryl Ether Containing Lignin Model Compounds. *Energy & Fuels*, **14**, 1314–1322 (2000).

47. P. F. Britt, M. K. Kidder, A. C. Buchanan I. Oxygen substituent effects in the pyrolysis of phenethyl phenyl ethers. *Energy & Fuels*, **21**, 3102–3108 (2007).
48. H. Kawamoto, T. Nakamura, S. Saka. Pyrolytic cleavage mechanisms of lignin-ether linkages: A study onp-substituted dimers and trimers. *Holzforschung*, **62**, 50–56 (2008).
49. H. Kawamoto, M. Ryoritani, S. Saka. Different pyrolytic cleavage mechanisms of β-ether bond depending on the side-chain structure of lignin dimers. *Journal of Analytical and Applied Pyrolysis*, **81**, 88–94 (2008).
50. H. Kawamoto, S. Saka. Role of Side-Chain Hydroxyl Groups in Pyrolytic Reaction of Phenolic β-Ether Type of Lignin Dimer. *Journal of Wood Chemistry and Technology*, **27**, 113–120 (2007).
51. H. Kawamoto, S. Horigoshi, S. Saka. Pyrolysis reactions of various lignin model dimers. *Journal of Wood Science*, **53**, 168–174 (2007).
52. G. W. Huber, J. N. Chheda, C. J. Barrett, *et al.* Production of Liquid Alkanes by Aqueous-Phase Processing of Biomass-Derived Carbohydrates. *Science*, **308**, 1446–1450 (2005).
53. N. Yan, C. Zhao, C. Luo, *et al.* One-Step Conversion of Cellobiose to C6-Alcohols Using a Ruthenium Nanocluster Catalyst. *Journal of the American Chemical Society*, **128**, 8714–8715 (2006).
54. D. K. Shen, S. Gu. The mechanism for thermal decomposition of cellulose and its main products. *Bioresource Technology*, **100**, 6496–6504 (2009).
55. S. Wang, X. Guo, K. Wang, *et al.* Influence of the interaction of components on the pyrolysis behavior of biomass. *Journal of Analytical and Applied Pyrolysis*, **91**, 183–189 (2011).
56. R. Alén, E. Kuoppala, P. Oesch. Formation of the main degradation compound groups from wood and its components during pyrolysis. *Journal of Analytical and Applied Pyrolysis*, **36**, 137–148 (1996).
57. W. Christian, Klampfl; Gerold Breuer; Schwarzinger CK, B. Investigations on the Effect of Metal Ions on the Products Obtained From the Pyrolysis of Cellulose. *Acta Chimica Slovenica*, **53**, 437–443 (2006).
58. Q. Lu, W. M. Xiong, W. Z. Li, *et al.* Catalytic pyrolysis of cellulose with sulfated metal oxides: a promising method for obtaining high yield of light furan compounds. *Bioresource Technology*, **100**, 4871–4876 (2009).
59. P. Rutkowski, A. Kubacki, Influence of polystyrene addition to cellulose on chemical structure and properties of bio-oil obtained during pyrolysis. *Energy Conversion and Management*, **47**, 716–731 (2006).
60. S. C. Moldoveanu, Pyrolysis GC/MS, present and future (recent past and present needs). *Journal of Microcolumn Separations*, **13**, 102–125 (2001).
61. P. R. Patwardhan, R. C. Brown, B. H. Shanks, Product distribution from the fast pyrolysis of hemicellulose. *ChemSusChem.*, **4**, 636–643 (2011).
62. Gao-Jin Lv, Shu-Bin Wu, Rui Lou. Characteristics of corn stalk hemicellulose pyrolysis in a tubular reactor. *BioResources*, **5**, 2051–2062 (2010).
63. Y. Peng, S. Wu, The structural and thermal characteristics of wheat straw hemicellulose. *Journal of Analytical and Applied Pyrolysis*, **88**, 134–139 (2010).

64. D. K. Shen, S. Gu, A. V. Bridgwater, Study on the pyrolytic behaviour of xylan-based hemicellulose using TG–FTIR and Py–GC–FTIR. *Journal of Analytical and Applied Pyrolysis*, **87**, 199–206 (2010).
65. P. J. Hernes, J. I. Hedges. Tannin signatures of barks, needles, leaves, cones, and wood at the molecular level. *Geochimica et Cosmochimica Acta*, **68**, 1293–1307 (2004).
66. Seiji Ohara, Y. Yasuta, H. Ohi. Structure elucidation of condensed tannins from barks by pyrolysis/gas chromatography. *Holzforschung*, **57**, 145–149 (2003).
67. H. Ben, A. Ragauskas Pyrolysis of biomass to biofuels, 243rd ACs National Meeting & Exposition, San Diego, CA.
68. M. Gaugler, W. Grigsby. Thermal Degradation of Condensed Tannins from Radiata Pine Bark. *Journal of Wood Chemistry and Technology*, **29**, 305–321 (2009).
69. B. Scholze, C. Hanser, D. Meier. Characterization of the water-insoluble fraction from fast pyrolysis liquids (pyrolytic lignin): Part II. GPC, carbonyl goups, and ^{13}C-NMR. *Journal of Analytical and Applied Pyrolysis*, **58–59**, 387–400 (2001).
70. A. Chaala, T. Ba, M. Garcia-Perez, *et al.* Colloidal Properties of Bio-oils Obtained by Vacuum Pyrolysis of Softwood Bark: Aging and Thermal Stability. *Energy & Fuels*, **18**, 1535–1542 (2004).
71. M. Kosa, H. Ben, H. Theliander, *et al.* Pyrolysis oils from CO_2 precipitated Kraft lignin. *Green Chemistry*, **13**, 3196–3202 (2011).
72. M. E. Boucher, A. Chaala, H. Pakdel, *et al.* Bio-oils obtained by vacuum pyrolysis of softwood bark as a liquid fuel for gas turbines. Part II: Stability and ageing of bio-oil and its blends with methanol and a pyrolytic aqueous phase. *Biomass and Bioenergy*, **19**, 351–361 (2000).
73. D. Mohan, C. U. Pittman, P. H. Steele. Pyrolysis of Wood/Biomass for Bio-oil: A Critical Review. *Energy & Fuels,* **20**, 848–889 (2006).
74. Y. Huang, Z. Wei, Z. Qiu, *et al.* Study on structure and pyrolysis behavior of lignin derived from corncob acid hydrolysis residue. *Journal of Analytical and Applied Pyrolysis*, **93**, 153–159 (2012).
75. J. Ke, D. Singh, X. Yang, *et al.* Thermal characterization of softwood lignin modification by termite Coptotermes formosanus (Shiraki). *Biomass and Bioenergy*, **35**, 3617–3626 (2011).
76. Q. Liu, S. Wang, Y. Zheng, *et al.* Mechanism study of wood lignin pyrolysis by using TG–FTIR analysis. *Journal of Analytical and Applied Pyrolysis*, **82**, 170–177 (2008).
77. Q. Liu, Z. Zhong, S. Wang, *et al.* Interactions of biomass components during pyrolysis: A TG-FTIR study. *Journal of Analytical and Applied Pyrolysis*, **90**, 213–218 (2011).
78. D. K. Shen, S. Gu, K. H. Luo, *et al.* The pyrolytic degradation of wood-derived lignin from pulping process. *Bioresources Technology*, **101**, 6136–6146 (2010).
79. S. Wang, K. Wang, Q. Liu, *et al.* Comparison of the pyrolysis behavior of lignins from different tree species. *Biotechnology Advances*, **27**, 562–567 (2009).
80. C. Lievens, D. Mourant, M. He, *et al.* An FT-IR spectroscopic study of carbonyl functionalities in bio-oils. *Fuel*, **90**, 3417–3423 (2011).

81. A. E. Pütün, N. Özbay, E. Apaydın Varol, *et al.* Rapid and slow pyrolysis of pistachio shell: effect of pyrolysis conditions on the product yields and characterization of the liquid product. *International Journal of Energy Research*, **31**, 506–514 (2007).
82. C. A. Mullen, G. D. Strahan, A. A. Boateng. Characterization of Various Fast-Pyrolysis Bio-Oils by NMR Spectroscopy†. *Energy & Fuels*, **23**, 2707–2718 (2009).
83. Z. Luo, S. Wang, X. Guo. Selective pyrolysis of Organosolv lignin over zeolites with product analysis by TG-FTIR. *Journal of Analytical and Applied Pyrolysis* 2012; 95: 112–117.
84. J. Joseph, C. Baker, S. Mukkamala, *et al.* Chemical Shifts and Lifetimes for Nuclear Magnetic Resonance (NMR) Analysis of Biofuels. *Energy & Fuels*, **24**, 5153–5162 (2010).
85. Gr. Gellerstedt, J. Li, I. Eide, *et al.* Chemical Structures Present in Biofuel Obtained from Lignin. *Energy & Fuels*, **22**, 4240–4244 (2008).
86. W. J. DeSisto, N. Hill, S. H. Beis, *et al.* Fast Pyrolysis of Pine Sawdust in a Fluidized-Bed Reactor. *Energy & Fuels*, **24**, 2642–2651 (2010).
87. K. David, M. Kosa, A. Williams, *et al.* ^{31}P-NMR analysis of bio-oils obtained from the pyrolysis of biomass. *Biofuels*, **1**, 839–845 (2010).
88. K. David, H. Ben, J. Muzzy, *et al.* Chemical characterization and water content determination of bio-oils obtained from various biomass species using ^{31}P NMR spectroscopy. *Biofuels*, **3**, 123–128 (2012).
89. H. Ben, A. J. Ragauskas. Torrefaction of Loblolly pine. *Green Chemistry*, **14**, 72–76 (2012).
90. H. Ben, A. J. Ragauskas. NMR Characterization of Pyrolysis Oils from Kraft Lignin. *Energy & Fuels*, **25**, 2322–2332 (2011).
91. H. Ben, A. J. Ragauskas. Heteronuclear Single-Quantum Correlation–Nuclear Magnetic Resonance (HSQC–NMR) Fingerprint Analysis of Pyrolysis Oils. *Energy & Fuels*, **25**, 5791–5801 (2011).
92. H. Ben, A. J. Ragauskas. In Situ NMR Characterization of Pyrolysis Oil during Accelerated Aging. *ChemSusChem*, **5**, 1687–1693 (2012).
93. H. Ben, A. J. Ragauskas. Pyrolysis of Kraft Lignin with Additives. *Energy & Fuels*, **25**, 4662–4668 (2011).
94. A. Oasmaa, E. Kuoppala, Y. Solantausta. Fast Pyrolysis of Forestry Residue. 2. Physicochemical Composition of Product Liquid. *Energy & Fuels*, **17**, 433–443 (2003).
95. S. Şensöz. Slow pyrolysis of wood barks from Pinus brutia Ten. and product compositions. *Bioresource Technology*, **89**, 307–311 (2003).
96. A. Oasmaa, Y. Solantausta, V. Arpiainen, *et al.* Fast Pyrolysis Bio-Oils from Wood and Agricultural Residues. *Energy & Fuels*, **24**, 1380–1388 (2010).
97. J. P. Diebold, S. Czernik. Additives To Lower and Stabilize the Viscosity of Pyrolysis Oils during Storage. *Energy & Fuels*, **11**, 1081–1091 (1997).
98. R. Fahmi, A. V. Bridgwater, I. Donnison, *et al.* The effect of lignin and inorganic species in biomass on pyrolysis oil yields, quality and stability. *Fuel*, **87**, 1230–1240 (2008).
99. W. Mu, H. Ben, A. Ragauskas, Y. Deng. Lignin Pyrolysis Components and Upgrading—Technology Review. *BioEnergy Research*, 1–22 (2013).

CHAPTER 9

UPGRADE OF BIO-OIL TO BIO-FUEL AND BIO-CHEMICAL

HAOXI BEN

School of Chemistry and Biochemistry, Institute of Paper Science & Technology Georgia Institute of Technology
500 10th St. NW, Atlanta, GA, 30332, United States
benhaoxi@gmail.com

Although bio-oil production from pyrolysis of biomass has been proven on a commercial scale to be a very promising option for production of renewable chemicals and fuels, there are still several drawbacks that have not been solved. The components of biomass pyrolysis oils are very complicated, and it is related to the properties of bio-oil. In this chapter, the details about pyrolysis oil components particularly from lignin pyrolysis processes will be discussed first. Due to the poor physical and chemical properties, the lignin pyrolysis oil has to be upgraded before usage. The most common method to upgrade bio-oil is to employ zeolite as the additives or hydrotreating. Catalysts have been widely used in petroleum industry for pyrolysis bio oil upgrading. In this chapter, the mechanism of the improved decomposition on the surface of zeolite and hydrodeoxygenation reaction will be discussed and the effects from the reaction condition will be summarized.

1. Introduction

With declining petroleum resources, increasing fuel demands and growing concerns about the effects of carbon dioxide emissions from fossil fuels, it is imperative to develop sustainable production of fuels and chemicals. Biomass is a renewable resource for the sustainable production of fuels and chemicals that, to date, have been made primarily from fossil resources. The U.S. Department of Agriculture and U.S. Department of

Energy established a vision to derive 25% of chemicals and materials and 20% of transportation fuels from biomass by 2030.[1] The use of biomass for renewable energy production is increasingly viewed as a promising alternative to reduce net carbon dioxide emissions and to gain long term energy security.[2] Among the various conversion technologies being investigated, pyrolysis has been reported as one of the most economic ways (i.e., low capital and operating costs) to utilize biomass for bio-fuels and bio-chemicals.[3] The major drawbacks towards commercialization of pyrolysis oil are several challenging properties, which have been summarized in the last chapter, including poor volatility, high oxygen content, acidity and viscosity, corrosiveness and cold flow problems.[4] Therefore, upgrading technologies that could convert bio-oils to a potential substitution of diesel and gasoline fuels are necessary. Among these technologies, in situ upgrading pyrolysis oil during the pyrolysis process appears to be pragmatic.[5]

Many researchers have examined the use of additives including, zeolite and metal salts to upgrade the properties of bio-oil during the pyrolysis of lignin. For example, Mullen *et al.*[6] used analytical pyrolysis methods (Py-GC-MS) to pyrolyze four different lignins at 650°C with 1/1 or 3/1 ($W_{additive}/W_{lignin}$) H-ZSM-5 zeolite and CoO/MoO_3. They suggested that H-ZSM-5 zeolite could improve the depolymerization of lignin and CoO/MoO_3 could facilitate the production of aromatic hydrocarbons through a direct deoxygenation of methoxyphenol units. Zhao *et al.*[7] upgraded the water-insoluble fraction (pyrolytic lignin) from pyrolysis oil of rice husk, with several additives, including H-ZSM-5, MCM-41, SBA-15 and β-zeolite from 500°C to 800°C. They found that compared to the results of pyrolysis without catalyst, most of oxygenates were converted to arenes and polycyclic aromatic hydrocarbons over zeolites. It indicated that this conversion was favored at higher temperatures. French *et al.*[8] used molecular-beam mass spectrometry (MBMS) to analyze the product vapor from pyrolysis of cellulose, straw lignin and ground aspen wood with forty different additives at 400°C, 500°C and 600°C. They found that the highest yield of hydrocarbons (~16 wt%) was achieved by using nickel, cobalt, iron, and gallium-substituted ZSM-5 zeolite during the pyrolysis. Being consistent with the results of Zhao *et al.*[7], they also suggested that the zeolites could improve deoxygenation reactions during pyrolysis. In addition, they also indicated that the best-performing catalysts belonged to ZSM-5 zeolite while larger-pore zeolites showed less deoxygenation activity. Jackson *et al.*[9] used sand, H-ZSM-5 zeolite, K-ZSM-5 zeolite, Al-MCM-41, solid phosphoric acid and Co/Mo/Al_2O_3 as additives in the pyrolysis of lignin at 600°C and analyzed the gas and liquid product by GC-MS. They found that H-ZSM-5 zeolite almost completely deoxygenated the liquid phase which produced simple aromatics and naphthalenics.

In addition, many researchers have examined the use of additives for the pyrolysis of biomass. Aho *et al.*[10] used H-Y, H-Beta, H-ZSM-5 and mordenite zeolites as additives to pyrolyze pine wood at 450°C. They concluded that the yield of the pyrolysis product was only slightly influenced by the different structures of the zeolite. However, the chemical components of bio-oil were dependent on the structures of zeolite, such that the content of ketones was higher and the amounts of acids and alcohols were lower in the bio-oil when ZSM-5 was used as an additive during the pyrolysis. Zhang *et al.*[11] studied the effects of the flow rate of the carrier gas, temperature and the particle size for the pyrolysis of corncob with H-ZSM-5 zeolite monitoring bio-oil composition and gas yield. They concluded that the optimal bio-oil yield was 56.8 wt% and the use of H-ZSM-5 zeolite caused a marked decrease of the heavy oil fraction and an increase in water and gas yields. They also indicated that the use of H-ZSM-5 zeolite with pyrolysis of corncob the contents of aromatic hydrocarbons in the bio-oil increased eight-fold whereas the oxygen content of the bio-oil decrease by 25%. Pan *et al.*[12] pyrolyzed *Nannochloropsis sp.* microalga residue with different amounts of H-ZSM-5 zeolites at 300°C, 350°C, 400°C, 450°C, and 500°C. They found that the optimal conditions for pyrolysis oil yield (~45 wt%) occurred with a pyrolysis temperature of 400°C and an additive-to-material ratio of 1:1. They also indicated that the use of zeolite during pyrolysis yielded a bio-oil that had a lower oxygen content, higher heating value and a higher content of aromatic hydrocarbons. The pyrolysis of several biomasses such as, corn stalks,[13] cassava rhizome,[14,15] hybrid poplar wood,[16] rice husks,[17] and pine[18,19] with ZSM-5 zeolites have also been studied in recent years. All of these studies reported that zeolites in would result pyrolysis oil with less oxygen content when mixed with biomass during pyrolysis.

In the last chapter, we have summarized that the average molecular weight of biomass pyrolysis oil by GPC and reported that the molecular weights are from 210–1700 g/mol.[20–23] In addition, there are significant amount of aromatics as well as high oxygen contained components in the biomass pyrolysis oils. Therefore, the pyrolysis oil has to be upgraded by hydrogenation before being used as biofuels. The hydrogenation with hydrotreating catalysts is one of the promising routes to upgrade pyrolysis oils.[5]

Many researchers have examined the hydrogenation of biomass components and pyrolysis oils. For example, the hydrogenation of phenolic pyrolysis oil model compounds to the corresponding aliphatic compounds has been reported recently.[24,25] Noble metals such as Ru, Rh, Pd and Pt have been used as catalysts and the conversion for several model compounds such as anisole, *4*-ethylphenol, 2-methoxy-4-n-propylphenol and 4-hydoxy-3-methoxyphenyl-acetone to the corresponding aliphatic compounds could be accomplished with yields approaching 100%. A two-step hydrogenation process has been reported[26] to upgrade the pyrolysis oils. In the first step, Ru/C was used as catalyst at 125°C with a H_2 pressure of 10 MPa and then Pt/C was used as catalyst at 250°C with a H_2 pressure of 10 MPa in the second step. Through a two-step hydrogenation process, the water-soluble fraction of pinewood pyrolysis oil was converted to gasoline-like compounds, which had boiling points range from 65–175°C with a carbon yield of 17%.[26] For the hydrogenation of whole pyrolysis oils produced from pinewood,[26] it has been reported that the carbon yield of the gas and coke is 26.5% and 34.6%, respectively. The hydrogenation of the whole fast pyrolysis oils produced from beech wood with Ru/C as the catalyst at 350°C, 20 MPa has also been examined.[27] On the basis of ^{1}H-NMR results, the aliphatic/aromatic proton ratio increased from 6.4:1 to 11.2–16.4:1 after hydrogenation, which suggested the hydrogenation of the benzene ring during the process, however, was not complete. Conversion of carbohydrate model compounds, including sorbitol and furoin,[28] and sugar solutions prepared by hydrolysis of maple wood[29] to the C_1-C_{15} alkanes has also been reported. All of these treatments involved hydrogenation process with Pt, Pd or Ru as catalysts at 120–265°C under a H_2 pressure of 5.2–6.2 MPa. The carbon yields of aqueous carbohydrates conversion derived from maple wood into gasoline range products were reported up to 57%.

In this chapter, we will first discuss the aging problem, which is another challenge for the usage of pyrolysis oil and another reason for upgrading pyrolysis oils. The second section will focus on the influence of Si/Al ratios and frameworks of zeolites on the properties of upgraded pyrolysis oils. The last section will summarize the catalyst behaviors and reaction mechanism of hydrogenation process of pyrolysis oil.

2. Aging Process of Pyrolysis Oils*

The liquid products of pyrolysis of biomass, known as pyrolysis oil, however, have several challenging properties including high oxygen and water content, high viscosity and acidity, poor volatility, corrosiveness, and aging problem.[4,5] Many researchers have investigated the aging process of pyrolysis oils. Some chemical and physical properties of pyrolysis oils have been examined during the aging process. It is well known that the viscosity of pyrolysis oils increases with storage of time, especially when the oils is stored or handled at high temperatures.[23,30–37] The average molecular weight of pyrolysis oils also increases during the storage[23,31,33,35] and it has been indicated that the formation of larger molecules is one of the reasons for the increasing viscosity of pyrolysis oils during the storage. Since the water content of the pyrolysis oils has been found to increase during the aging process, it is considered one of the by-products of the process.[23,31,32,34,36,37] By using small-angle neutron scattering (SANS), pyrolysis oils has been shown to be nanostructured fluids, consisting of a complex continuous phase and nanoparticles mainly formed by the association of units of pyrolytic lignin components from biomass. The aggregation of these units producing branched structures is reported to be responsible for the aging.[38] Compared to the well known properties including viscosity, water content and molecular weight of aged pyrolysis oils, only a few of researchers have characterized aged pyrolysis oils primarily by FT-IR and GC/MS to identify changes in chemical structures. By using FT-IR, esterification of pyrolysis oils has been proposed as one of the major reactions during the

*The full data of this section was accepted for publication in ChemSusChem, 2012. It is entitled as "In situ NMR characterization of pyrolysis oil during accelerated aging". The other author is Arthur J. Ragauskas. Reproduced with permission of John Wiley and Sons.

aging process.[32] Compared with freshly generated pyrolysis oil, the contents of hydroxyacetaldehyde,[34,35] furfurals[34] and guaiacols[34] in the aged pyrolysis oils have been detected to decrease by GC/MS.

To fully understand the fundamental chemistry of the aging process of pyrolysis oils, our previous work investigated chemical structural changes by NMR. For the aged pyrolysis oils, which have very high molecular weights (~500–1710 g/mol), the traditional analysis methods such as FT-IR and GC/MS, could not provide the insight into the chemical structures.[39] In contrast, NMR has a much higher resolution and provides detailed structural information of pyrolysis products.[22,40–42] In addition, the heating function of the NMR equipment provides an opportunity for in situ characterization of pyrolysis oil during the accelerated aging process. Our previous work[43] examined the relationship between the structures of various pyrolysis oils at room temperature and various time points during the accelerated aging process (at 80°C) by ^{1}H-NMR, ^{13}C-NMR and Heteronuclear Single Quantum Correlation (HSQC)-NMR.

The ^{1}H-NMR integration results shown in Table 1 indicates that the content of carbonyl and carboxyl protons slightly increases during accelerated aging process. Ortega *et al.*[36] indicated that the pH value of mixed hardwood pyrolysis oils decreased during the aging, which supports these results. The protons in the aliphatic groups also increased during the aging process, which indicated the formation of aliphatic functional groups. In contrast, the content of aliphatic groups protons containing C-O bonds decreased during aging, which was the evidence of the decomposition of ether and methoxyl groups. The content of aromatic protons also slightly decreased, which suggested that the aryl condensation reactions were also occurring during aging. For the room temperature (RT) aged pyrolysis oil, the same chemical trends have been found. Compared with the 60 h accelerated aged pyrolysis oil, the one year RT aged oil has a more serious aging phenomena.

To verify whether these chemical trends during the aging process will work for the other types of pyrolysis oil, the fresh and aged pyrolysis oils produced from two major biomass components — lignin and cellulose and a hardwood — poplar wood have also been investigated. The ^{1}H and ^{13}C-NMR results are shown in Tables 1 and 2. Compared with the cellulose and whole biomass pyrolysis oil, lignin pyrolysis oil has much more aromatic, carbonyl and carboxyl protons but relatively lower amount of protons in aliphatic groups containing C-O bonds. For the two types of aliphatic protons, lignin pyrolysis oil has a much higher aging rate than the cellulose pyrolysis oil. Therefore, for the whole biomass pyrolysis oils such as pine and poplar wood pyrolysis oils, the pyrolysis components from lignin in the whole biomass may have more responsibilities for the aging. Compared with the second 30 h accelerated aging process, there are relatively more differences between each types of protons at the first 30 h treatment, which indicates the aging rate is not constant. Nolte *et al.*[33]

Table 1. Functional group contributions for the fresh, one year room temperature aged and accelerated aged at 80°C for 60 h lignin, cellulose, pine residue and poplar wood pyrolysis oils, detected by ^{1}H-NMR.[43]

	Lignin		Cellulose	
Type of protons	Fresh	1 y @ RT	Fresh	1 y @ RT
-C<u>H</u>O, -COO<u>H</u>	5	5.2	0.3	0.7
Ar<u>H</u>, <u>H</u>C=C-	29.2	27.1	2.3	1.5
-C$\underline{H}_n$-O- C$\underline{H}_3$-O-	34.3	22.8	66.1	63
-C$\underline{H}_3$, -C$\underline{H}_n$-	31.5	44.9	31.3	34.8

	Poplar wood			Pine wood		
		30 h	60 h		60 h	
Type of protons	Fresh	@ 80°C		Fresh	@ 80°C	1 y @ RT
-C<u>H</u>O, -COO<u>H</u>	0.4	0.6	0.6	0.2	0.4	0.8
Ar<u>H</u>, <u>H</u>C=C-	11.2	10.5	10	12	11.2	10.8
-C$\underline{H}_n$-O- C$\underline{H}_3$-O-	43.4	36.6	35.9	45.7	43.2	39.2
-C$\underline{H}_3$, -C$\underline{H}_n$-	45	52.3	53.5	42.1	45.2	49.2

found a similar phenomena of the viscosity changes of an oak pyrolysis oil during the accelerated aging process. For all the tested pyrolysis oils, the chemical trends in the aging process were the same, which indicated that similar aging mechanisms might be involved in all aging processes.

To fully characterize the changes of functional groups in the aged pyrolysis oils, detailed analyses of ten different types of carbons presented in the pyrolysis oils was accomplished using ^{13}C-NMR (Tables 2 and 3). Similar with ^{1}H-NMR results, the contents of aromatic C-H bonds and aliphatic C-O bonds (including levoglucosan and methoxyl groups) decrease during the aging. Shen *et al.*[44] studied the mechanism for the thermal decomposition of cellulose and indicated that the anhydrosugars such as levoglucosan could further decompose to small molecule or

Table 2. ^{13}C-NMR functional group contributions for the pyrolysis oils produced from pine residue and poplar wood during the accelerated aging process at 80°C and aged at room temperature for one year. The results are shown as percentage of carbon.[43]

	Pine residue			Poplar wood		
Structure	0 h	60 h	One year	0 h	30 h	60 h
R–C(=O)–R'	7.1	6.8	6.0	7.3	6.9	6.9
Ar–C–O–R	9.8	12.3	12.1	11.5	13.3	14.3
Ar–C–R	12.0	12.3	12.9	12.6	13.5	14.3
Ar–C–H	21.9	20.2	18.6	22.2	21.3	20.4
Levoglucosan (C1–C6, OH)	5.2	4.6	4.3	4.3	3.8	3.5
R–CH2–O–R	22.1	19.2	18.7	16.7	13.8	12.9
Ar–O–CH3	3.1	2.7	2.2	6.3	5.4	4.6
R–CH2–R	24.0	26.5	29.5	23.4	25.8	26.6
Ar–CH3	2.0	1.9	1.7	5.3	4.6	4.2
Ar(–CH3)(–O–R)	0.2	0.4	0.9	0.7	0.8	0.9

Table 3. ^{13}C-NMR functional group contributions for the pyrolysis oils produced from lignin and cellulose aged at room temperature for one year. The results are shown as percentage of carbon.[43]

	Lignin		Cellulose	
Structure	Fresh	Aged	Fresh	Aged
R-C(=O)-R'	2.6	1.0	3.0	2.4
Ar-C-O-R	16.6	18.7	6.7	9.8
Ar-C-R	17.2	21.5	0.3	0.5
Ar-C-H	34.5	30.8	16.1	14.4
Anhydrosugar (C1–C6, OH)	0.0	0.0	53.0	50.1
R-CH$_2$-O-R	4.5	1.5	66.3	66.1
Ar-O-CH$_3$	6.4	5.3	1.9	1.1
R-CH$_2$-R	18.2	21.2	5.7	5.7
Ar-CH$_3$	3.3	3.2	2.0	1.5
Ar(-O-R)-CH$_3$	2.5	2.9	0.0	0.0

furfural. Figure 1 shows the tentative chemical pathways for the decomposition. The condensation reactions between two aliphatic C-O bonds could also be completed by release of water,[31,32,37] which would produce an ether bond and lead to reduce the content of aliphatic C-O bonds. In contrast, the contents of aliphatic C-C bonds, aromatic C-O and C-C bonds increased during the aging process. The condensation reactions to form aromatic C-O and C-C bonds during the thermal treatment of lignin and lignin model compounds have been reported in the literature.[45,46] During the aging, such polymerization reactions could be initiated, in part, by organic peroxides presented in pyrolysis oils, which could

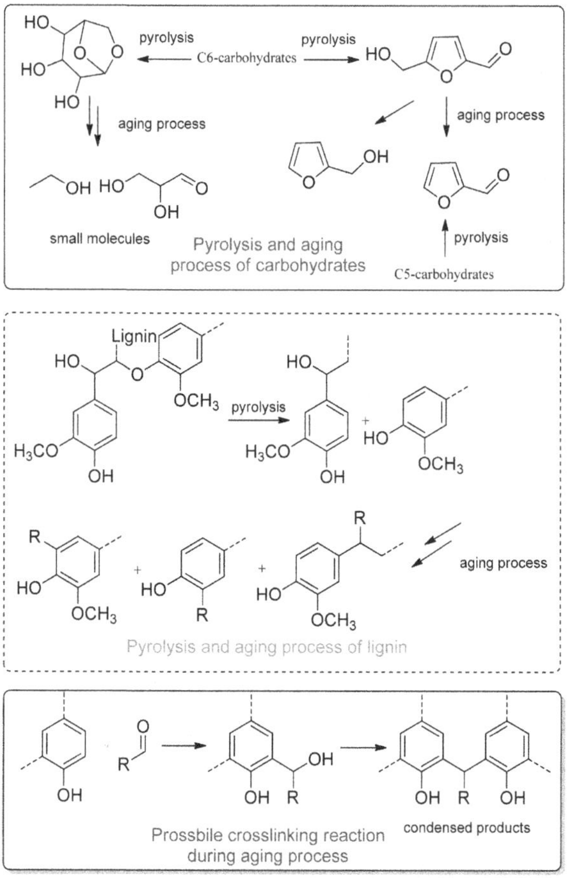

Fig. 1. Tentative chemical pathways for the reactions occurred during the aging process.[40,43,46–48]

spontaneously decompose, generating radicals.[31] The possible reaction pathways are shown in Fig. 1 and such type of condensation leads to reduce the contents of aromatic C-H bonds, which is supported by the ^{1}H and ^{13}C-NMR. Other than radical initiated condensation, Binder *et al.*[47] reported an acid-catalyzed self-condensation of organosolv lignin and lignin model compounds in ionic liquid. The α-ethers in lignin could undergo acid-catalyzed elimination to form electrophilic methide intermediates. These can react with lignin nucleophiles to from condensed structures linked by C-C bonds. Brosse *et al.*[48] also proposed a cross linking reaction (see the Fig. 1 for the detail) between formaldehyde, which is the decomposed products of carbohydrates and aromatic C-H bonds during the thermal treatment of beech wood. Such cross linking reaction will form new aromatic C-C bonds and consume aromatic C-H bond and carbonyl groups. All the condensation reactions summarized in Fig. 1 consume aliphatic C-O bonds, including levoglucosan and methoxyl groups, and form new aromatic C-C and C-O bonds, are consistent with our NMR results. In addition, such condensation reactions also increased the molecular weight of the pyrolysis components, which could explain the well known aging phenomena — increasing average molecular weight of pyrolysis oil with the storage of time. For instance, the molecular weight of poplar pyrolysis oil increased from 260 g/mol to 928 g/mol after 60 h accelerated aging at 80°C.

Similar with ^{1}H-NMR result, the one year RT aged pine wood pyrolysis oil shows more serious aging properties. Cellulose pyrolysis oil is relatively more stable than the lignin pyrolysis oil under the same aging condition, which may be due to the less complexity, since about 50% of carbon signals came from levoglucosan. Compared with the pine, there are more methoxyl groups in the poplar pyrolysis oil, which can be readily attributed to the fact that the hardwood contains syringyl type lignin[49,50] with two methoxyl groups per benzene ring. For this pyrolysis oil, the differences between the fresh and aged oil are more significant than that from pine wood. It has been reported[51,52] that methoxyl-aromatic bonds are one of the first groups to decompose during thermal treatment. Hence, the greater content of methoxyl groups in the poplar pyrolysis oil may initiate further aging reactions and cause more serious aging issues than pine wood pyrolysis oil. Being consistent with ^{1}H-NMR results, the chemical trends in the aging process detected by ^{13}C-NMR are the same for all the tested pyrolysis oils.

In summary, the NMR results indicate that the content of aliphatic C-O bonds, aromatic C-H bonds decrease during the aging process. In contrast, the contents of aliphatic C-C bonds, aromatic C-C and C-O bonds will increase. The condensation reactions between two aliphatic C-O bonds could be completed by release of water, which will produce an ether bond and reduce the content of aliphatic C-O bonds. The condensation reaction to form aromatic C-O and C-C bonds could be initiated by instable organic peroxides presented in pyrolysis oils, which can spontaneously decompose, and generate radicals. The cross linking reaction between formaldehyde, the decomposed products of carbohydrates, and aromatic C-H bonds could also form new aromatic C-C bonds and consume aromatic C-H bond and carbonyl groups. Contents of the aromatic C-H bonds in the ortho and para positions of hydroxyl groups, were found to decrease more quickly than the aromatic C-H bonds in the meta positions of hydroxyl groups, which indicates that the former two types of aromatic C-H bonds are more favored for the condensation reactions. All of condensation reactions mentioned above consume aliphatic C-O bonds, including levoglucosan and methoxyl groups, and form new aromatic C-C and C-O bonds. In addition, such condensation reactions also increase the molecular weight of the pyrolysis components. The contents of levoglucosan decreased during the aging process, which indicates that the anhydrosugars such as levoglucosan could be further decomposed. These chemical trends also work for other types of pyrolysis oil, such as the pyrolysis oils produced from two major biomass components — lignin and cellulose, and a hardwood — poplar wood. In addition, these results suggest that the pyrolysis components result from lignin in the whole biomass may have more responsibilities for the aging effects. The hardwood pyrolysis oil has a higher aging rate than the softwood pyrolysis oil, which may be due to the syringyl type products in the hardwood pyrolysis oil.

3. Upgrade Pyrolysis Oil with Zeolites

Zeolites are a promising class of additives to improve the properties of pyrolysis oils. ZSM zeolite is one of the most famous types of zeolite, which has been used in literature. Investigations of other types of zeolite including Beta (BEA), Y (FAU), ZSM-5 (MFI), Mordenite (MOR) and Ferrierite (FER) zeolites also have been reported in literatures. In the meanwhile, different Si/Al ratios of zeolite have been shown to affect the cracking reactions of model compounds during thermal treatment.[53–59] However, only a few researchers have investigated the influence of different Si/Al ratios of zeolites on the properties of pyrolysis products. In this section, influences of Si/Al ratios and frameworks of zeolites on the properties of upgraded pyrolysis oils will be discussed.

3.1. *Influences of Si/Al ratios of zeolites on the properties of upgraded pyrolysis oils**

Compared to the model compounds studies, there are only limited reported studies about the influences from Si/Al ratios. For example, Aho *et al.*[60] pyrolyzed pine wood with various H-Beta zeolites at 450°C and they indicated that zeolites with lower Si/Al ratios formed less organic oil, char and gas, but more water and polyaromatic hydrocarbons. Compared with the non-catalytic pyrolysis of pine, the yield of the organic oil decreases, but there will be more char and water. They also found that the contents of different groups of compounds were slightly affected by zeolites. There will be more aldehydes by employing more acidic zeolites. The contents of ketones and phenols decreased after the use of zeolites. The amounts of acids and alcohols have only been slightly affected by using zeolites, but PAHs was only found in the zeolite upgraded pyrolysis oils and the contents decreased with elevated Si/Al ratios of H-Beta zeolites. The authors also indicated that zeolites might convert/eliminate levoglucosan during the upgrading process, since the content of which decreased significantly after the use of zeolites. The used zeolite catalysts also have been regenerated by calcination in air and it has been found that the surface area and the acidity could be recovered.

Our previous work[61] examined the influences of different Si/Al ratios of H-ZSM-5 zeolites on the properties of lignin pyrolysis products. The elemental analysis, GPC, HHV, energy, carbon and mass yields from pyrolyzing a softwood kraft lignin with different ZSM-5 zeolites at 600°C were examined. The upgraded pyrolysis oils have been characterized by ^{13}C, ^{31}P and HSQC NMR. The results of yields indicated that the yields of char slightly increased after the use of additives and the char yield was almost constant for all zeolites, which suggests that the additives have very limited effects on the primary decomposition of lignin. In contrast, the yield of heavy oil decreased after using zeolite. Correspondingly, the

*The full data of this section was accepted for publication in ACS Sustainable Chemistry & Engineering, 2013. Reprinted with permission from "Ben, H.; Ragauskas, A. J., Influence of Si/Al Ratio of ZSM-5 Zeolite on the Properties of Lignin Pyrolysis Products. ACS Sustainable Chemistry & Engineering 2013, 1, (3), 316-324."

yield of light oil and gas increased with used zeolites. In addition, pyrolysis samples with lower Si/Al ratio zeolites formed relatively less heavy oil but more light oil. HHV for the heavy oil upgraded by Z280 (SiO_2/Al_2O_3 mole ratio is 280) are similar with the control pyrolysis oil, which indicates that the H-ZSM-5 zeolite with a very large SiO_2/Al_2O_3 mole ratio only limitedly affects the properties of pyrolysis oil. Nevertheless, both energy and carbon yields show that Z50 upgraded pyrolysis oil retained most energy and carbon from the SW kraft lignin, which indicates that this type of zeolite is a promising candidate to upgrade the properties of pyrolysis oil. The GPC analysis indicated that the molecular weight decreased by 8–16% after the use of H-ZSM-5 zeolite. The lower molecular weight of upgraded pyrolysis oil is the evidence of enhanced decomposition of methoxyl group, carboxyl acid and dehydration of aliphatic hydroxyl groups. After the use of zeolites, the carbon contents of heavy oils increased up to 5%. In contrast, there is up to 13% less oxygen in the upgraded pyrolysis oils, which indicates that zeolite could improve the deoxygenation during the pyrolysis process. Most interestingly, the sulfur content also decreased by 50% after using Z50 as the additive, which exhibits a potential method to decrease the sulfur content in the kraft lignin pyrolysis oils.

For all the heavy oils (water insoluble parts of pyrolysis oil), the ^{31}P NMR integration results show that there are significant amount (from 2 to 3 mmol/g of heavy oil) of guaiacyl and catechol types of hydroxyl groups. If there is only one of these two types of functional groups in each pyrolysis oil components, there should be >11–17 wt% of catechol and its derivatives as well as >24–37 wt% of guaiacol and its derivatives in the heavy oils. After the use of H-ZSM-5 zeolite as an additive the aliphatic hydroxyl groups and carboxylic acid in the heavy oils are nearly completely decomposed. With the exception of those two hydroxyl groups, the heavy oil produced from pyrolysis of lignin with Z280 is similar with the control pyrolysis oil, which indicates that the H-ZSM-5 zeolite with very large SiO_2/Al_2O_3 mole ratio has limited effects on the phenolic hydroxyl groups of pyrolysis oil during the pyrolysis process. Other than Z280, after the use of H-ZSM-5 zeolites, the contents of C_5 substituted and normal guaiacyl phenolic hydroxyl groups decreased with the increasing SiO_2/Al_2O_3 mole ratio of zeolites. In contrast, the contents of catechol type and *p*-hydroxy-phenyl hydroxyl groups increased when a high SiO_2/Al_2O_3 mole ratio zeolite was used. Since those two types of hydroxyl groups are the decomposition products of methoxyl-aromatic bonds and ether bonds in the lignin structure,[40,41] it indicates that a zeolite with a relatively higher SiO_2/Al_2O_3 mole ratio could improve the cleavage of aromatic C-O bonds including methoxy groups and ether bonds more efficiently. Some possible degradation pathways of ether bonds and methoxyl groups on the surface of zeolite have been proposed in literatures[62–64] and these are summarized in Fig. 2. The limited amount of aliphatic OH also decreased after the use of a high SiO_2/Al_2O_3 mole ratio zeolite, which indicated an enhanced dehydration of aliphatic C-O bonds. In contrast, the content of carboxylic acid-OH increased with the increasing SiO_2/Al_2O_3 mole ratio, which suggested that zeolite with lower SiO_2/Al_2O_3 mole ratio was more effective for the decomposition of carboxylic acid. Several researchers

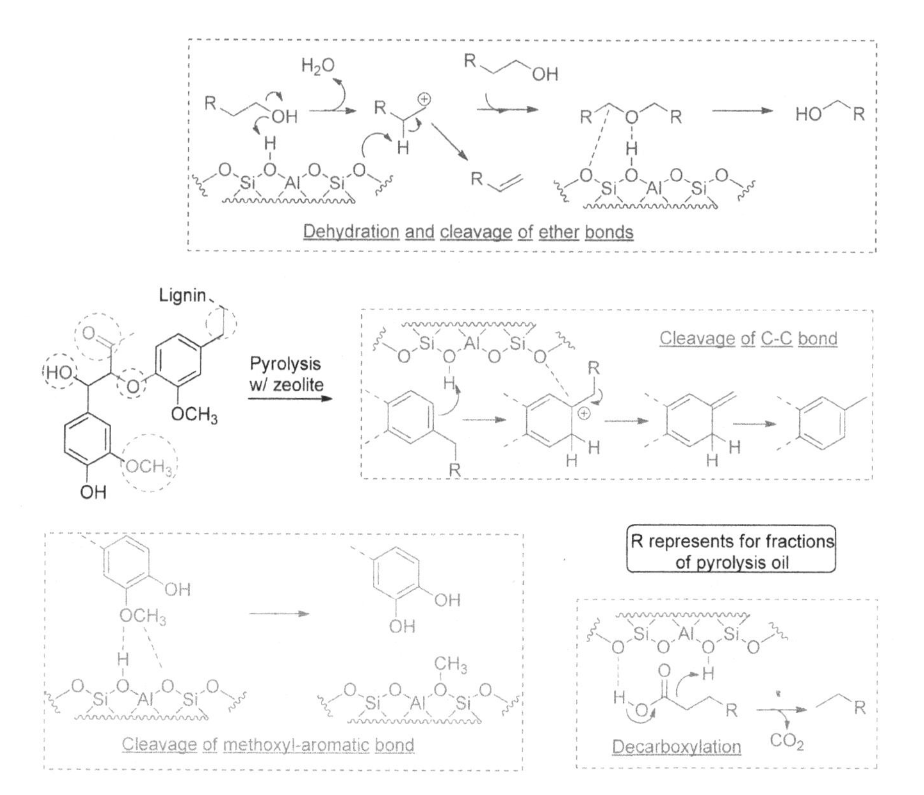

Fig. 2. The primary decomposed functional groups in lignin during the pyrolysis (circled by dash line in lignin model structure) and the possible cracking pathways on zeolite.[61–72]

have examined the effects of zeolite on the dehydration and decarboxylation of model compounds.[62,65–69] The tentative mechanisms in the literatures have also been shown in Fig. 2.

The ^{13}C NMR integration results of this analysis for the heavy oils showed that compared to the use of other zeolites, the effects of Z280 on the heavy oil were very limited and the product was more similar with the control heavy oil, which was consistent with ^{31}P-NMR results. The carbonyl groups were nearly completely eliminated and the content increased with the increasing SiO_2/Al_2O_3 mole ratio of H-ZSM-5 zeolite, which is additional evidence that the upgraded pyrolysis oil has lower acidity. After the use of zeolite, the content of methoxyl groups in the heavy oil decreased by ~45% and the decomposition of this functional group was improved when a higher SiO_2/Al_2O_3 mole ratio of H-ZSM-5 zeolite was used. This effect could explain the increasing contents of catechol type hydroxyl groups in the heavy and light oils detected by ^{31}P-NMR. The percentage of aromatic C-C bonds decreased upon the use of a higher SiO_2/Al_2O_3 mole ratio zeolite, which suggests that the zeolite with a higher SiO_2/Al_2O_3 mole ratio favors to cleave aromatic C-C bonds or prevent the formation of such bonds. There were also less aliphatic C-C bonds after the use of zeolite. Zeolite has been reported[70–72] to improve the cleavage of aliphatic C-C bonds, which could explain the reduced content of such bonds in the upgraded pyrolysis oil. The possible pathways reported in the literatures have been summarized in Fig. 2. The two types of methyl aromatic bonds in the heavy oils are the rearrangement products of methoxyl groups.[40] Since the cleavage of methoxyl groups has been enhanced, the content of those two methyl aromatic bonds also increased after the use of zeolite. The increased aromatic C-O bonds in the upgraded heavy oils indicated a higher phenol hydroxyl groups, which was supported by ^{31}P-NMR results.

HSQC-NMR results showed that after the use of zeolite as additive, the pyrolysis oils contained some polyaromatic hydrocarbons (PAH) and the content decreased with the increasing SiO_2/Al_2O_3 mole ratio of zeolite, which were supported by the GC-MS analysis. The formation of PAH on the surface of zeolites was also observed by several model compounds studies.[73–76] It has been reported that the Brønsted acid sites of H-ZSM-5 zeolites could be calculated by the unit cell formula of the zeolite

($H_nAl_nSi_{96-n}O_{192}$ for H-ZSM-5 zeolite).[55] Therefore, the Brønsted acid sites for H-ZSM-5 zeolites used in this work linearly decreased from 0.35 mmol/g (Z23) to 0.03 mmol/g (Z280). Figure 2 shows that the improved cracking pathways of pyrolysis oil on the zeolites always involve protons from Brønsted acid sites, which indicate that more acidic zeolites should perform better upgrading results. However, both ^{31}P and ^{13}C-NMR results showed that the H-ZSM-5 zeolite with a relatively higher SiO_2/Al_2O_3 (~50–80) was more effective to upgrade the properties of pyrolysis oil. Since the PAH has been reported as the precursor of coke[75,76] that will deactivate the zeolite, the higher contents of PAH in more acidic zeolite upgraded pyrolysis oils provide insight into the reason why Z50 performs better upgrading results than Z23.

3.2. *Influences of frameworks of zeolites on the properties of upgraded pyrolysis oils**

Zeolites are a promising class of additives to improve the properties of pyrolysis oils, which have been examined by several researchers. For example, Murzin's group investigated pyrolysis of pine wood with various zeolites, including Beta (BEA), Y (FAU), ZSM-5 (MFI), Mordenite (MOR) and Ferrierite (FER) zeolites.[10,60,77–79] They concluded that the mass yield of pyrolysis product was only slightly influenced by the types of zeolite, whereas the chemical compositions of bio-oil were dependent on the structure of the zeolite employed. The content of ketones was higher and the amount of acids and alcohols was lower in the bio-oil when ZSM-5 was used as an additive during the pyrolysis. By using a Mordenite zeolite, the content of polyaromatic hydrocarbons was relatively low. Beta, Y and Ferrierite zeolites were more effective at catalyzing dehydration and decarbonylation reactions. Huber's group[80–82] also examined the influence of zeolite on the properties of pyrolysis products produced from pine wood, cellulose and glucose. They reported that the thermal conversion of

---

*The full data of this section was accepted for publication in RSC Advances, 2012. It is entitled as "One step thermal conversion of lignin to the gasoline range liquid products by using zeolites as additives". The other author is Arthur J. Ragauskas. Reproduced with permission of the Royal Society of Chemistry.

glucose to aromatics was a function of the pore size of the zeolite catalyst. Small pore zeolites did not produce aromatics. In contrast, medium pore size (5.2–5.9 Å) zeolites produce mostly aromatic products. Higher coke, lower aromatics and reduced oxygenated species were observed with the use of large pore zeolites. Zhao *et al.*[7] upgraded the water-insoluble fraction (pyrolytic lignin) from pyrolysis oil of rice husks with several additives, including ZSM-5 and Beta zeolites. They found that compared to the Beta zeolite, ZSM-5 produced more aromatics and less coke. Most of oxygenates were found to be converted to arenes and polycyclic aromatic hydrocarbons with zeolites and this conversion was favored at higher temperatures. French *et al.*[8] used molecular-beam mass spectrometry (MBMS) to analyze the product vapor from pyrolysis of cellulose, straw lignin and ground aspen wood with forty different additives at 400°C, 500°C and 600°C. They found that the highest yield of hydrocarbons (~16 wt%) was achieved by using metal-substituted ZSM-5 zeolite during pyrolysis and the best-performing catalyst belonged to ZSM-5 zeolite while larger-pore zeolites showed less deoxygenation activity.

The pyrolysis of several biomasses such as corn stalks,[11,13] cassava rhizome,[14,15] maple,[83,84] poplar,[16,85] rice husks,[7,17] bamboos,[86] empty palm fruit bunch,[87] oak,[88] beech[58,89] and pine[18,19]with various zeolites have also been studied in recent years. Most of the reported additives studies have focused on the pyrolysis of biomass, while only a few have examined the behavior of lignin in the presence of additives during pyrolysis. Lignin is a natural aromatic polymer and a main constituent of lignocellulosics biomass; however, it has received much less biorefining efforts than plant polysaccharides[2,90,91] due to its complexity, biological recalcitrance and relative thermal stability, as its weight loss occurs in a very wide temperature range of 160–900°C.[92] Furthermore, it is well known that the pyrolysis of lignin yields more char and the least amount of bio-oil compared to cellulose or whole biomass.[4,93] It has been reported that most of the water insoluble and the aromatic C-H groups from the pyrolysis of whole biomass pyrolysis oils could be attributed to the pyrolysis of lignin.[93] Therefore, upgrading technologies converting lignin to a potential replacement for diesel and gasoline could provide an attractive biofuel technology and insight into the conversion of whole biomass to green fuels. One-step thermal conversion of lignin to gasoline range (molecular weight is ~105 g/mol) liquid products or simple petrochemicals such as benzene, toluene, xylene, phenol and catechol appears to be very pragmatic.

Our previous work[94] investigated this pragmatic process and successfully converted lignin to gasoline molecular weight range liquid products by one step pyrolysis process. Pyrolysis of lignin with various different framework zeolites including MFI (Z), FAU (Y), BEA (B), FER (F) and MOR (M) has been examined. The number-average and weight-average molecular weights (M_n and M_w) and polydispersity values indicated that the molecular weight decreased to the gasoline range (80–120 g/mol) after the use of Y and B zeolites, which represented simple aromatic molecules and could be used as the precursor of gasoline and possible substitution of petrochemicals. The GC-MS analysis of Y and B zeolites upgraded pyrolysis oils indicated that phenol, methyl phenols, dimethyl phenols, catechol, methyl catechols, naphthalene, methyl naphthalenes and guaiacol were the major components. In contrast, after adding F and M zeolites the molecular weights were almost intact.

The upgraded pyrolysis oils have been characterized by ^{31}P, ^{13}C and HSQC NMR. The ^{31}P NMR integration results for the heavy oils indicated that after the use of Y and B zeolites as additives, the aliphatic hydroxyl groups in the heavy oils were completely decomposed. With the other three types of zeolite, the aliphatic hydroxyl groups also decreased by 70–95%. The results showed that for the zeolites studied the resulting pyrolysis oils were significantly dehydrated. Compared to the Z, F and M zeolites, after the use of Y and B zeolites, the heavy oils contained more C_5 substituted guaiacyl phenolic, catechol and *p*-hydroxy-phenyl types of hydroxyl groups but less normal guaiacyl phenolic hydroxyl group. In contrast, by adding F and M zeolites, lignin pyrolysis yielded more normal guaiacyl phenolic but less *p*-hydroxy-phenyl and C_5 substituted guaiacyl phenolic types of hydroxyl groups. Since catechol and *p*-hydroxy-phenyl types of hydroxyl groups are the decomposition products of methoxyl-aromatic and ether bonds in the lignin structure,[40,41] this indicates that the Y and B zeolites could perform the cleavage of aromatic C-O bonds more effectively. After the use of zeolites, the content of carboxylic acid decreased by 44–85%, which was anticipated to be a much more suitable biofuel precursor. Compared to the other studied zeolites, Z and M zeolites could induce decarboxylation reactions more efficiently. Several researchers have reported similarly enhanced dehydration and decarboxylation of model compounds on the surface of zeolite.[62,65–69]

Our previous studies indicated that the light oils from pyrolysis of lignin contained more than 80 wt% of water and ~10 wt% of methanol, catechol and acidic acid.[22,40,41] For all the upgraded light oils, the yields of methanol and acetic acid significantly decreased. In contrast, the content of catechol, increased after the use of zeolites. Compared with the Z, F and M zeolites, the use of Y and B zeolites yielded a light oil with increased catechol which indicates a more efficient decomposition of ether bond and methoxyl group. The content of catechol increased by 320% after the use of Y zeolite, this pure catechol in the light oil could be used as an alternative pathway for its synthesis. As anticipated, the content of the major dehydration product, water increased with the use of zeolites.

The ^{13}C NMR integration results of this analysis for the heavy oils show that after the use of Y and B zeolites, the heavy oils contained ~80% less methoxy groups than the native pyrolysis oil, which indicated a very efficient decomposition of methoxyl groups. The two types of methyl aromatic bonds in the heavy oils are reported to be the rearrangement products of methoxyl groups.[40] After the use of these two zeolites, the heavy oils contained relatively large amount of methyl aromatic bonds, which was evidence for the enhanced cleavage of methoxyl groups. The ^{13}C-NMR results also showed that there were only 13–56% carbonyl groups remaining in the upgraded pyrolysis oils, which was consistent with ^{31}P-NMR results. It has been reported that the acidity of pyrolysis oils is mainly (60–70%) derived from carboxylic acids and their content detrimentally impacts the stability of pyrolysis oils.[95] The significantly reduced amount of carboxylic acids in the zeolite-upgraded pyrolysis oils represents a more stable pyrolysis oil which was expected to cause less corrosion problems.[95] With the use of F and M zeolites, the heavy oils contained relatively larger amount of aromatic C-O bonds but the least amount of aromatic C-C bonds, which indicated that those two zeolites prefer to cleave aromatic C-C bonds or prevent the formation of such bonds. For the Y and B zeolites upgraded pyrolysis oils, the contents of oxygen substituents including aromatic C-O and aliphatic C-O bonds, carbonyl and methoxyl groups decreased by more than 40%. In addition, almost all the remaining oxygen functionality (up to ~87%) in these two upgraded pyrolysis oils belongs to the phenolic hydroxyl groups.

HSQC-NMR results showed that after the use of Z, Y and B zeolites, the pyrolysis oils contained some polyaromatic hydrocarbons (PAH). In contrast, there were only very limited amount of PAH in the F and M zeolites upgraded pyrolysis oils and almost no PAH in the native pyrolysis oil. The native methoxyl groups (with a hydroxyl group or ether bond in the ortho position) have been completely eliminated after adding Y zeolite during the pyrolysis. The improved degradation of methoxyl groups on the surface of zeolite have been reported in literatures.[62–64] Since Y type zeolite has a relatively larger pore size and a three dimensional channel system[7,10,77,78,81,84,87,89] that could let small aromatics such as phenol, naphthalene, and xylene[63,81] go through the channels it thereby improves the upgrading effect more efficiently. For the other pyrolysis samples, the decomposition of native type of methoxyl group has also been improved. The content of rearranged methoxyl groups (no hydroxyl group or ether bond in the ortho position) also decreased after the use of Y and B zeolites, which was consisted with the significant reduced amount of methoxyl groups detected by ^{13}C-NMR. Compared with the rearranged methoxyl groups, the zeolites prefer to cleave the native methoxyl groups, which may be due to the hydroxyl group in the ortho position facilitates the cleavage. The reported possible degradation pathway of methoxyl groups on the surface of zeolite have been summarized in Fig. 3.[62–64] After

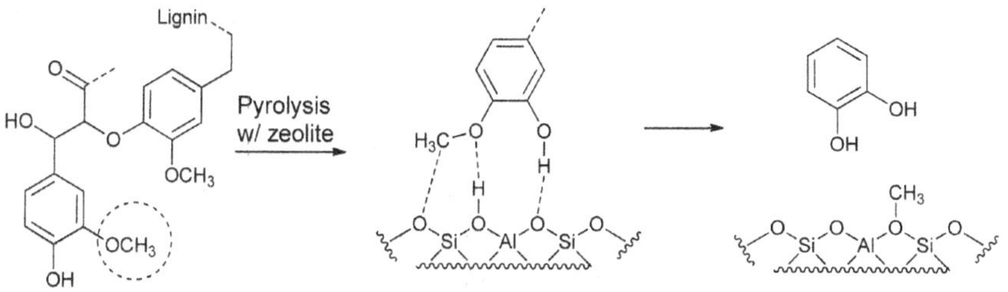

Fig. 3. The reported possible degradation pathway of methoxyl groups on the surface of zeolite.[62–64,94]

the use of zeolites, there were much more methyl aromatic bonds in the pyrolysis oil, which was consistent with our ^{13}C-NMR result. Compared with the native pyrolysis oil, the upgraded pyrolysis oils have relatively lower amount of long chain aliphatic C-C bonds. Zeolite has been reported[70–72] to improve the cleavage of aliphatic C-C bonds in the model compounds, which could explain the reduced contents of such bonds in the upgraded pyrolysis oils.

4. Hydrodeoxygenation of Pyrolysis Oils

Hydrodeoxygenation (HDO), as the name indicates, is composed of hydrogenation and deoxygenation parts. The common features of key model compounds for pyrolysis oil are oxygen atoms and aromatic structure. The purpose for HDO is to hydrogenate the unstable unsaturated bonds and to reduce the oxygen in the pyrolysis oil. Research showed that the pyrolysis oil tends to repolymerize under 175°C without catalyst or hydrogen, followed by char formation within a couple of minutes.[96] However, in the presence of catalyst and hydrogen, the pyrolysis oil will convert to stable compounds first under the same condition. When the temperature goes higher than 250°C, the HDO reaction occurs with the existence of hydrogen and catalyst.

In 1983, Furimsky *et al.*[97] published the first review paper on the catalyst, mechanism and kinetics study related to HDO process in crude oil upgrading. Very limited information was available at that time. Afterwards, numerous papers appeared in HDO study on coal-derived liquid and biomass-derived oil. In 2000, Choudhary *et al.*[98] reviewed the catalytic hydrodeoxygenation again. This time much more detailed information was included such as the phenol, furan, ether and other bio-oil compounds. Huber and Corma *et al.*[5] reviewed the synthesis of transportation fuel from biomass in 2006. Most recently (2012), Bu *et al.*[99] reviewed the catalytic hydrodeoxygenation of lignin-derived phenols. Phenol compound HDO was discussed intensively in the paper. Our recent review[100] focuses on the catalyst behaviors and reaction mechanism study of lignin derived bio-oil HDO process.

4.1. *Catalysts used in hydrodeoxygenation process**

Two types of catalysts are commonly used in HDO process. The first type is sulfide catalysts, such as $NiMoS/Al_2O_3$, $CoMoS/Al_2O_3$, etc. This type of catalyst has been widely used in the petroleum industry for HDO purpose for decades.[5, 66] The technique is mature and the reaction mechanisms are very well studied. The cost for this type of catalysts is much lower than the second type described below. Oxygen atoms in phenolic compounds can be effectively removed by these sulfided catalysts with high yields of aromatic and saturated products. However, bio-oil is different from conventional fossil feedstock that the oxygen content in bio-oil is much larger than that in the fossil oil and it has inherently low amount of sulfur. Although these catalysts are good at oxygen removal, the high amount of oxygen can cause rapid catalyst deactivation during the HDO. Second, the water in raw bio-oil would also induce deactivation to the catalysts; therefore, most of the reactions catalyzed by sulfide catalysts were conducted in gas phase. The high coking formation would also reduce the life of the catalysts. Some of the catalysts that need sulfur during the reaction would cause sulfur contamination.[100]

The second type of catalysts is transition metal catalysts, including platinum, palladium, ruthenium, rhodium, etc. This type of catalyst is tolerant to the solvent that it can perform HDO reaction with the existence of water or even in water phase. Generally, noble metals have higher reactivity for hydrogenation and require less severe reaction condition than sulfided catalyst. One of the disadvantages of the experiment is that it is sensitive to sulfur. The feedstock with certain amount of sulfur requires special treatment to remove the sulfur before the HDO process.

Part of this section was accepted for publication in Fuel, 2012. It is entitled as "Production of renewable gasoline from aqueous phase hydrogenation of lignin pyrolysis oil". The other authors are Arthur J. Ragauskas, Wei Mu and Yulin Deng. Reproduced with permission of Elsevier.

The other part of this section was accepted for publication in BioEnergy Research, 2013. It is entitled as "Lignin Pyrolysis Components and Upgrading – Technology Review". The other authors are Arthur J. Ragauskas, Wei Mu and Yulin Deng. Reproduced with permission of Springer Science+Business Media.

4.2. *Sulfided catalyst*

In the study of Ryymin *et al.*,[101] the HDO of phenol over sulfided NiMo/γ-Al_2O_3 was evaluated under 250°C and 7.5 MPa hydrogen. They found that the competition on active site greatly affected the selectivity, i.e., phenol was sensitive to the presence of the other reactant such as methyl heptanoate. The presence of sulfur induced catalyst deactivation and decreased the fraction of cyclohexane, because the active sites for alkene saturation were occupied by sulfur. The sulfur-containing intermediates were also observed.

The temperature effect on guaiacol HDO was also evaluated by Lin *et al.*[102] It was found that between 300 and 400°C, the yield of conversion was increased and the coke formation was decreased. The NiMo had high selectivity towards cyclohexane (~80%) and CoMo produced more phenol and methyl-phenol. Cokes and sulfur stripping from catalyst were found on both NiMo and CoMo catalysts.

4.3. *Noble metal catalyst*

4.3.1. *Platinum*

Platinum is the most extensively studied catalyst in this category. It is widely used in various HDO reactions and shows robust reactivity.[100] Very detailed reaction networks for HDO conversion of guaiacol catalyzed by Pt/γ-Al_2O_3 have been proposed by Gates' group.[103,104] Phenol has been detected as major HDO product from guaiacol and anisole is the minor HDO product. In addition, transalkylation, bimolecular transalkylation, hydrogenolysis and bimolecular transalkylation have also been found during the HDO process of guaiacol, which will produce catechol, veratrole and methylguaiacol, respectively. The authors also indicated that both HDO reaction and hydrogenolysis reaction were observed under the condition. The author defined the former as oxygen removal from the aromatic ring and the latter meant that oxygen still attached to the aromatic ring after the C-O cleavage (Fig. 4). The kinetics data was calculated for all four model compounds that hydrogenolysis reaction dominated in two of them (anisole, 4-methylanisole). For guaiacol, the hydrogenolysis reaction rate was still higher than the rate of HDO. After the aromatic ring was saturated, the Pt did not effectively deoxygenate the cyclohexanone and cyclohexanol under this condition.

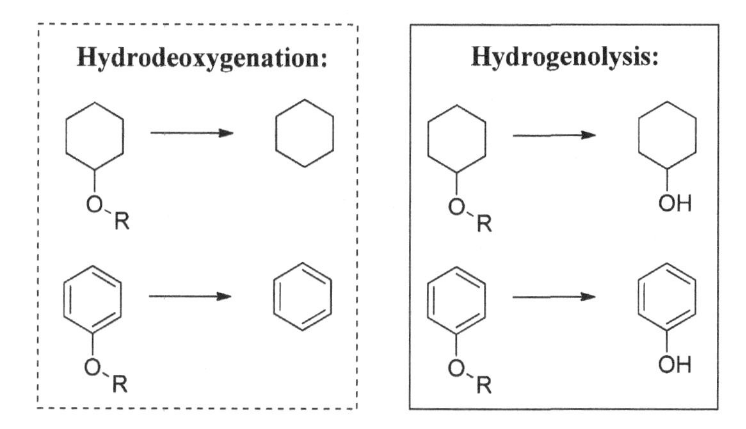

Fig. 4. Difference between Hydrogenolysis and Hydrodeoxygenation.[100]

Ohta *et al.*[105] used platinum supported on activated carbon (AC) to catalyze the 4-propyl-phenol HDO reaction in water phase at 280°C under acid-free condition. The aromatic structure was initially hydrogenated on the metal surface, and then the deoxygenated products were produced. In the NH_3-TPD (temperature programmed desorption) profile, Pt/AC showed no peak, which meant that there was no acidity site on the surface. Therefore, the acid-site catalyzed deoxygenation was not possible in the reaction. The hydrogenolysis reaction occurred on platinum surface. This catalyst was reused for three times and no deactivation was observed.

How the model compounds interacted with catalyst surface directly determines the behavior of catalyst. Pt (111) are the most stable facets of the Pt crystal.[106] Adsorption of anisole and its derivatives on Pt [111] surface was studied.[107] The anisole was less strongly bonded to the surface compared to parent molecule benzene, probably due to the steric hindrance of the methoxy group. The most stable configuration was that the molecule adsorbed parallel to the surface with both aromatic ring and oxygen above the bridge sites. The binding energy for this configuration was 2.23 eV for Pt (111). Another vertical configuration resulted in a much weaker adsorption. The binding energy was only 1.09 eV. The adsorption on stepped surface was also studied because it was often considered preferable sites for catalysis. The binding energy for Pt (211) was only 0.64 eV. The dissociation of phenol into phenoxy was endothermic on Pt (111) with reaction energy equal to 0.26 eV.

4.3.2. *Palladium*

Palladium is another widely used HDO catalyst. The behavior of palladium is similar to platinum; therefore, they were compared in some studies. The activity of these two catalysts highly depended on the reaction condition. In Liu *et al.*'s study,[108] metal dispersion of Pd was only half of the value on Pt, but the CO uptake amount was 50% higher than that on Pt. Pd catalyst also showed higher activity than Pt in the both hydrogenation and HDO of benzofuran. The study indicated that the hydrogenation (HYD) of benzofuran and the cleavage of C-O were easier over Pd than Pt. In some other cases, Pt showed better performance than Pd. In the HDO of 4-propyl-phenol under 280°C in acid-free aqueous solution, Pd on active carbon was used as catalyst.[105] Pd was good at hydrogenation but not at hydrogenolysis of C-O bond compared with Pt, Ru and Rh. More than half of the aliphatic-OH was left uncleaved. Another example was that under 100°C and 8 MPa H_2, Pd was not effective in either HYD or HDO of guaiacol. Half of the methoxy group was removed under this condition.[109]

Orita and Itoh studied phenol the formation of phenol from benzene on the surface on Pd (111).[110] Pd (111) was the most stable facet of palladium. Even in polycrystalline Pd, it was still most abundant. The deoxygenation was basically the reverse of this reaction. The adsorption of phenol on Pd surface was less stable than benzene. The adsorption energy of phenol on Pd (111) was -7.85 eV. When one hydrogen atom attached to the C-1, the adsorption energy of the intermediate increased to -6.85 eV. Then the O-atom and benzene were both adsorbed on one unit cell and the co-adsorption system became more stable (-7.06 eV). The last step was C-O bond cleavage. The sum of adsorption energies of oxygen atom and benzene in separate domains was -7.26 eV, with benzene adsorption energy at -1.43 eV and oxygen adsorption energy at -5.83 eV.

4.3.3. *Rhodium*

Rhodium was a relatively newly developed catalyst compared with platinum and palladium catalysts. However, it showed excellent performance in some cases. Under 280°C in acid-free aqueous solution, Rhodium on activated carbon successfully catalyzed the HDO reaction of

4-propyl-phenol.[105] The temperature effect on guaiacol HDO was also studied for Rh catalyst.[102] The temperature range studied was 300–400°C. The increasing temperature led to higher cyclohexane yield. The coke formed on Rh decreased with the increasing temperature. The $C_xH_yO_2$ compound decreased dramatically in the temperature range (especially increased from 350°C to 400°C). Both mono-oxygen compound and aliphatic compound increased. At 400°C, the aliphatic yield was almost 50%. RhPt and RhPd were both synthesized for comparison and monometallic Rh catalyst had higher HDO reactivity than any of them.

Rh (111) is the most stable facet of Rhodium.[106] Phenol showed strong adsorption on Rh (111) with binding energy of 2.79 eV under horizontal configuration. Under vertical configuration, the binding energy was 1.15 eV, which was much weaker than that under horizontal configuration. On the stepped surface Rh (211), the binding energy reduced to 1.79 eV. The dissociation of phenol to phenoxy was -0.27 eV, which indicated an exothermic reaction.

4.3.4. *Ruthenium*

Ruthenium on activated carbon was used to catalyze the HDO of 4-propyl-phenol under 280°C in acid-free aqueous solution.[105] In the study, Ru catalyst achieved 100% hydrogenation but the deoxygenation capability was weaker than Rh and Pt that only 50% aliphatic-OH bond was cleaved.

The performance of Ru is highly affected by the support. Although it is also observed for other metal catalysts, the effect is stronger for Ru. In the study of Park *et al.*,[111] platinum, rhodium, palladium and ruthenium were compared in guaiacol HDO on three different support materials. When using γ-Al_2O_3/SiO_2-Al_2O_3/nitric acid treated carbon black as supports, the highest product yield were cyclohexanol/cyclohexane/2-methoxy cyclohexanol respectively. Furthermore, the selectivity of the Ru was the best among noble metals.

Guo *et al.*[112] studied the hydrotreating of eugenol by using Ru catalyst. The major crystal facet of Ru was Ru (101) and the major facet of Pd is Pd (111). Under similar condition, ruthenium showed much higher activity than palladium. The reactivity of catalyst had no loss after two runs with model compounds. When using distillated fraction from bio-oil as reactant (major components were phenolic compounds, such as phenol, 2-methoxyphenol, 4-ethylphenol, 4-methyl-2-methoxyphenol, 4-ethyl-2-methoxyphenol), the catalyst lost activity rapidly. The BET study showed the reduction in both surface area and pore volume, which indicated the coke or tar formation in pore channels. The study further revealed that Ru^0 were the major role in HYD reaction.

Wildschut *et al.*[96] studied the hydrotreatment of fast pyrolysis oil using Ru/C at 623K and 20 MPa H_2 pressure. The ruthenium catalyst was very effective in deoxygenation in the beginning. After 1 hour, the O/C molar ratio decreased from 0.45 to 0.02. However, with the time going, it increased from 0.02 to 0.07 in 1–6 hours of recaction time. Probably because the low O/C ratio compounds transferred from oil phase to gas phase. The H/C ratio was 1.35 for the pyrolysis oil. At 1 hour, it dropped to 1.05 at first, and then increased to 1.32 after 6 hours. The initial drop was probably caused by loss of hydrogen for dehydration.

Our previous work[113] investigated the two step hydrogenation (HYD) of the heavy oil produced from pyrolysis of pine wood ethanol organosolv lignin (EOL) at 600°C for 30 min. Ru/C was used as the catalyst and water was used as the dispersant for the heavy oil and the hydrogenation products. The first hydrogenation step was accomplished at 300°C with a H_2 pressure of 14 Mpa for 4 h and Ru/C as catalyst. This treatment converted the water insoluble dark brown heavy pyrolysis oil to a transparent product. Further hydrogenation of this product was accomplished at 250°C and a H_2 pressure of 14 Mpa with Ru/C as catalyst for 2 h to produce the final product. The carbon yields for the first and second step hydrogenation are 35% and 33% (overall molar% of carbon content in the heavy oil) respectively.

By employing ^{1}H-NMR, compared with the EOL heavy oil and first step HYD product, the aromatic protons were completely eliminated in the second HYD product and 85% of protons belonged to the aliphatic protons with no oxygen attached to the α-carbon. It indicates that the second HYD product contains only aliphatic carbons and has relatively lower oxygen content, which represents a potential resource for bio-gasoline. Compared with the EOL heavy oil, first step HYD product contains less aromatic protons but more aliphatic protons, which indicates that the hydrogenation of benzene ring also occurs during first HYD process. The ^{13}C-NMR integration results show that there is no aromatic carbon in the second step HYD product, which is consistent with ^{1}H-NMR result. There are less methoxy groups but much more aliphatic C-O bonds in the second step HYD product. This is attributed to the hydrogenation of the aromatic ring that converts aromatic C-O bonds to aliphatic C-O bonds. Compared with the EOL heavy oil, there are relatively lower amounts of carbonyl C=O bonds, aromatic C-C bonds and C-H bonds but more aliphatic C-C bonds in the first step HYD product, which is also consistent with ^{1}H-NMR result and is the evidence that the hydrogenation of carbonyl C=O bonds and benzene ring occurred during the first HYD process. To fully characterize the HYD products, DEPT-135 ^{13}C-NMR and HSQC-NMR were used. Compared with the complexity of the EOL heavy oil, the first step HYD product is relatively easier to identify. On the basis of GC-MS result and HSQC-NMR, the major components in the first step HYD are guaiacol, 4-methylguaiacol and catechol, which indicate that the EOL heavy oil (weight-average molecular weight (M_w) is 265 g/mol) has been upgraded to the simple aromatic molecule after first step HYD process. The DEPT-135 and quantitative ^{13}C-NMR also show that after first step HYD process, the carbonyl C=O bonds and the aliphatic C-O bonds in the EOL heavy oil have been reduced consistent with a hydrodeoxygenation (HDO) process. Chen *et al.*[114] reported that the hydrodeoxygenation and the cleavage of C-C bond of carboxylic acids to alcohols and alkanes over supported Ru in the aqueous phase were favored at high temperature and with Ru/C as the catalyst, which supported our results. There are also more aliphatic CH_2 peaks in the first step HYD products, which is due to the hydrogenation of benzene ring in the EOL heavy oils. The NMR results also show that all the aliphatic C-O bonds in the second step HYD product contain a tertiary carbon, which is formed from hydrogenation of phenols and represent to the α-carbon in the alcohols. A significant amount of the secondary carbons in the second step HYD products have a chemical shift of 30–40 ppm, which represent β-substituted alcohols.

On the basis of our previous work[93,115] and GC-MS results, most peaks in the HSQC-NMR of the second step HYD products could be assigned to cyclohexanol and its derivatives. The carbon yield from first to second step HYD products is very high (~95 wt%), which is similar to literature reports about the hydrogenation of several phenolic pyrolysis oil model compounds.[24,25] It is also the evidence that the products of first step HYD process are simple aromatic molecules. Several researchers have reported the tentative reaction pathways of HYD and HDO process of phenolic model compounds.[104,114,116–118] On the basis of those pathways and our results, the first step HYD process of EOL heavy oil mainly involves the cleavage of ether bonds and methoxyl groups in the heavy oils, which produces simple aromatic molecules, such as guaiacol. The second step HYD process will further upgrade those aromatics to the aliphatic rings, such as cyclohexanol. The tentative pathways are shown in Fig. 5.

There are some reports showing that the tar or coke will be formed during the HYD or HDO process of phenolic model compounds[104,118] and water soluble pyrolysis oils.[26,28] We also find the formation of tar after the first step HYD process. The M_w (462 g/mol) of the tar is almost as twice as EOL heavy oil (M_w=265 g/mol). The quantitative ^{13}C-NMR for the tar shows that there are much more condensed aromatic C-C bonds than the EOL heavy oil, which represent condensation reactions during the HYD process. Comparable condensation reactions have been reported during the HYD process of model compounds in the literatures[104,118] and the possible pathways are shown in Fig. 5. The SEM pictures for the catalysts show that after first step HYD of EOL heavy oil, the catalyst was coated by tar, which

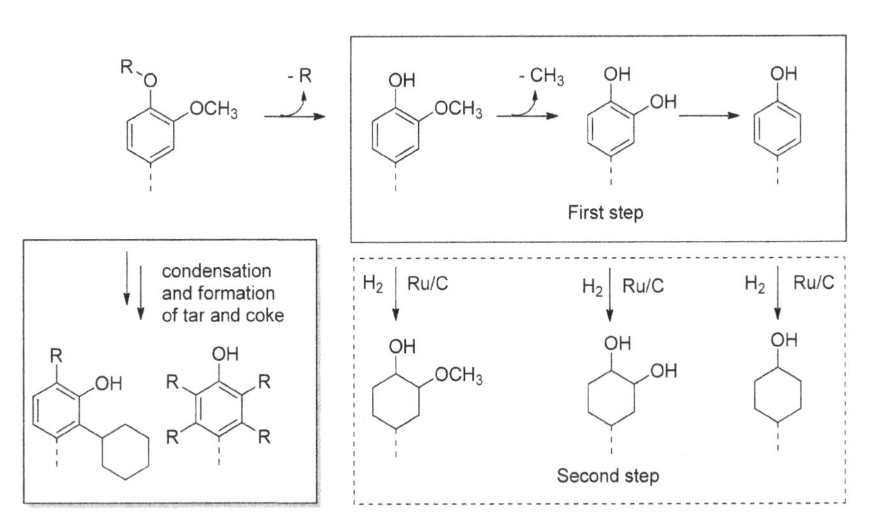

Fig. 5. Tentative reaction pathways of HYD and HDO process of EOL heavy oil.[104,113,114,116–118]

may affect the hydrogen transfer and the HYD reactions. The surface of catalyst for the second step HYD was almost intact after the process, which was consistent with the high carbon yield from the first to the second step HYD process. Thermogravimetric analysis (TGA) was also employed to further investigate the spent catalyst after the first step HYD process. The result indicating a significant weight loss (~40 wt %) from 300–400°C for the spent catalyst was also the evidence of the catalyst being coated by tar.

In summary, a two-step hydrogenation of water insoluble pyrolysis oil (heavy oil) produced from pyrolysis of pine wood ethanol organosolv lignin (EOL) at 600°C for 30 min was examined. Ru/C was used as the catalyst while water was used as the dispersant for the heavy oil and hydrogenation products. The carbon conversion yields for the first and second step hydrogenation were 35% and 33% (overall molar% of carbon content in the heavy oil), respectively. The products of first step of hydrogenation were primarily aromatic molecules, which were produced from the hydrolytic cleavage of ether bond and methoxy groups in the heavy oils. Further hydrogenation was shown to covert the insoluble heavy oils (weight average molecular weight was 265 g/mol) to the aliphatic alcohols and other aliphatic components, which could be used as renewable gasoline. As far as we know, this is the first reported effort to upgrade water insoluble parts of lignin pyrolysis oil to the total aliphatic components by aqueous phase via hydrogenation.

Acknowledgments

The authors are grateful for financial support from the Paper Science & Engineering (PSE) Fellowship program at Institute of Paper Science & Technology (IPST) and the School of Chemistry and Biochemistry at Georgia Institute of Technology.

References

1. R. D. Perlack, L. L. Wright, A. F. Turhollow, *et al.* Biomass as Feedstock for a bioenergy and bioproducts Industry: The technical feasibility of a billion-ton annual supply. *Biomass as Feedstock for a bioenergy and bioproducts Industry: The technical feasibility of a billion-ton annual supply* (2005).

2. A. J. Ragauskas, C. K. Williams, B. H. Davison, *et al.* The path forward for biofuels and biomaterials, *Science,* **311**, 484–489 (2006).
3. R. P. Anex, A. Aden, F. K. Kazi, *et al.* Techno-economic comparison of biomass-to-transportation fuels via pyrolysis, gasification, and biochemical pathways, *Fuel,* **89**, Supplement 1:S29–S35 (2010).
4. S. Czernik and A. V. Bridgwater, Overview of Applications of Biomass Fast Pyrolysis Oil, *Energy & Fuels,* **18**, 590–598 (2004).
5. G. W. Huber, S. Iborra and A. Corma, Synthesis of transportation fuels from biomass: Chemistry, catalysts, and engineering, *Chemical Reviews,* **106**, 4044–4098 (2006).
6. C. A. Mullen and A. A. Boateng, Catalytic pyrolysis-GC/MS of lignin from several sources, *Fuel Processing Technology,* **91**, 1446–1458 (2010).
7. Y. Zhao, L. Deng and B. Liao, Aromatics Production via Catalytic Pyrolysis of Pyrolytic Lignins from Bio-Oil, *Energy & Fuels,* **24**, 5735–5740 (2010).
8. R. French and S. Czernik, Catalytic pyrolysis of biomass for biofuels production, *Fuel Processing Technology,* **91**, 25–32 (2010).
9. M. A. Jackson, D. L. Compton and A. A. Boateng, Screening heterogeneous catalysts for the pyrolysis of lignin, *Journal of Analytical and Applied Pyrolysis,* **85**, 226–230 (2009).
10. A. Aho, N. Kumar and K. Eranen, Catalytic pyrolysis of woody biomass in a fluidized bed reactor: Influence of the zeolite structure, *Fuel,* **87,** 2493–2501 (2008).
11. H. Zhang, R. Xiao and H. Huang, Comparison of non-catalytic and catalytic fast pyrolysis of corncob in a fluidized bed reactor, *Bioresource Technology,* **100**, 1428–1434 (2009).
12. P. Pan, C. Hu and W. Yang, The direct pyrolysis and catalytic pyrolysis of Nannochloropsis sp. residue for renewable bio-oils, *Bioresource Technology,* **101**, 4593–4599 (2010).
13. B. B. Uzun and N. Sarioǎlu, Rapid and catalytic pyrolysis of corn stalks, *Fuel Processing Technology,* **90**, 705–716 (2009).
14. A. Pattiya, J. O. Titiloye and A. V. Bridgwater, Evaluation of catalytic pyrolysis of cassava rhizome by principal component analysis, *Fuel,* **89**, 244–253 (2010).
15. A. Pattiya, J. Titiloye and A. Bridgwater, Fast pyrolysis of cassava rhizome in the presence of catalysts, *Journal of Analytical and Applied Pyrolysis,* **81**, 72–79 (2008).
16. F. A. Agblevor, S. Beis, O. Mante, *et al.* Fractional Catalytic Pyrolysis of Hybrid Poplar Wood, *Industrial & Engineering Chemistry Research,* **49**, 3533–3538 (2010).
17. P. T. Williams and N. Nugranad, Comparison of products from the pyrolysis and catalytic pyrolysis of rice husks, *Energy,* **25**, 493–513 (2000).
18. B. Valle, A. G. Gayubo, A. S. T. Aguayo, *et al.* Selective Production of Aromatics by Crude Bio-oil Valorization with a Nickel-Modified HZSM-5 Zeolite Catalyst, *Energy & Fuels,* **24**, 2060–2070 (2010).
19. A. Atutxa, R. Aguado, A. G. Gayubo, *et al.* Kinetic Description of the Catalytic Pyrolysis of Biomass in a Conical Spouted Bed Reactor, *Energy & Fuels,* **19**, 765–774 (2005).

20. B. Scholze, C. Hanser and D. Meier, Characterization of the water-insoluble fraction from fast pyrolysis liquids (pyrolytic lignin): Part II. GPC, carbonyl goups, and ^{13}C-NMR, *Journal of Analytical and Applied Pyrolysis,* **58–59**, 387–400 (2001).
21. A. Chaala, T. Ba, M. Garcia-Perez, *et al.* Colloidal Properties of Bio-oils Obtained by Vacuum Pyrolysis of Softwood Bark: Aging and Thermal Stability, *Energy & Fuels,* **18**, 1535–1542 (2004).
22. M. Kosa, H. Ben, H. Theliander, *et al.* Pyrolysis oils from CO_2 precipitated Kraft lignin, *Green Chemistry,* **13**, 3196–3202 (2011).
23. M. E. Boucher, A. Chaala, H. Pakdel, *et al.* Bio-oils obtained by vacuum pyrolysis of softwood bark as a liquid fuel for gas turbines. Part II: Stability and ageing of bio-oil and its blends with methanol and a pyrolytic aqueous phase, *Biomass and Bioenergy,* **19**, 351–361 (2000).
24. N. Yan, Y. Yuan, R. Dykeman, *et al.* Hydrodeoxygenation of lignin-derived phenols into alkanes by using nanoparticle catalysts combined with Bronsted acidic ionic liquids, *Angewandte Chemie International Edition,* **49**, 5549–5553 (2010).
25. C. Zhao, Y. Kou, A. A. Lemonidou, *et al.* Highly selective catalytic conversion of phenolic bio-oil to alkanes, *Angewandte Chemie International Edition,* **48**, 3987–3990 (2009).
26. T. P. Vispute, H. Zhang, A. Sanna, *et al.* Renewable Chemical Commodity Feedstocks from Integrated Catalytic Processing of Pyrolysis Oils, *Science,* **330**, 1222–1227 (2010).
27. J. Wildschut, M. Iqbal, F. H. Mahfud, *et al.* Insights in the hydrotreatment of fast pyrolysis oil using a ruthenium on carbon catalyst, *Energy & Environmental Science,* **3**, 962 (2010).
28. G. W. Huber, J. N. Chheda, C. J. Barrett, *et al.* Production of Liquid Alkanes by Aqueous-Phase Processing of Biomass-Derived Carbohydrates, *Science,* **308**, 1446–1450 (2005).
29. N. Li, G. A. Tompsett and T. Zhang, Renewable gasoline from aqueous phase hydrodeoxygenation of aqueous sugar solutions prepared by hydrolysis of maple wood, *Green Chemistry,* **13**, 91–101 (2011).
30. S. Czernik, D. K. Johnson and S. Black, Stability of wood fast pyrolysis oil, *Biomass and Bioenergy,* **7**, 187–192 (1994).
31. A. Chaala, T. Ba and M. Garcia-Perez, Colloidal properties of bio-oils obtained by vacuum pyrolysis of softwood bark: Aging and thermal stability, *Energy & Fuels,* **18**, 1535–1542 (2004).
32. C. D. Naske, P. Polk and P. Z. Wynne, Postcondensation Filtration of Pine and Cottonwood Pyrolysis Oil and Impacts on Accelerated Aging Reactions, *Energy & Fuels,* **26**, 1284–1297 (2012).
33. M. W. Nolte and M. W. Liberatore, Real-Time Viscosity Measurements during the Accelerated Aging of Biomass Pyrolysis Oil, *Energy & Fuels,* **25**, 3314–3317 (2011).
34. A. Oasmaa and E. Kuoppala, Fast Pyrolysis of Forestry Residue, 3. Storage Stability of Liquid Fuel, *Energy & Fuels,* **17**, 1075–1084 (2003).

35. R. Fahmi, A. V. Bridgwater and I. Donnison, The effect of lignin and inorganic species in biomass on pyrolysis oil yields, quality and stability, *Fuel,* **87**, 1230–1240 (2008).
36. J. V. Ortega, A. M. Renehan and M. W. Liberatore, Physical and chemical characteristics of aging pyrolysis oils produced from hardwood and softwood feedstocks, *Journal of Analytical and Applied Pyrolysis,* **91**, 190–198 (2011).
37. A. Oasmaa, E. Kuoppala and D. C. Elliot, Development of the Basis for an Analytical Protocol for Feeds and Products of Bio-oil Hydrotreatment, *Energy & Fuels,* **26**, 2454–2460 (2012).
38. E. Fratini, M. Bonini and A. Oasmaa, SANS analysis of the microstructural evolution during the aging of pyrolysis oils from biomass, *Langmuir,* **22**, 306–312 (2006).
39. M. Garcia-Perez, A. Chaala and H. Pakdel, Characterization of bio-oils in chemical families, *Biomass and Bioenergy,* **31**, 222–242 (2007).
40. H. Ben and A. J. Ragauskas, NMR Characterization of Pyrolysis Oils from Kraft Lignin, *Energy & Fuels,* **25**, 2322–2332 (2011).
41. H. Ben and A. J. Ragauskas, Pyrolysis of Kraft Lignin with Additives, *Energy & Fuels,* **25**, 4662–4668 (2011).
42. E. D. Christensen, G. M. Chupka and J. Luecke, Analysis of Oxygenated Compounds in Hydrotreated Biomass Fast Pyrolysis Oil Distillate Fractions, *Energy & Fuels,* **25**, 5462–5471 (2011).
43. H. Ben and A. J. Ragauskas, *In Situ* NMR Characterization of Pyrolysis Oil during Accelerated Aging, *ChemSusChem.,* **5**, (9), 1687–1693 (2012).
44. D. K. Shen and S. Gu, The mechanism for thermal decomposition of cellulose and its main products, *Bioresource Technology,* **100**, 6496–6504 (2009).
45. P. F. Britt, M. K. Kidder and A. C. Buchanan, I. Oxygen substituent effects in the pyrolysis of phenethyl phenyl ethers, *Energy & Fuels,* **21**,3102–3108 (2007).
46. J. B. Binder, M. J. Gray and J. F. White, Reactions of lignin model compounds in ionic liquids, *Biomass and Bioenergy,* **33**, 1122–1130 (2009).
47. T. Nakamura, H. Kawamoto and S. Saka, Condensation Reactions of Some Lignin Related Compounds at Relatively Low Pyrolysis Temperature, *Journal of Wood Chemistry and Technology,* **27**, 121–133 (2007).
48. N. Brosse, R. El Hage and M. Chaouch, Investigation of the chemical modifications of beech wood lignin during heat treatment, *Polymer Degradation and Stability,* **95**, 1721–1726 (2010).
49. K. David, H. Ben and J. Muzzy, Chemical characterization and water content determination of bio-oils obtained from various biomass species using ^{31}P NMR spectroscopy, *Biofuels,* **3**, 123–128 (2012).
50. F. G. Calvo-Flores and J. A. Dobado, Lignin as renewable raw material, *ChemSusChem.,* **3**, 1227–1235 (2010).
51. R. W. Coughlin and F. Davoudzadeh, Coliquefaction of lignin and bituminous coal, *Fuel,* **65**, 95–106 (1986).
52. T. Watanabe, H. Kawamoto and S. Saka, Radical chain reactions in pyrolytic cleavage of the ether linkages of lignin model dimers and a trimer, *Holzforschung,* **63**, 424–430 (2009).

53. P. Sazama, J. Dedecek and V. Gabova, Effect of aluminium distribution in the framework of ZSM-5 on hydrocarbon transformation. Cracking of 1-butene, *Journal of Catalysis,* **254**, 180–189 (2008).
54. J. Huang, W. Long and P. K. Agrawal, Effects of Acidity on the Conversion of the Model Bio-oil Ketone Cyclopentanone on H–Y Zeolites, *The Journal of Physical Chemistry C,* **113**, 16702–16710 (2009).
55. J. F. Haw, Zeolite acid strength and reaction mechanisms in catalysis, *Physical Chemistry Chemical Physics,* **4**, 5431–5441 (2002).
56. D. J. Mihalcik, C. A. Mullen and A. A. Boateng, Screening acidic zeolites for catalytic fast pyrolysis of biomass and its components, *Journal of Analytical and Applied Pyrolysis,* **92**, 224–232 (2011).
57. A. M. Azeez, D. Meier, and J. Odermatt, Effects of zeolites on volatile products of beech wood using analytical pyrolysis, *Journal of Analytical and Applied Pyrolysis,* **91**, 296–302 (2011).
58. E. F. Iliopoulou, E. V. Antonakou and S. A. Karakoulia, Catalytic conversion of biomass pyrolysis products by mesoporous materials: Effect of steam stability and acidity of Al-MCM-41 catalysts, *Chemical Engineering Journal,* **134**, 51–57 (2007).
59. N. Fonseca, F. Lemos and S. Laforge, Influence of acidity on the H-Y zeolite performance in n-decane catalytic cracking: evidence of a series/parallel mechanism, *Reaction Kinetics, Mechanisms and Catalysis* 2010.
60. A. Aho, N. Kumar and K. Eranen, Catalytic Pyrolysis of Biomass in a Fluidized Bed ReactorInfluence of the Acidity of H-Beta Zeolite, *Process Safety and Environmental Protection,* **85**, 473–480 (2007).
61. H. Ben and A. J. Ragauskas, Influence of Si/Al Ratio of ZSM-5 Zeolite on the Properties of Lignin Pyrolysis Products, *ACS Sustainable Chemistry & Engineering* **1**(3), 316–324 (2013).
62. C. A. Mullen, A.A. Boateng and D. J. Mihalcik, Catalytic Fast Pyrolysis of White Oak Wood in a Bubbling Fluidized Bed, *Energy & Fuels,* **25**, 5444–5451 (2011).
63. G. Fogassy, N. Thegarid and Y. Schuurman, From biomass to bio-gasoline by FCC co-processing: effect of feed composition and catalyst structure on product quality, *Energy & Environmental Science,* **4**, 5068 (2011).
64. P. D. Chantal, S. Kaliaguine, and J. L. Grandmaison, Reactions of phenolic compounds over HZSM-5, *Applied Catalysis,* **18**, 133–145 (1985).
65. D. J. Mihalcik, A. A. Boateng and C. A. Mullen, Packed-Bed Catalytic Cracking of Oak-Derived Pyrolytic Vapors, *Industrial & Engineering Chemistry Research,* **50**, 13304–13312 (2011).
66. P. M. Mortensen, J. D. Grunwaldt, P. A. Jensen, *et al.* A review of catalytic upgrading of bio-oil to engine fuels, *Applied Catalysis A: General,* **407**, 1–19 (2011).
67. M. A. Peralta, T. Sooknoi and T. Danuthai, Deoxygenation of benzaldehyde over CsNaX zeolites, *Journal of Molecular Catalysis A: Chemical,* **312**, 78–86 (2009).
68. H. Chiang and A. Bhan, Catalytic consequences of hydroxyl group location on the rate and mechanism of parallel dehydration reactions of ethanol over acidic zeolites, *Journal of Catalysis,* **271**, 251–261 (2010).

69. C. Pereira, G. T. Kokotailo, R. J. Gorte, Adsorption and reaction of 2-propen-1-ol in H-ZSM-5, *The Journal of Physical Chemistry,* **94**, 2063–2067 (1990).
70. H-K, Min and S. B. Hong, Mechanistic Investigations of Ethylbenzene Disproportionation over Medium-Pore Zeolites with Different Framework Topologies, *The Journal of Physical Chemistry C,* **115**, 16124–16133 (2011).
71. S. J. Collins and P. J. Omalley, A Theoretical Description for the Monomolecular Cracking of C-C Bonds over Acidic Zeolites, *Journal of Catalysis,* **153**, 94–99 (1995).
72. T. Jin, D. H. Xia and Y. Z. Xiang, The Effect of Metal Introduced Over ZSM-5 Zeolite for C_9-Heavy Aromatics Hydrodealkylation, *Petroleum Science and Technology,* 27, 1821–1835 (2009).
73. S. M. Holmes, A. Garforth and B. Maunders, A solvent extraction method to study the location and concentration of coke formed on zeolite catalysts, *Applied Catalysis A: General,* **151**, 355–372 (1997).
74. S. Laforge, m-Xylene transformation over H-MCM-22 zeolite, 1. Mechanisms and location of the reactions, *Journal of Catalysis,* **220**, 92–103 (2003).
75. P. Magnoux, A. Rabeharitsara and H. S. Cerqueira, Influence of reaction temperature and crystallite size on HBEA zeolite deactivation by coke, *Applied Catalysis A: General,* **304**, 142–151 (2006).
76. P. Magnoux, H. S. Cerqueira and M. Guisnet, Evolution of coke composition during ageing under nitrogen, *Applied Catalysis A: General,* **235**, 93–99 (2002).
77. A. Aho, N. Kumar and K. Eränen, Catalytic pyrolysis of woody biomass, *Biofuels,* **1**, 261–273 (2010).
78. A. Aho, N. Kumar and A. V. Lashkul, Catalytic upgrading of woody biomass derived pyrolysis vapours over iron modified zeolites in a dual-fluidized bed reactor, *Fuel,* **89**, 1992–2000 (2010).
79. A. Aho, A. Tokarev and P. Backman, Catalytic Pyrolysis of Pine Biomass Over H-Beta Zeolite in a Dual-Fluidized Bed Reactor: Effect of Space Velocity on the Yield and Composition of Pyrolysis Products, *Topics in Catalysis,* **54**, 941–948 (2011).
80. T. R. Carlson, Y-T. Cheng and J. Jae, Production of green aromatics and olefins by catalytic fast pyrolysis of wood sawdust, *Energy & Environmental Science,* **4**, 145 (2011).
81. J. Jae, G. A. Tompsett, and A. J. Foster, Investigation into the shape selectivity of zeolite catalysts for biomass conversion, *Journal of Catalysis,* **279**, 257–268 (2011).
82. T. R. Carlson, G. A. Tompsett and W. C. Conner, *et al.* Aromatic Production from Catalytic Fast Pyrolysis of Biomass-Derived Feedstocks, *Topics in Catalysis,* **52**, 241–252 (2009).
83. J. D. Adjaye and N. N. Bakhshi, Production of hydrocarbons by catalytic upgrading of a fast pyrolysis bio-oil. Part II: Comparative catalyst performance and reaction pathways, *Fuel Processing Technology,* **45**, 185–202 (1995).
84. J. D. Adjaye and N. N. Bakhshi, Production of hydrocarbons by catalytic upgrading of a fast pyrolysis bio-oil. Part I: Conversion over various catalysts, *Fuel Processing Technology,* **45**, 161–183 (1995).

85. D. Fabbri, A. Adamiano and C. Torri, GC-MS determination of polycyclic aromatic hydrocarbons evolved from pyrolysis of biomass, *Analytical and Bioanalytical Chemistry,* **397**, 309–317 (2010).
86. W. Y. Qi, C. W. Hu and G. Y. Li, Catalytic pyrolysis of several kinds of bamboos over zeolite NaY, *Green Chemistry*, **8**, 183 (2006).
87. M. Misson, R. Haron and M. F. Kamaroddin, Pretreatment of empty palm fruit bunch for production of chemicals via catalytic pyrolysis, *Bioresource Technology,* **100**, 2867–2873 (2009).
88. S. Vitolo, M. Seggiani and P. Frediani, Catalytic upgrading of pyrolytic oils to fuel over different zeolites, *Fuel,* **78**, 1147–1159 (1999).
89. S. Stephanidis, C. Nitsos, and K. Kalogiannis, Catalytic upgrading of lignocellulosic biomass pyrolysis vapours: Effect of hydrothermal pre-treatment of biomass, *Catalysis Today,* **167**, 37–45 (2011).
90. M. Nagy, M. Kosa and H. Theliander, Characterization of CO_2 precipitated Kraft lignin to promote its utilization, *Green Chemistry,* **12**, 31 (2010).
91. P. Tomani, The lignoboost process, *Cellulose Chemistry and Tgechnology,* **44**, 53–58 (2010).
92. H. Yang, R. Yan and H. Chen, Characteristics of hemicellulose, cellulose and lignin pyrolysis, *Fuel,* **86**, 1781–1788 (2007).
93. H. Ben and A. J. Ragauskas, Heteronuclear Single-Quantum Correlation–Nuclear Magnetic Resonance (HSQC–NMR) Fingerprint Analysis of Pyrolysis Oils, *Energy & Fuels,* **25**, 5791–5801 (2011).
94. H. Ben, A. J. Ragauskas, One step thermal conversion of lignin to the gasoline range liquid products by using zeolites as additives, *RSC Advances,* **2**, (33), 12892–12898 (2012).
95. A. Oasmaa, D. C. Elliott and J. Korhonen, Acidity of Biomass Fast Pyrolysis Bio-oils, *Energy & Fuels,* **24**, 6548–6554 (2010).
96. J. Wildschut, M. Iqbal and F. H. Mahfud, Insights in the hydrotreatment of fast pyrolysis oil using a ruthenium on carbon catalyst, *Energy & Environmental Science,* **3**, 962–970 (2010).
97. E. Furimsky, Chemistry of Catalytic Hydrodeoxygenation, *Catalysis Reviews,* **25**, 421–458 (1983).
98. T. V. Choudhary and C. B. Phillips, Renewable fuels via catalytic hydrodeoxygenation, *Applied Catalysis A: General,* **397**, 1–12 (2011).
99. Q. Bu, H. Lei, A. H. Zacher, A review of catalytic hydrodeoxygenation of lignin-derived phenols from biomass pyrolysis, *Bioresource Technology,* **124**, 470–477 (2012).
100. W. Mu, H. Ben, A. Ragauskas and Y. Deng, Lignin pyrolysis components and upgrading — Technology review, *BioEnergy Research* 1–22 (2013).
101. E-M. Ryymin, M. L. Honkela and T-R. Viljava, Competitive reactions and mechanisms in the simultaneous HDO of phenol and methyl heptanoate over sulphided NiMo/γ-Al2O3, *Applied Catalysis A: General,* **389**, 114–121 (2010).
102. Y-C. Lin, C-L. Li, H-P. Wan, Catalytic Hydrodeoxygenation of Guaiacol on Rh-Based and Sulfided CoMo and NiMo Catalysts, *Energy & Fuels,* **25**, 890–896 (2011).

103. R. C. Runnebaum, T. Nimmanwudipong and D. E. Block, Catalytic conversion of compounds representative of lignin-derived bio-oils: a reaction network for guaiacol, anisole, 4-methylanisole, and cyclohexanone conversion catalysed by Pt/[gamma]-Al2O3, *Catalysis Science & Technology,* **2**, 113–118 (2012).
104. T. Nimmanwudipong, R. C. Runnebaum, D. E. Block, Catalytic Conversion of Guaiacol Catalyzed by Platinum Supported on Alumina: Reaction Network Including Hydrodeoxygenation Reactions, *Energy & Fuels,* **25**, 3417–3427 (2011).
105. H. Ohta, H. Kobayashi and K. Hara, Hydrodeoxygenation of phenols as lignin models under acid-free conditions with carbon-supported platinum catalysts, *Chemical Communications,* **47**, 12209–12211 (2011).
106. M. L, Honkela, J. Bjork and M. Persson, Computational study of the adsorption and dissociation of phenol on Pt and Rh surfaces, *Physical Chemistry Chemical Physics,* **14**, 5849–5854 (2012).
107. N. Bonalumi, A. Vargas, and D. Ferri, Theoretical and Spectroscopic Study of the Effect of Ring Substitution on the Adsorption of Anisole on Platinum, *The Journal of Physical Chemistry, B* 110, 9956–9965 (2006).
108. C. Liu, Z. Shao and Z. Xiao, Hydrodeoxygenation of Benzofuran over Silica–Alumina-Supported Pt, Pd, and Pt–Pd Catalysts, *Energy & Fuels,* **26**, 4205–4211 (2012).
109. A. Gutierrez, R. K. Kaila and M. L. Honkela, Hydrodeoxygenation of guaiacol on noble metal catalysts, *Catalysis Today,* **147**, 239–246 (2009).
110. H. Orita and N. Itoh, Simulation of phenol formation from benzene with a Pd membrane reactor: ab initio periodic density functional study, *Applied Catalysis A: General,* **258**, 17–23 (2004).
111. C. R. Lee, J. S. Yoon, Y-W. Suh, Catalytic roles of metals and supports on hydrodeoxygenation of lignin monomer guaiacol, *Catalysis Communications,* **17**, 54–58 (2012).
112. J. Guo, R. Ruan and Y. Zhang, Hydrotreating of Phenolic Compounds Separated from Bio-oil to Alcohols, *Industrial & Engineering Chemistry Research,* **51**, 6599–6604 (2012).
113. H. Ben, W. Mu Y. Deng and A. J, Ragauskas, Production of renewable gasoline from aqueous phase hydrogenation of lignin pyrolysis oil, *Fuel,* **103**, 1148–1153 (2013).
114. L. Chen, Y. Zhu, H. Zheng, Aqueous-phase hydrodeoxygenation of carboxylic acids to alcohols or alkanes over supported Ru catalysts, *Journal of Molecular Catalysis A: Chemical,* **351**, 217–227 (2011).
115. H. Ben, A. Rogauskas, Pyrolysis of biomass to biofuels., 243rd ACS National Meeting & Exposition, San Diego, CA.
116. C. Zhao, J. He, and A. A. Lemonidou, Aqueous-phase hydrodeoxygenation of bio-derived phenols to cycloalkanes, *J Catal.,* **280**, 8–16 (2011).
117. V. N. Bui, D. Laurenti, P. Delichère, Hydrodeoxygenation of guaiacol, *Applied Catalysis B: Environmental,* **101**, 246–255 (2011).
118. V. N. Bui, D. Laurenti and P. Afanasiev, Hydrodeoxygenation of guaiacol with CoMo catalysts. Part I: Promoting effect of cobalt on HDO selectivity and activity, *Applied Catalysis B: Environmental,* **101**, 239–245 (2011).

CHAPTER 10

CORROSION ISSUES IN BIOFUELS

LINDSEY R. GOODMAN*, PREET M. SINGH†

**GATE, Inc. 16360 Park Ten Pl. Suite 206, Houston, TX 77084*
lgoodman@gateinc.com
†Professor, School of Materials Science and Engineering,
Georgia Institute of Technology, Atlanta, GA 30332
Preet.singh@mse.gatech.edu

In this chapter, the compatibility of metals and alloys with different biofuels is discussed. As the use of renewable biofuels becomes more and more widespread, issues of material compatibility with these biofuels arise. Corrosion behavior of industrial metals and alloys, such as carbon and stainless steels, is well understood in traditional fossil fuel environments, and methods of mitigating and preventing corrosion of steel tanks, pipes, engine parts, and other items have been developed. When the environment in contact with these metal surfaces is changed from fossil fuel to biofuel, the nature of the metal-environment interaction can change, leading to corrosion or degradation of expensive industrial equipment, sometimes leading to catastrophic failures. Failures of steel tanks and pipes have been attributed to stress corrosion cracking caused by ethanol fuels. Research has shown that corrosion and cracking of steels in fuel grade ethanol (FGE) are mostly caused by minor chemical constituents and impurities within the fuels. Specifically, water and oxygen contained in FGE play large roles in their corrosivity. Similarly, pyrolysis oil or bio-oil fuels produced from biomass are significantly more corrosive compared to traditional fuels. As-produced pyrolysis oils contain a substantial amount of oxygen, primarily as component of water, carboxylic acids, phenols, ketones and aldehydes. Depending on the biomass sources, production processes, and resulting chemical constituents, bio-oils can be significantly more corrosive than fossil fuels. Extent and type of corrosion in different biofuels depend on their chemical constituents and other factors. Water, found in all biofuels as an impurity, plays an important role in their overall corrosivity. It is necessary to understand the detrimental interactions of new and existing biofuels with industrial materials so that existing fuel infrastructure and equipment in contact with these fuels can be properly maintained and new structures can be designed with appropriate materials. Known mechanisms of material–biofuel interactions for FGE and pyrolysis oil fuels are discussed in this chapter.

1. Introduction

Liquid biofuels are important alternatives for petroleum products as transportation fuels. Most common types of biofuels used in the transportation sector are FGE and biodiesel. However, there are other biofuels, like pyrolysis oils or bio-oils that are either used as such to produce energy or can be further upgraded in refineries to produce transportation fuel. The corrosion behavior of alloys commonly used for production and transport of fossil fuels is well known, however in many instances these new alternative fuels interact differently, sometimes detrimentally, with these materials. Familiarity with the compatibility of these biofuels with common metallic industrial materials, like steels, is extremely important, as unmitigated corrosive interactions of these chemicals with production equipment can cause expensive and catastrophic equipment failures, leading to plant shutdowns or human casualties. An industry survey compiled in 2007 brought to light a number of failures of carbon steel equipment used in the storage and handling of FGE. Similarly, methanol, a renewable fuel often used to power fuel cells, leads to pitting corrosion of iron and carbon steels. There have been numerous reports of localized corrosion due to microbial activity and stress corrosion cracking (SCC) of steel storage tanks in bio-diesel fuel applications.[2,3] With an increase in the production and utilization of these biofuels, associated corrosion issues are of major concern. Bio-oils or Pyrolysis oils are being developed and produced by heating biomass at a very high heating rate. Resulting pyrolysis oil is a complex mixture of chemical species, which depends on the feedstock and processes used. Due to significant amounts of acetic and formic acids in raw pyrolysis oils, these bio-oils may have a total acid number (TAN) that can reach levels as high as 100 mgKOH/g,[4] and exhibit a pH in the range of 2.0 to 3.0. This is the main reason why biomass pyrolysis oils can be significantly more corrosive than traditional fuels.

Much of the industrial equipment used in ethanol service is comprised of carbon or low-alloy steels due to their high strengths and low costs. Carbon steels generally have good corrosion resistance in non-aqueous environments, but are highly susceptible to general corrosion and stress corrosion cracking when in contact with neutral pH or mildly acidic aqueous environments. Currently, an entire industry exists around mitigating and preventing corrosion in the petroleum industry and thus most corrosion issues and mechanisms in petroleum are well-defined and have been studied much longer than those in renewable biofuels.

Within this chapter, the focus will be biofuel product-related corrosion issues rather than production-related corrosion issues. Corrosion issues discussed in this chapter are related to transportation, handling, and storage of these biofuels. In this chapter, common industrial metals and corrosion behavior of alloys in ethanol and methanol will be discussed and exemplified with instances of corrosion-related equipment failures in the ethanol and methanol biofuel industry. The continued study of corrosion in biofuels is imperative for the development of solutions to corrosion issues.

2. Corrosion

Corrosion of a metal or alloy is generally due to its electrochemical interaction with the surrounding environment. As a result, the metal may undergo an oxidation reaction. For example, iron or steel may undergo oxidation to produce ferrous or ferric ions. Corroding metal undergoes an anodic reaction, producing or liberating electrons according to:

$$M \rightarrow M^{n+} + ne^-$$

For the corrosion or anodic reaction to occur, a reaction which consumes the liberated electrons, or a cathodic reaction, must also occur simultaneously. The nature of the cathodic reaction depends on composition of the surrounding electrolyte. The most common cathodic reactions in aqueous environments[5] are hydrogen evolution, O_2 reduction, and metal ion reduction, as shown by the reactions below:

$$2H^+ + 2e^- \rightarrow H_2$$

$$2H_2O + O_2 + 4e^- \rightarrow 4OH^-$$

$$M^{3+} + ne^- \rightarrow M^{2+}$$

Corrosion reactions in aqueous environments are well-known and have been more thoroughly studied. Reduction reaction in a given fuel environment will also depend on the reactants in that fuel. Lou and Singh[6] studied cathodic reactions in ethanol fuel environments and found that the reduction reactions involving dissolved oxygen are more efficient in these environments. Reduction reactions in other biofuels will also depend on the composition of that fuel and the presence of other contaminants. However, this has not been studied systematically for some of the bio-oils.

Corrosion reactions can contribute to failure of metals in several ways. General corrosion leads to the degradation of the entire surface of a metal part of piece of equipment, thinning the material until it no longer retains its integrity. This type of corrosion attack can typically be seen on carbon steel equipment or other alloys where the surface film may not be very protective under given conditions. However, localized corrosion may still occur in alloys, which generally form a protective surface film under given environmental conditions. Stress corrosion cracking (SCC) is such a phenomenon wherein a crack initiates and grows within a metal, assisted by both corrosive action and mechanical stresses, until it becomes large enough for the material to fail. Crevice and pitting corrosion are other corrosion processes that remain highly localized, in which visible corrosion occurs over a small area but penetrates deeply into an object, often traveling completely through the wall of a pipe or tank, causing leaks and failures. Localized microbial activity in biofuels like bio-diesel can cause localized corrosion, called microbiologically influenced corrosion (MIC) in areas where these microbial colonies thrive. These colonies have typically been found near the water-fuel boundary.[7,8] All of these modes of corrosion have been encountered in the biofuels industry.

3. Constituents of Biofuels and Their Potential Relationships to Corrosiveness to Steels

Biofuels may vary widely in their chemical compositions, which may depend on the feedstock or processes used to produce them. Although the ethanol used as fuel contains mostly ethanol molecules, or C_2H_5OH, there are a number of other constituents that are present in FGE, which may be intentionally added or may come from the production process. Similarly, biodiesel is comprised mainly of fatty acids like those in vegetable oil or animal fats. Bio-oils or pyrolysis oils have been reported to contain over 400 chemical constituents.[9,10] Commercial biofuels are almost never *pure* substances; i.e., in general, biofuels contain not just ethanol molecules or fatty acids, they also may contain small amounts of water, organic acids, chemical additives, or impurities. Generally, these impurities are what affect corrosion and stress corrosion cracking of metals and alloys in these biofuel environments.

There are numerous alloys utilized in industrial biofuel manufacturing. Carbon steels as well as stainless steels comprise a large portion of production equipment, storage tanks, piping, and other equipment used for biofuels. Carbon steel, specifically, is an inexpensive high-strength material widely used in many industrial processes. For this reason, carbon steel will be the focused material for much of this chapter. Alloying elements in carbon steel include small amounts of carbon, manganese, sulfur, and phosphorous to improve strength, ductility, and weldability. Carbon steels are not known for their great corrosion resistance in low pH water-based environments, but they here good general corrosion resistance to many organic solvents, petroleum products, and erosive slurries and high pH aqueous environments at low temperatures.

Stainless steels are also used for the piping and high-temperature components in biofuel production plants. Stainless steels are more expensive than carbon steels but generally have much greater corrosion resistance in aqueous environments due mainly to their higher chromium content. Corrosion and stress corrosion cracking of some stainless grades of steel will also be discussed in this chapter.

3.1 *Corrosion issues in ethanol and methanol biofuels*

Ethanol biofuel can be produced from many feedstocks, including corn, wood cellulose, sugarcane, and fruit. Ethanol from corn feedstocks comprises 3.5% of motor vehicle gasoline in the United States," and sugar cane ethanol is the source of more than 45% of automotive fuel in Brazil.[12] Increasing demand of alternative liquid fuels has resulted in a push for more production of ethanol from non-corn feedstocks.

Development of ethanol production from cellulosic sources is already occurring.[13,14] In light of the multiple feedstocks and processes used to manufacture ethanol, FGE is a multi-component system that may contain trace constituents based on the distillation process and feedstock.[1,15,16] Furthermore, ethanol is an excellent solvent, thus FGE may also absorb contaminants, like water, during production, processing, and transportation. It has been shown by many[15,17–24] that these trace constituents and impurities are the major causal factors of the different forms of corrosion in steel equipment used in ethanol service. In electrochemical studies on general corrosion of iron and steels in ethanol,[15,18,19,25,26] it was found that ethanol containing diminutive concentrations (on the order of 10^{-4} M) of acid, water, or chloride causes the presence of localized corrosion and the significant increase of corrosion rate. A similar phenomenon was observed in methanol environments.[18,19,27,28] These results are intriguing because generally such small concentrations of acids, water, or chloride ions do not appreciably affect carbon steel corrosion in aqueous environments.

3.1.1. *Chloride contamination of FGE*

Chloride is an example of a common FGE contaminant, entering the process stream via process water or chlorine-based pesticides on the feedstocks.[16] The presence of chloride exacerbates corrosion and stress corrosion cracking of many steels, including some stainless steels, in aqueous environments. The exceptions are the costly duplex stainless steels, which are significantly less sensitive to chloride SCC.[16] Studies have proven that susceptibility of carbon steels to localized corrosion and stress corrosion cracking in simulated FGE environments depends on chloride ion concentration,[20,21,24] and Cl^- quantities as low as 5 ppm can lead to SCC of carbon steel in FGE environments. In stress corrosion cracking tests of X65 carbon steel in corn-based FGE, incremental Cl^- additions were made. At Cl^- concentrations below 5 ppm, no cracking was observed, whereas at 5 ppm and above, crack density increased. Figure 1 shows the increase in crack density of X65 carbon steel as the Cl^- concentration of the FGE environment increases.

In the U.S., FGE composition is regulated by a written standard compiled by the ASTM.[29] ASTM D 4806 dictates that FGE must contain less than 10 ppm Cl^-. However, other countries possess their own regulatory standards for FGE composition, so that, depending on locale, FGE may contain varying amounts of chlorides as well as acids, water, denaturants, and dissolved oxygen. These standards are mainly for purposes of ensuring clean fuel combustion and environmental protection, and are not geared toward corrosion prevention. A comparison table of regulated constituents in the USA, Brazil, and mainland Europe is shown in Table 1.

Chloride ion concentrations below the levels regulated by ASTM[29] can worsen stress corrosion cracking of carbon steels and lead to failures.

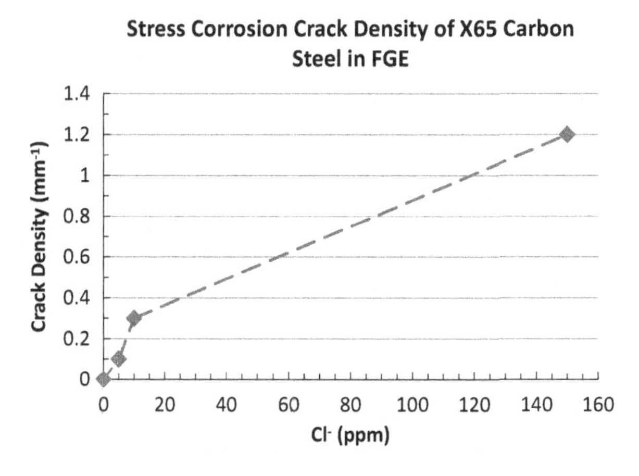

Fig. 1. Increase in occurrence of environmental cracking of X65 carbon steel in FGE when Cl^- content is increased.

Table 1. Comparison of regulated constituents of FGE globally.[30]

Contaminant	U.S. (ASTM D 4806-10) Maximum	Brazil (Hydrated) Maximum	Europe (Anhydrous) (prEN 15376) Maximum
Ethanol, volume %	92.1 min	95.1 min	98.7 min
Methanol, volume %	0.5	—	1.0
Solvent-washed gum, mg/100L	5.0	5.0	—
Water, volume %	1.0 max	—	0.3 max
Acidity (as acetic acid) mg/L	56	30	56
pHe	6.5–9.0	6.5–8.0	6.5–9.0
Inorganic chloride ppm	10		20
Sulfur, mg/L	30		10
Sulfate, mg/L	4	4	—
Copper, mg/kg	0.1		0.1
Denaturant	1.96–5.0		10

3.1.2. *Effects of water concentration on corrosion and SCC of steels in methanol and ethanol biofuels*

Water is another constituent that is always present in FGE, as ethanol is hygroscopic and will absorb available atmospheric water until it reaches its azeotropic water concentration of approximately 5 vol%.

In Brazil, two types of ethanol are used for fuel: hydrated and anhydrous. The main difference in the composition of these ethanol fuels is their water contents. Anhydrous ethanol is regulated to have a maximum water content of 1 vol%, whereas hydrated ethanol normally contains between 4–6 vol% of water. Though the difference in water content may not seem significant, laboratory studies[15,18–20,25] and field experience[1,31] demonstrate the contrary.

For instance, in Brazil, much of the fuel ethanol is hydrated, and general corrosion of carbon steel Brazilian hydrated ethanol storage tanks is a known issue.[32] This general corrosion is mitigated through the use of corrosion inhibiting chemicals, or corrosion inhibitors. Curiously, in the United States, carbon steel failures described in an ethanol industry survey' were attributed to stress corrosion cracking (SCC) indstead of general corrosion. Electrochemical measurements of X65 carbon steel in a 1%-water-content commercial FGE distilled from corn gave an estimated general corrosion rate of 0.002 mils per year (mpy),[33] which is considered negligible. Similarly, in general corrosion tests of carbon steel in simulated FGE solutions, no appreciable corrosion occurred in solutions containing 1 vol% water (Fig. 2(a)), whereas severe high density pitting occurred in the FGE containing 5 vol% water Fig. 2(b).[22,26] In SCC tests of carbon steel in simulated FGE environments, Lou *et al.*[26] verified that SCC of carbon steels occurs when water content of the FGE is 2 vol% or less. SCC rarely occurred above this water concentration.

3.1.3. *Understanding the mechanisms of effects of water on corrosion and stress corrosion cracking in methanol and ethanol biofuels*

Because corrosion is an electrochemical process, electrochemical measurements of metals in FGE can help researchers understand the corrosion phenomena occurring. Measurements of current density flowing from a metal in solution can indicate the rate of corrosive processes (corrosion rate). Electrochemical studies of the effects of added water on the

(a) (b)

Fig. 2. Images of carbon steel surfaces after exposure for 14 days to simulated FGE solutions containing (a) 1 vol% water, (b) 5 vol% water (Reproduced from Ref. 26).

corrosion behavior of pure iron in acidic methanol solution were performed. These particular species were chosen because of their presence in commercially available methanol fuels;[28] they are similarly found in other alcohol-based biofuels, particularly ethanol.[29] Studies showed that the iron and carbon steel remained in a state of active dissolution when no water was present, but when water was added to the solution the iron surface formed a stable protective (passive) film.[27,28] Potentiodynamic scans of pure Fe metal in a methanol – 0.5 M H_2SO_4 – X M H_2O environment indicated that when X = 0.1M, the iron corrosion current density remains high as potential is incrementally increased.[27] When X = 0.5M, Fe dissolution current drops sharply when potential reaches approximately 600 mV, indicating that a protective film is formed under this condition. The protective film formation is facilitated with further water increase, indicated by the decreasing film formation potential and widening of the region in which decreased current density is evident. In a similar study, researchers concluded[28] that water decreases the mobility of H^+ ions due to preferential protonation of water over methanol. This data supported the conclusion that water produces a corrosion inhibiting effect in organic solvent-based fuels containing acid.

Since the presence of seemingly minute concentrations of water can produce appreciable changes in the corrosion behavior of iron and carbon steels in methanol and ethanol, it was proposed[19] that in small C-chain organic solvents such as methanol and ethanol, solute water forms a chemisorbed monolayer on the electrode surface, producing a difference between the concentration of water available at the metal/solution interface and that in the bulk solution. This phenomenon of inhomogeneous distribution of water in ethanol was indeed observed and reported in a letter to Nature in 2002.[34] Researchers commented on their observations of incomplete mixing in water/alcohol solutions, stating that neutron diffraction experiments interpreted by the empirical potential structure refinement (EPSR) model showed that local ordering of methanol molecules was modified by small additions of water. In a 70% methanol–30% water mixture. It was expected that most water would exist as isolated water molecules, but radial distribution function data showed three-molecule clusters or strings of two to twenty molecules. Results of a low-frequency Raman spectroscopy study on clustering behavior in ethanol-water binary solutions by Egashira *et al.*[35] supported the Nature article[34] findings that, on a molecular level, ethanol-water binary solutions are not ideally mixed but are in a state of microscopic phase separation. This clustering effect (Fig. 3) allows even the presence of minute water quantities in FGE to influence the corrosion behavior of carbon steel.

3.1.4. *Role of dissolved oxygen in corrosion and stress corrosion cracking of carbon steels in ethanol biofuels*

Dissolved oxygen has a significant presence in FGE environments. Solubility of dissolved oxygen in ethanol is far greater than that in water (Table 2). Oxygen also plays a substantial role in corrosion of iron and steels. Because corrosion is an electrochemical process, the anodic

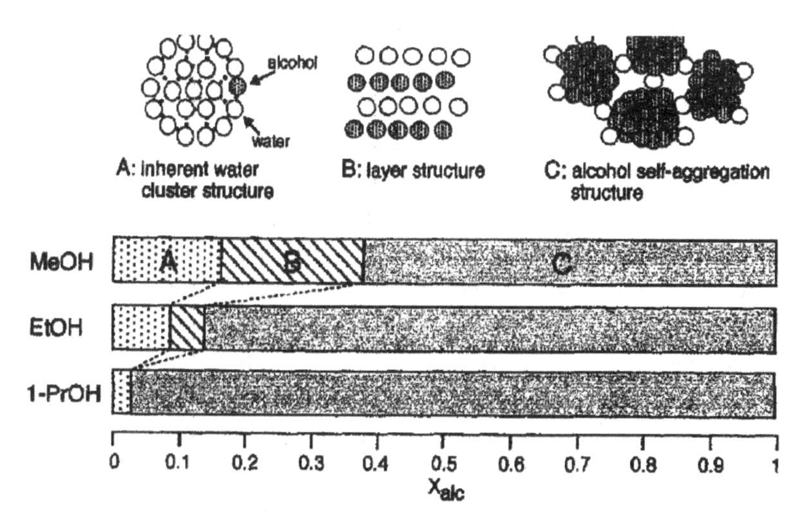

Fig. 3. Schematic illustrating the structure of alcohol-water solutions based on the mole fraction of alcohol (X_{alc}) present. Clustering decreases as carbon chain length increases (Reproduced from Ref. 36).

Table 2.

Solvent	O_2 solubility (mM fraction, 298K, 1atm)[67]
Ethanol	0.56
Water (DI)	0.0228

reaction must be accompanied by a complimentary cathodic reaction. Anodic reaction rate is controlled by the rate of cathodic reaction and vice versa. Dissolved oxygen gas within a solution can become a cathodic reactant, leading to increased quantity and rate of dissolution of iron metal into the solution, thus increasing corrosion rate. The cathodic reaction of oxygen when water is present is:

$$O_2 + 2H_2O + 4e^- \rightarrow 4OH^- \tag{1}$$

The following cathodic reaction has also been proposed when oxygen is present in a FGE environment.[6]

$$O_2 + 2C_2H_5OH + 4e^- \rightarrow 2C_2H_5O^- + 2OH^-, \tag{2}$$

where C_2H_5OH are, of course, ethanol molecules.

Dissolved oxygen can also aid in formation of a passive oxide film on the steel surface. This can affect the stress corrosion cracking process. Corrosion researchers have proposed that formation of a passive film at a specific rate is necessary to maintain the geometry of a stress corrosion crack.[37–39] So, if no film can be formed, the crack walls will corrode laterally, and the crack will blunt. However, if a film is formed at an intermediate rate along the walls of the crack, the shape of the crack will be maintained, and stress will continue to concentrate on the crack tip leading to further crack growth. Interestingly, in ethanol fuel environments, several researchers have illustrated[15,20,23,40,41] that carbon steel will not undergo SCC in any ethanol fuel solution if oxygen is absent. Figure 4 compares two carbon steel samples after being slowly strained to failure in oxygen-containing (Fig. 4(a)) and de-oxygenated (Fig. 4(b)) simulated FGE solutions.

It was previously mentioned that dense pitting and general corrosion occurred on pipeline steel surfaces in simulated FGE solutions containing 5 vol% water.[26] Within the same study, it was also demonstrated that when oxygen is completely removed from solutions by deaeration, pit density and pit size become drastically reduced. This illustrates the significant effect of oxygen on the severity and morphology of corrosion of pipeline steel in FGE solutions.

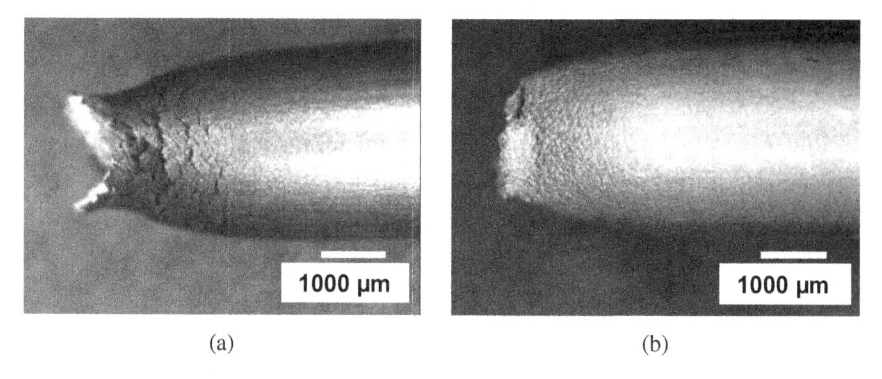

(a) (b)

Fig. 4. Carbon steel specimens to which stress has been applied in the presence of (a) oxygenated simulated FGE, and (b) de-oxygenated simulated FGE (Figure adopted from Ref. 20).

Dissolved oxygen is nearly impossible to control or regulate, thus it is not usually included in FGE standards. However, due to the stress corrosion cracking issues in ethanol plants and transport piping, ethanol facilities have begun to use oxygen scavengers and de-aeration techniques to extend the life of their steel infrastructure.

3.1.5. *Corrosive effects of organic impurities in fuel grade ethanol*

Aside from water and chloride impurities, researchers in Brazil identified and characterized organic impurities in a Brazilian FGE from sugarcane[42] using a gas chromatography-mass spectrometry (GC-MS) technique. Compounds such as saturated linear hydrocarbons, ketones, aldehydes, and alcohols were detected. However, many of the detected compounds are oxidation products of ethanol, implying that they may not have been present from distillation, but might have formed during production. Additionally, presence of these compounds varies qualitatively as well as quantitatively, depending on ethanol batch.

A survey on FGE SCC[1] established that corn-based "producer" ethanol (un-denatured ethanol from an ethanol production plant before transport), caused less severe stress corrosion cracking than "user ethanol" (fuel-ready ethanol from a holding tank or blending station) from the same corn feedstock. These observations imply that during transportation and processing, ethanol might either undergo compositional changes affecting its corrosiveness or absorb impurities during transport from a production facility to a holding tank. The specific alterations to the FGE environment that lead to SCC of carbon steel in contact with user ethanol but not producer ethanol have not yet been determined. Possible corrosive impurities include a range of organic and inorganic compounds, such as water, carboxylic acids, ketones, sulfates, chlorides, and others.

3.1.6. *Corrosion of materials in ethanol/gasoline blended fuels*

In the US, FGE is not used full-strength as automotive fuel; FGE is blended with gasoline in a ratio of 10%/90% ethanol/gasoline. Certain cars are specifically engineered to burn 85%/15% ethanol/gasoline blends, called E85. One of the major reasons why higher-ethanol-content gasolines cannot be utilized for fuel in regular autos is its corrosive effect on many auto parts and its tendency to degrade many of the polymeric materials present in automobile engines. Aluminum alloys, carbon steel, and stainless steels, all present within automobile engines, have been shown to be susceptible to corrosion and pitting in ethanol/gasoline blends. Copper and copper alloys, used primarily in automobile wiring, are also susceptible to corrosion in certain ethanol/gasoline blend ratios.

For both aluminum and carbon steel, the corrosive effect of ethanol blended with gasoline increases as the percentage of ethanol increases. Temperature also plays a major role in corrosion rate of aluminum and carbon steel alloys. In experiments performed at room temperature in 5%/95%, 10%/90%, and 15%/85% ethanol/gasoline blends, the aluminum alloy Al 6061 did not exhibit appreciable mass loss due to corrosion over a 128-day exposure,[47] though mass loss did increase with an increase in ethanol concentration. Mass loss rates were on the order of 1.5 g/cm^3. For plain carbon steel (0.12% C, 0.3 S, 0.11 P, 1.15 Mn, 0.03 Si, bal Fe) and 304 stainless steel, mass loss after 128 days was less than 1 g/cm^3 in the 5%/95% ethanol/gasoline solution but rose to 12–14 g/cm^3 in the 15%/85% ethanol/gasoline environment. In 10%/90% ethanol/gasoline with very low water content, aluminum alloys exhibited severe pitting and general corrosion at 120°C and 130°C.[43] The severity of this aluminum corrosion has been shown to decrease however with minor (0.2%) increases in water content of the blended fuels.[44] Magnesium and lead have also been found to be susceptible to this type of corrosion in ethanol-containing fuel blends.[45]

Electrochemical measurements in a 20%/80% ethanol/gasoline mixture showed that general corrosion of copper occurs in that environment.[46] Immersion of a copper fuel plug component in 5%/95%, 10%/90%, and 15%/85% ethanol/gasoline mixtures showed increase in mass loss with increase in ethanol fraction.[47] Additions of Cl^- and acetic acid to a 10%/90% ethanol/gasoline blend increased the severity of de-zincification, or preferential corrosion of zinc, from the brass alloy CuZn39Pb3.[43]

3.2. *Corrosion issues in biodiesel*

Corrosion behavior of biodiesel is different from the traditional diesel or petro-diesel due to the differences in chemical makeup. Biodiesels typically have high cetane numbers and low fuel densities compared to the petrodiesels and do not contain high sulfur. Similarly to ethanol-based biofuels, biodiesels contain chemical substances other than fatty acids. In the US, biodiesel chemical composition is regulated by ASTM D 6751.[20] Biodiesels are produced by esterification of feedstocks including vegetable oils from plants and seeds, animal fats, recycled cooking oil, and other sources of fatty acids. Corrosivity of biodiesels is typically enhanced due to the impurities or chemical constituents formed by the deterioration of biodiesels through oxidation. Corrosivity of biodiesels is different for different types of alloys. In general, copper alloys have been found to be more corrosive than the ferrous alloys.[48] Geller *et al.*[49] also reported that copper and brass are prone to corrosion and localized pitting corrosion under deposits covering the surface. Aluminum alloys have also been reported to show pitting corrosion in biodiesel. The surface morphology of copper and aluminum strips showed increased pitting corrosion compared to that in diesel.[50] Stainless steel, however, did not exhibit any measurable general corrosion or localized corrosion attack after exposure to biodiesel.

Corrosion has been directly linked to presence of water in diesel storage tanks produced by microbial growth in the tank. Sorensen *et al.*[51] reported that biodiesels are able to hold about 20–30 times more water than mineral diesel. The saturation level of water in the tested biodiesel was found to be 1395 ppm as compared to 62 ppm in mineral diesel. Work by Videla *et al.*[52] has shown that microbial activity in hydrocarbon/water mixtures can increase the local proton concentration due to organic acid metabolites as well as change the oxidizing characteristics of the solution, which may produce favorable conditions for localized corrosion. Microbial activity can also decrease the surface energy of the passive film/electrolyte interface and enhance metal dissolution.

3.2.1. *Microbial corrosion in biodiesel*

Problems arising from microbial growth in the water fraction of stored fuels are well documented with a long history.[8,52–55] A number of researchers have also reported MIC associated with the microbial activity in bio-derived hydrocarbon fuels and fuel blends.[2,53,56,57] Klofutar and Golob[55] showed that the presence of water and adequate temperature are important for microbial activities in the diesel or biodiesel and the resulting microbiologically influenced corrosion of metals.[8,52] Given the high water content of biofuels, serious problems with microbial growth and resulting MIC can be anticipated. However, the amount of water required for microbial growth in these environments is very small. Water may also be produced by the microbial activity. Bosecker[53] has shown that in kerosene, *Cladosporium resinae* could produce 80 mg of water per liter of fuel during a four week incubation period. However, this effect may vary with the microbial species as well as other environmental conditions. In tanks, typically the water accumulates at the bottom but some bacteria may also produce surfactants and cause the formation of stable water-oil emulsions.[8] The role of such emulsion on MIC as well non-microbial corrosion is less clear.

There is a push to use low sulfur containing biodiesel. However, Londry and Suflita[54] have reported that lowering concentration of sulfur compounds in diesel may also promote bio-activity as some of the sulfur compounds such as thiophenes, thiols, organic sulfides, and others may act as biocides. A secondary effect of sulfur removal may be the production of other organic compounds during processing that may promote microbial activity. Lee *et al.* have observed a difference of nearly 1 V in the corrosion potential of carbon steel C10200 in different diesel fuel mixtures, showing the most active corrosion potential in ultra-low-sulfur diesel (ULSD).[8] Variations in the microbial activity in different fuel blends were attributed to this difference in the corrosion behavior.

3.2.2. *Stress corrosion cracking in biodiesel*

There have been numerous reports of steel storage tank stress corrosion cracking (SCC) and non-uniform corrosion in diesel fuel applications.[2,3] The corrosion propensity of biodiesel with regard to various grades of steel seems to increase with increasing water content, decreasing pH, increasing Cl^- and SO_4^{-2} levels.[57]

Lee *et al.*[8] have illustrated that profound changes in microbial consortia can occur in diesel fuel exposed to seawater, including activation of sulfate reducing bacteria (SRB). Sulfate reducing bacteria, which can develop under anaerobic conditions, are well-documented contributors to steel stress corrosion cracking[2,3] and likely exhibit complex microbial ecology with other microorganisms in biofuels, some of which may worsen corrosion whereas others may mitigate corrosion. Depending on the seawater composition, biodiesel grades, and fuel-blends used, different microbial activities are expected on the metal surfaces of fuel/ballast tank and related equipment. Javaherdashti *et al.*[2] reported SCC of stainless steel in the presence of SRB whereas hydrogen embrittlement of low alloy steel in the presence of SRB was reported by Raman *et al.*[3]. Steels that appear to be resistant to MIC and pitting may still embrittle and become susceptible to forms of environmental cracking such as SCC or corrosion fatigue. Kane and Papavinasam[7] have reported that dynamic loading conditions (specifically slow fatigue mode) are required to promote SCC.

3.3. ***Corrosion issues in bio-oils or pyrolysis oils***

Pyrolysis is one of the numerous possible paths by which biomass can be converted into higher value products. Pyrolysis oil technology can be used in biorefineries to expand the suite of product options available from biomass. Pyrolysis oils or "Bio-oils" are produced by heating biomass at a very high heating rate with a corresponding high heat flux.[4,9,10,58–60]

The application of heat in the absence of oxygen breaks down the complex polymeric constituents of biomass (cellulose, hemicellulose, and lignin) to simpler oxygenated molecular fragments of the starting biomass. The lower molecular weight compounds remain as permanent gases at ambient temperature while the majority of compounds condense to collectively make up what is called bio-oil or pyrolysis oil. Raw pyrolysis oil contains significant amounts of water, parts of which are formed during pyrolysis and other parts are due to original moisture in the biomass feed that ends up as water in bio-oil.[9,10,58] Biomass-derived pyrolysis oils are rich in carbon and can be used as such or can be refined in ways similar to crude oils to upgrade them for use in production of fuel and chemicals in petroleum refineries.

Corrosivity of pyrolysis oils depends on the feedstock as well as the process parameters used to produce bio-oil. Pyrolysis oils are complex mixtures where their typical constituents can be classified into five broad categories: hydroxyaldehydes, hydroxyketones, sugars, carboxylic acids, and phenolics that are primarily derived from the lignin component of biomass.[59] A more detailed classification of compounds[9,10,58] includes the following categories: acids, alcohols, aldehydes, esters, ketones, phenols, guaiacols, syringols, sugars, furans, alkenes, aromatics, nitrogen compounds, and misc. oxygenates. While there is a rich mixture of known compounds in bio-oils, the vast majority of these compounds are found in low concentrations. The highest concentration of any single chemical compound (after water) is hydroxyacetaldehyde at levels up to 10 wt%. This is followed by acetic and formic acids, at about 5 wt% and 3 wt%, respectively. This is the primary reason why bio-oils may have a total acid number (TAN) that can reach levels as high as 100 mgKOH/g,[4] and exhibit a pH in the range of 2.0 to 3.0. Corrosivity of bio-oils or pyrolysis oils has been attributed to high TAN values. Most of the upgraded fuels have lower TAN values than the raw pyrolysis oils. Due to large variations in the starting material and the high temperature reactive environment to which the biomass vapor fragments are exposed during pyrolysis, there are variations in many of the physio-chemical properties of bio-oil, causing their corrosivity to vary.

Apart from the source and procedures used to produce pyrolysis oil, the extent and mode of corrosion attack also depends on the alloy composition, microstructure, and environmental parameters. Pyrolysis oil produced from rice-husk has been shown to cause significant corrosion to mild steel, even at room temperature.[61] Lu *et al.*[61] reported that the corrosion rates for aluminum alloys were lower than those of mild steel, and brass was found to be relatively resistant to the general corrosion at room temperature. Corrosion rates for all alloys increased with an increase in the test temperature up to 70°C. Mild steel exposed to pyrolysis oil was covered with loosely attached corrosion product, which was easy to remove by washing and wiping. However, brass underwent dezincification attack in pyrolysis oil.[61] Aubin and Roy[62] investigated the effect of temperature, water content and acetic/formic acid concentration on the corrosivity of pyrolysis oils towards different alloys.

Pyrolysis oils were reported not to be very corrosive to the cast iron at ambient temperature when tested in bio-oils with low water content for all concentrations of acids used.[62] However, increasing the test temperature to 45°C dramatically increased the iron's corrosion activity in the pyrolysis oil. Aubin *et al.*[62] also found that the corrosion rate of bio-oil became enhanced with the increase of water content. Our previous work on pine and oak pyrolysis oils has shown that carbon steel, 516-Gr70, undergoes significant corrosion.[63] Although the oak pyrolysis oil in this study had lower water content, it exhibited higher corrosivity than the pine pyrolysis oils tested. Corrosion rate of the carbon steel increased with an increase in the pyrolysis oil temperature up to 80°C. Selective corrosion of the ferrite phase in pearlitic steel microstructure was observed, whereas attack on the cementite phase of this steel was relatively less. Pyrolysis oils are unstable above 80°C[61] so the effect of temperature above 80°C has not been studied systematically.

Stainless steels are resistant to general corrosion in pyrolysis oils up to 80°C. Darmstadt *et al.*[64] tested corrosivity of softwood bark bio-oil with significant amounts of water, produced by vacuum pyrolysis, at 80°C and showed that the aluminum alloys and, to a much smaller degree, copper were corroded by the bio-oil, whereas 316 stainless steel was not affected. A compact protective layer of Cr_2O_3 and NiO was detected on the surface of austenitic stainless steels, as detected through X-ray-photoelectron spectroscopy (XPS). Respective oxide and/or hydroxide layers were formed on all three metals. However, in the case of aluminum and copper, these layers were not compact and adherent and did not protect the underlying metal against further oxidation. It was speculated that for 316 SS, after initial modification of the surface (e.g., leaching of Fe species), subsequent oxidation is prevented by the formation of a chromium oxide layer. Kirk *et al.*[65] tested three types of biofuels for their compatibility with a variety of polymeric materials and metal alloys at 80°C to represent aggressive field usage conditions. 304L, 316L, 430 and 20MO4 stainless steels had corrosion rates of less than 0.007 mm/y, whereas the non-traditional low chromium alloy steel, MASH, was found to be highly susceptible to all fuels and corroded at rates up to 3.7 mm/y at 80°C.

Keiser *et al.*[66] tested metallic samples immersed in the oil as well as in the vapor space above the oil, with the bio-oil at a temperature of 50°C. Results showed significant corrosion of carbon steel and the 2¼ Cr-1 Mo steel samples for all tested pyrolysis oils, except for the samples exposed in the hydro-processed bio-oil. A hydrated iron formate corrosion product was identified on the surface of these specimens, which also showed significant weight loss. Calculated corrosion rates for these alloys were much higher than that could be tolerated in any system with a design life of several years. The stainless steel specimens showed minimal weight change.

3.3.1. *Effect of water content and temperature on pyrolysis oil corrosivity*

Pyrolysis oils separate into two distinct fractions with time. These fractions have different amounts of water as well as other constituents. Singh *et al.*[63] tested different alloys in the two separate fractions of pine based pyrolysis oil. The top layer of pyrolysis oil was a light-brown layer with more water (~69% water) than bottom layer, which was dark-brown in color and more viscous (~19% water). Carbon steel corrosion behavior showed significant differences when tested in each separated layer. Corrosion rates for the carbon steel 516-G70 from these tests are shown in Fig. 5. Carbon steel coupons had a uniform corrosion at all tested temperatures. At 50°C the corrosion rate of carbon steel coupons was almost 75 mils per year (mpy) in the light-brown oil but only about 14 mpy in the dark-brown oil. However, at 80°C the corrosion rate in light-brown oil was lower (~50 mpy) than the equivalent test at 50°C. This was attributed to the evaporation of water at higher temperature, but may also be due to some degradation of pyrolysis oil at the higher temperature.

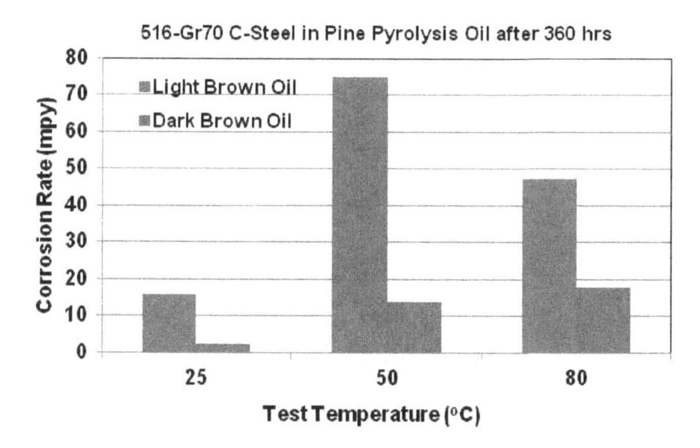

Fig. 5. Corrosion rate of carbon steel, 516-Gr70, in two different fractions of pine pyrolysis oil. Light brown fraction had ~69% water whereas the dark brown fraction had ~20% water.

Similar trends in corrosion rates with temperature were seen for aluminum, Al-1100 alloy, although corrosion rates of Al-1100 were significantly lower than that of the carbon steel under all tested conditions. However, Al-1100 showed localized corrosion in the form of random pits at the surface when tested in the higher-water-content light-brown pine-pyrolysis oil fraction. Pit density and size was higher at higher temperature. The light-brown oil fraction led more pits to form on the metal's surface than the dark-brown oil fraction. Limited EDS analysis of pit areas suggested that the intermetallic particles containing Fe and Si were associated with pit initiation sites. Austenitic stainless steel, 316L, and two duplex stainless steels, 2205 and 2101, were also exposed to the pine pyrolysis oils at three temperatures. Results in Fig. 6 show that the corrosion rates for all stainless steels up to 50°C were less than 0.05 mpy. No indication of localized corrosion was seen on these steels under tested conditions.

Copper and Brass samples tested in pine-pyrolysis oil fractions had higher corrosion rate than stainless steels tested, as shown in Fig. 7. However, corrosion rates for Cu and brass were significantly lower than the aluminum and carbon steel samples under equivalent conditions. No dezincification of brass was observed under tested conditions in tested pine pyrolysis oil.

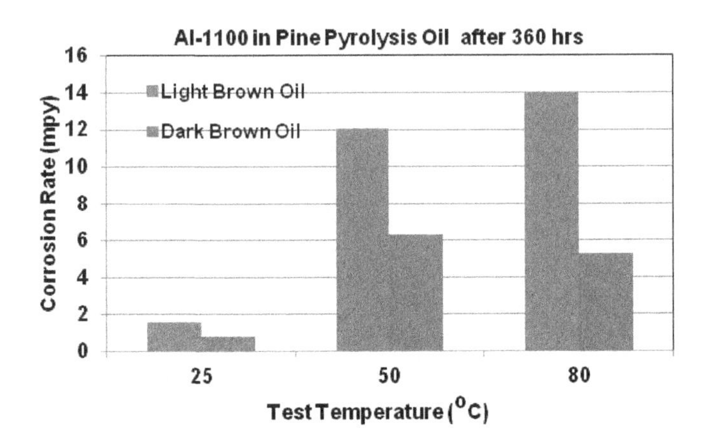

Fig. 6. Corrosion rate of Pure Al, Al-1100, in pine pyrolysis oils, where the light brown oil had ~69% water and dark brown fraction had ~20% water.

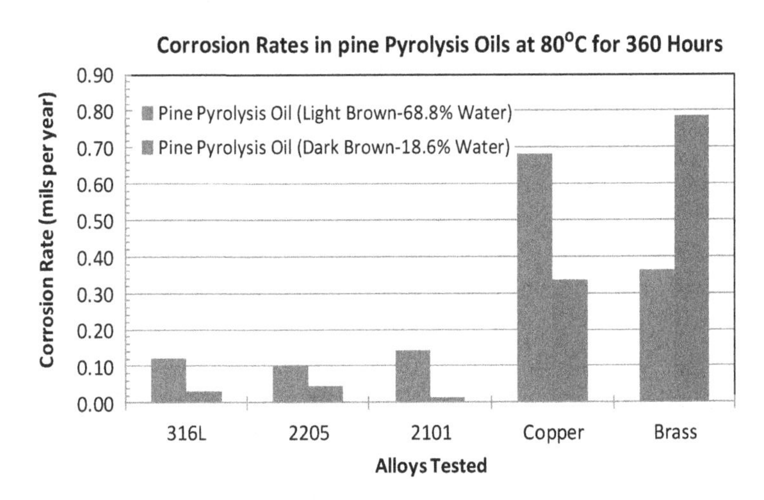

Fig. 7. Corrosion rate of three stainless steels pure copper and brass samples in pine pyrolysis oils at 80°C.

4. Conclusions

In this chapter, corrosive effects of several renewable biofuels, namely ethanol, biodiesel, and pyrolysis oil, were discussed. Corrosion issues differ for each bio-fuel, and corrosion behavior of different alloy types varies widely in a given bio-fuel environment. Carbon steel does not undergo general corrosion in very low-water-content ethanol fuel, but pitting and general corrosion when water concentration rises near 5%. Aluminum alloys become less susceptible to general corrosion when water is added to ethanol fuel. Pyrolysis oil does not lead to de-zincification of brass alloys whereas ethanol fuel does. Other factors like temperature or flow rate may also affect extent and form of corrosion but have not been discussed in this chapter.

Though the list of fuels discussed is not exhaustive, a common theme exists: inorganic as well as organic impurities and water content of biofuels are the significant factors affecting the corrosion behavior of metals and alloys.

References

1. API, Stress Corrosion Cracking of Carbon Steel in Fuel-grade Ethanol: Review, Experience Survey, Field Monitoring, and Laboratory Testing *API Technical report 939-D*. Washington, D.C., American Petroleum Institute, (2007).
2. R. Javaherdashti, R. K. S. Raman, C. Panter and E. V. Pereloma, "Microbiologically assisted stress corrosion cracking of carbon steel in mixed and pure cultures of sulfate reducing bacteria." *International Biodeterioration & Biodegradation,* (2006).
3. R. K. S. Raman, R. Javaherdashti, C. Panter and E. V. Pereloma, "Hydrogen embrittlement of a low carbon steel during slow strain testing in chloride solutions containing sulphate reducing bacteria." *Materials Science and Technology*, (2005).
4. D. Meier and B. Schottze, Fast Pyrolysis Liquid Characteristics. *Biomass Gasification and Pyrolysis, State of the Art and Future Prospects*. M. Kaltschmitt and A. V. Bridewater. Newbury, CPL Scientific, 431–441 (1997).
5. M. G. Fontana, *Corrosion Engineering*. New York:, McGraw-Hill, (1986).
6. X. Lou, and P. M. Singh, "Cathodic Activities of Oxygen and Hydrogen on Carbon Steel in Simulated Fuel-grade Ethanol." *Electrochimica Acta.,* **56(5)**, 2312–2320 (2011).
7. R. D. Kane, and S. Papavinasam, Corrosion and SCC Issues in Fuel Ethanol and Biodiesel. *Corrosion 2009*. Atlanta, GA, NACE International (2009).
8. J. S. Lee, R. I. Ray and B. J. Little, *Microbiological and corrosivity characterizations of biodiesels and advanced diesel fuels*. Corrosion, Houston, TX, NACE International (2009).
9. J. Piskorz, D. S. Scott and D. Radlien, Composition of Oils Obtained by Fast Pyrolysis of Different Woods. *Pyrolysis Oils from Biomass: Producing Analyzing and Upgrading*. E. J. Soltes and T. A. Milne. Washington, D.C., ACS: 167–178 (1980).

10. T. A. Milne, F. Agblevor, M. Davis, S. Deutch and D. Johnson, A Review of the Chemical composition of Fast Pyrolysis Oils. *Developments in Thermochemical Biomass Conversion*. A. V. Bridgewater and D. G. B. Booncock. London, Blackie Academic & Professional: 409–424 (1997).
11. P. C. Westcott, Ethanol Expansion in the United States; How Will the Agricultural Sector Adjust? USDA. Washington, D.C, (2007).
12. M. Iootty, H. P. Jr. and F. Ebeling, Automotive Fuel Consumption in Brazil: Applying Static and Dynamic Systems of Demand Equations. *New Trends and Developments in Automotive System Engineering*. M. Chiaberge. Online: http://www.intechopen.com/books/new-trends-and-developments-in-automotive-system-engineering/automotive-fuel-consumption-in-brazil-applying-static-and-dynamic-systems-of-demand-equations, InTech: 29–44 (2011).
13. L. R. Lynd, J. H. Cushman, R. J. Nichols and C. E. Wyman, "Fuel Ethanol from Cellulosic Biomass." *Science,* **251(4999)**, 1318–1323 (1991).
14. Y. Pu, D. Zhang, P. M. Singh and A. J. Ragauskas, "The New Forestry Biofuels Sector." *Biofuels, Bioproducts and Biorefining,* **2(1)**, 58–73 (2008).
15. N. Sridhar, K. Price, J. Buckingham and J. Dante, "Stress Corrosion Cracking of Carbon Steel in Ethanol." *Corrosion,* **62(8)**, 687–702 (2006).
16. E. Torsner, "Solving corrosion problems in biofuels industry." *Corrosion Engineering, Science and Technology,* **45(1)**, 42–48 (2010).
17. J. T. Demo, "Effect of low concentrations of acid and water on the corrosion of metals in organic solvents." *Chemical Engineering World,* VII**(5),** (1972).
18. E. Heitz, Corrosion of Metals in Organic Solvents. *Advances in Corrosion Science and Technology*, New York, Plenum Press: 149–243 (1974).
19. P. L. de Anna, "The effects of Water and Chloride Ions on the Electrochemical Behaviour of Iron and 304L Stainless Steel in Alcohols." *Corrosion Science,* **25(1)**, 43–53 (1985).
20. X. Lou, D. Yang and P. Singh, "Effect of Ethanol Chemistry on Stress Corrosion Cracking of Carbon Steel in Fuel-grade Ethanol." *Corrosion.,* **65(11)**, 785–797 (2009).
21. F. Gui and N. Sridhar, "Conducting Electrochemical Measurements in Fuel-Grade Ethanol Using Microelectrodes." *Corrosion,* **66(4)**, 045005–045001–045005–045008 (2010).
22. X. Lou and P. M. Singh, "Role of Water, Acetic Acid, and Chloride on Corrosion and Pitting Behaviour of Carbon Steel in Fuel Grade Ethanol." *Corrosion Science,* **52(7)**, 2303–2315 (2010).
23. J. A. Beavers, F. Gui and N. Sridhar, "Effects of Environmental and Metallurgical Factors on the Stress Corrosion Cracking of Carbon Steel in Fuel Grade Ethanol." *Corrosion,* **67(2)**, 025005–025001–025005–025015 (2011).
24. L. R. Goodman and P. M. Singh, "Repassivation Behavior of X65 Pipeline Steel in Fuel Grade Ethanol and its Implications for the Stress Corrosion Cracking Mechanism." *Corrosion Science,* **65**, 238–248 (2012).

25. E. Cavalcanti, V. G. Wanderley, T. R. V. Miranda and L. Uller, "The Effect of Water, Sulphate and pH on the Corrosion Behaviour of Carbon Steel in Ethanolic Solutions." *Electrochimica Acta.,* **32(6)**, 935–937 (1987).
26. X. Lou, L. R. Goodman and P. M. Singh, Pitting Corrosion of Carbon Steel in Fuel Grade Ethanolic Environment. *Corrosion 2009*. Atlanta, GA, NACE International, (2009).
27. J. Banas, "Passivity of Iron and Nickel in a CH_3OH-H_2O-H_2SO_4 System." *Electrochimica Acta.,* **32(6)**, 871–875 (1987).
28. C. S. Brossia, E. Gileadi and R. G. Kelly, "The Electrochemistry of Iron in Methanolic Solutions and its Relation to Corrosion." *Corrosion Science,* **37(9)**, 1455–1471 (1995).
29. ASTM International, Standard Specification for Denatured Fuel Ethanol for Blending with Gasolines for Use as Automotive Spark-Ignition Engine Fuel. West Conshohocken, PA, USA, ASTM International. ASTM 4806-10 (2010).
30. Netherlands Standardization Institute, T, Worldwide Fuels Standards. *Overview of specifications and regulations on (bio)fuels*. Delft, NL, NEN, (2006).
31. Notes from PHMSA Trip to Brazil. PHMSA, (2007).
32. A. R. Moreira, Z. Panossian, C. A. L. Santos, G. M. Bragagnolo, M. C. Gandur and E. M. De Souza *Corrosivity of anhydrous ethanol, hydrated ethanol and fuel E25 (25% ethanol/75% gasoline)*. Corrosion 2010, March 14, 2010–March 18, 2010, San Antonio, TX, United states, National Assoc. of Corrosion Engineers International, (2010).
33. L. R. Goodman, Stress corrosion cracking of x65 pipeline steel in fuel grade ethanol environments. *Materials Science and Engineering*. Atlanta, Georgia Institute of Technology. Ph.D., (2012).
34. S. Dixit, J. Crain, W. C. K. Poon, J. L. Finney and A. K. Soper, "Molecular Segregation Observed in a Concentrated Alcohol-Water Solution." *Nature,* **416**, 829–832 (2002).
35. K. Egashira and N. Nishi, "Low-Frequency Raman Spectroscopy of Ethanol-Water Binary Solution: Evidence for Self-Association of Solute and Solvent Molecules." *Journal of Physical Chemistry B.,* **102**, 4054–4057 (1998).
36. A. Wakisaka, S. Komatsu and Y. Usui, "Solute-Solvent and Solvent-Solvent Interactions Evaluated through Clusters Isolated from Solutions: Preferential Solvation in Water-Alcohol Mixtures." *Journal of Molecular Liquids,* **90(1–3)**, 175–184 (2001).
37. J. C. Scully, "Kinetic Features of Stress Corrosion Cracking." *Corrosion Science.,* **7**, 197–207 (1967).
38. G. J. Bignold, "Electrochemical Aspects of Stress Corrosion of Steels in Alkaline Solutions." *Corrosion,* **28(8)**, 307–312 (1972).
39. *Stress Corrosion Cracking and Hydrogen Embrittlement of Iron Base Alloys*. Houston, Tx, NACE International, (1977).
40. F. Gui, N. Sridhar and J. A. Beavers, "Localized Corrosion of Carbon Steel and its Implications on the Mechanism and Inhibition of Stress Corrosion Cracking in Fuel-grade Ethanol." *Corrosion,* **66(12)**, 125001–125001–125001–125012 (2010).
41. X. Lou, D. Yang and P. M. Singh, "Film breakdown and anodic dissolution during stress corrosion cracking of carbon steel in bioethanol." *Journal of the Electrochemical Society,* **157(2)**, C86–C94 (2010).

42. R. Vilar and R. da Silva, "Preliminary characterization of anhydrous ethanol used in Brazil as automotive fuel." *Journal of Chromatography A.*, **985**, 367–373 (2003).
43. K. Eppel, M. Scholz, T. Trossmann and C. Berger, "Corrosion of metals for automotive applications in ethanol fuel blends." *Energy Materials,* **3(4)**, 227–231 (2008).
44. L. Krüger, F. Tuchscheerer, M. Mandel, S. Müller and S. Liebsch, "Corrosion behaviour of aluminium alloys in ethanol fuels." *Journal of Materials Science,* **47(6)**, 2798–2806 (2012).
45. O. E. Company, Report to the department of the environment and heritage to the environment. Australia (2002).
46. L. M. Baena, M. Gómez and J. A. Calderón, "Aggressiveness of a 20% bioethanol–80% gasoline mixture on autoparts: I behavior of metallic materials and evaluation of their electrochemical properties." *Fuel,* **95(0)**, 320–328 (2012).
47. H. Jafari, M. H. Idris, A. Ourdjini, H. Ramini and B. Ghobadian, "Effect of ethanol as gasoline assitive on vehicle fuel delivery system corrosion." *Materials and Corrosion,* **61(5)**, 432–440 (2010).
48. M. A. Fazal, A. S. M. A. Haseeb and H. H. Masjuki, "Biodiesel feasibility study: an evaluation of material compatibility; performance; emission and engine durability." *Renewable and Sustainable Energy Reviews,* **15**, 1314–1324 (2011).
49. D. P. Geller, T. T. Adams, J. W. Goodrum and J. Pendergrass, "Storage stability of poultry fat and diesel fuel mixtures: specific gravity and viscosity." *Fuel,* **87**, 92–102 (2008).
50. M. A. Fazal, A. S. M. A. Haseeb and H. H. Masjuki, "Comparative corrosive characteristics of petroleum diesel and palm biodiesel for automotive materials." *Fuel Process Technology,* **91**, 1308–1315 (2010).
51. G. Sorensen, D. V. Pedersen, A. K. Norgaard, K. B. Sorensen and S. D. Nygaard, "Microbial growth studies in biodiesel blends." *Bioresource Technology,* **102**, 5259–5264 (2011).
52. H. A. Videla, P. S. Guiamet, S. DoValle and E. H. Reinoso, *Effects of fungal and bacterial contaminants of kerosene fuels on the corrosion of storage and distribution systems.* Corrosion, Houston, TX, NACE International, (1988).
53. K. Bosecker, Deterioration of hydrocarbons. *Microbiologically Influenced Corrosion of Materials.* E. Heitz, H. C. Flemming and W. Sands. Berlin, Springer-Verlag: 439–444 (1996).
54. K. L. Londry and J. M. Suflita, "Toxicity effects of organosulfur compounds on anaerobic microbial metabolism." *Environmental Toxicology and Chemistry,* **17**, 1199–1206 (1998).
55. B. Klofutar and J. Golob, "Microorganisms in Diesel and in Biodiesel Fuels." *Acta Chimica Solvenica,* **54(4)**, 744–748 (2007).
56. M. E. May and R. A. Neihof, Microbial Deterioration of Marine Diesel Fuel from Oil Shale. N. R. Lab. Washington, D.C., (1981)
57. R. D. Kane and S. Campbell, Real-Time Corrosion Monitoring of Steel Influenced by Microbial Activity (SRB) in Simulated Seawater Injection Environments. *Corrosion.* Houston, TX, NACE, International, (2004).

58. F. Shafizadeh, "Pyrolysis and Combustion of Cellulose Materials." *Advances in Carbohydrate Chemistry*, 23 (1968).
59. J. P. Diebold, A Review of the Toxicity of Biomass Pyrolysis Liquids Formed at Low Temperatures. Boulder, CO, National Renewable Energy Laboratory, (1997).
60. A. Oasmaa and S. Czernik, "Fuel Oil Quallity of Biomass Pyrolysis Oils — State-of-the-Art for End Users." *Energy & Fuels* **13**, 914–921 (1999).
61. Q. Lu, J. Zhang, X. Zhu and D. Mohan, "Corrosion properties of bio-oils and its emulsions with diesel." *Chinese Science Bulletin,* **53(23)**, 3726–3734 (2008).
62. H. Aubin and C. Roy, "Study on the corrosiveness of wood pyrolysis oils." *Fuel Science & Technology International,* **8(1)**, 77–86 (1990).
63. P. M. Singh, A. J. Ragauskas, J. Mahmood, F. Huang and H. K. Dong (to be published). Corrosion and Stress Corrosion Cracking Susceptibility of Different Alloys in Pyrolysis Oils, (2013).
64. H. Darmstadt, M. Garcia-Perez, A. Adnot, A. Chaala, D. Kretschmer and C. Roy, "Corrosion of Metals by Bio-Oil Obtained by Vacuum Pyrolysis of Softwood Bark Residues. An X-ray Photoelectron Spectroscopy and Auger Electron Spectroscopy Study." *Energy Fuels,* **18(5)**, 1291–1301 (2004).
65. D. W. Kirk, Z. R. Li, D. Fuleki and P. C. Patnaik, Materials Compatibility with Pyrolysis Biofuel. *ASME Turbo Expo*. New Orleans, LA, ASME, (2001).
66. J. R. Keiser, M. A. Bestor, S. A. S. Lewis and R. M. Connatser, Corrosion Studies Of Raw and Treated Biomass-Derived Pyrolysis Oils. *Corrosion.* Salt Lake City, UT, NACE International, (2012).
67. R. Battino, T. R. Rettich and T. Tominaga, "The Solubility of Oxygen and Ozone in Liquids." *Journal of Physical and Chemical Reference Data,* **12(2)**, 163–178 (1983).

CHAPTER 11

INCORPORATION OF BIOFUELS TECHNOLOGY INTO A PULP MILL

MARKO HAKOVIRTA

Alabama Center for Paper and Bioresource Engineering,
Auburn University
Auburn, AL 36849, U.S.A.
marko.hakovirta@auburn.edu

This chapter gives a brief overview of the landscape and the biofuels technology related concepts incorporated into pulp mills. This chapter will also give a general overview of some commercialization related key issues for biorefineries in the U.S. and globally.

1. Introduction

It is not unorthodox to state that the biorefinery concepts are quite old in the pulp and paper industry as kraft and sulfite mills have always been working as biorefineries. The pulp mills have been designed to produce wood fiber for paper-product manufacturing, commercial chemicals such as tall oil turpentine and necessary energy for the manufacturing, process to make the manufacturing commercially viable. In the most modern mills the energy produced is also sold to the market as an additional product of the "biorefinery".

The focus of new biorefinery concepts during the recent years has been looking at the global need for more sustainable energy sources and the need of the pulp and paper industry to reinvent itself in the global competition. Of course key drivers have also been the political incentives for biofuels and national interests for self-sufficiency in energy.

Increasing energy demand is one of the major global megatrends that threatens the industrial growth engine and globalization as such. The magnitude of this increasing energy need is enormous; the current forecast is that from the 2004 primary energy demand of 11,200 Mtoe there is an increase of 53% to the level of 17,100 Mtoe by the end of 2030. During that period the electricity consumption will increase even more from 14,400 TWh to 28,100 TWh.[1]

The solution to the pressing energy crises will be a more balanced and sustainable portfolio of different energy sources. No single solution will be sufficient to address the existing gap of energy needs of this magnitude. One of the most discussed and internationally agreed solutions to ever increasing need for energy is biofuels. Its many benefits include the mitigation of greenhouse gases and the national security of industrial nations that are greatly dependent on oil-producing countries and more recent extremely volatile fuel prices.

Biofuels in general are defined as solid, liquid or gaseous fuels derived from biological matter. In principle one can produce biofuels from any carbon based biological materials. There are numerous different sources to produce biofuels (crops, forest, waste and residues) and the biofuel can be produced using first or second generation production techniques. The first generation technique is already in production and used to make biodiesel from vegetable oils by for example transesterification using methanol[2] and producing fuel that is almost similar to fossil diesel. First generation technology is also used for producing bioethanol from sugar and starch crops. The second generation techniques are more advanced and enable to use low-value agricultural products and residues and lignocellulosics from for example wood. Its carbon footprint is much lower than that of the first generation approach and its feedstock mix is more flexible, making it a more sustainable source for biofuels in the future.[3] The fuels that can be produced with this technique include Bioethanol, Fischer-Trops Diesel and biohydrogen. The second generation technology is however not yet commercially ready for extensive large-scale production. Currently there exist several pilot sized projects pioneering commercialization of the new technologies.[3–8]

2. Biofuels Landscape in the United States

Looking at the role of mitigation of greenhouse gas, the total role of transportation sector in the global overall GHG emissions is some 15%.[9–13] A major emission source being the energy industry's coal fired power plants. For example coal-fired power plants release over 40% of total U.S. carbon dioxide emission.[14–18] Although coal-fired power plants account for just over half of the electricity produced in the U.S. each year, they have been responsible for over 83% of the CO_2 emission since 1990.[19–22]

The self-sufficiency of transportation fuels is a totally different story. U.S. imports some 13 million barrels of oil per day, costing up to $350 billion per year in 2010.[23] The exports come from various countries globally but the top 5 are Canada (2,324 million barrels per day), Saudi Arabia (1,465 million barrels per day), Mexico (1099 million barrels per day), Venezuela (759 million barrels per day), and Nigeria (529 million barrels per day). Looking at the list further, one can see that the oil imports are widely diversified in order to have flexibility from a security and supply-demand perspective.[23–26]

The current biofuel's production targets passed by Congress require U.S. refineries to deliver 36 billion gallons a year by 2022 (Fig. 1.).[28] This means some 2.4 million barrels (1 barrel is 42 gallons) of biofuels per day. Of that amount, corn based ethanol production is targeted to be some 15 billion gallons per year in 2015 (or 1 million barrels a day). The rest of the demand (some 60%) is expected to be provided by the second generation — more advanced biofuels, the majority of which are cellulosic biofuels. It is estimated that by 2022, the production of cellulosic biofuels should be in a higher level than corn ethanol.[29]

With such ambitious and some may argue over positive targets for biofuels growth in the U.S. and globally (Fig. 1), the main questions to be addressed are: In what energy price levels and subsidy price levels will a biorefinery concept be profitable? What are the right feedstock basis, level of logistics cost, technology platform selection and the targeted product mix that make biorefinery profitable? What is the effect of different policy measures to the prices and above choices and what is the foreseen timeline and roadmap for the developed business (feedstock and logistics, production and investment subsidies)?

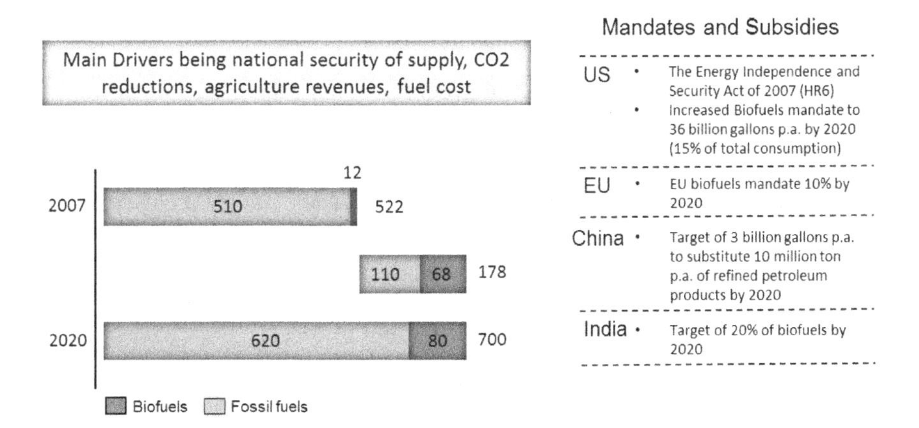

Fig.1. Biofuels are expected to capture a significant share of global transportation fuel growth.[1] All numbers are global and in quadrillion Btu's.

Looking more closely at the current progress to reach the ambitious strategic second generation biofuels goals, it may be argued that biofuels research is very much technology driven and lack of the true linkage to the pulp and paper markets and industry dynamics. The reality in this is that the existing pulp and paper and related industry will be the only major player that already has the feedstock issues, supply chain and logistic issues solved. Additionally pulp and paper industry is the only industry that has technology platforms into which the new biorefinery concepts may be readily integrated.

As a general statement in research, the investment opportunity aspects for different biorefinery platforms are many times missing. Also, local and global policy impact studies for the biofuels business should be addressed in more detail. Holistic sustainability studies on the feasibility and sustainability of biorefinery business (socio, environmental and economic views) needs to be evaluated and investigated thoroughly as well. This is important since biofuels will become a significant end use for biomass in parallel with power and heat production.

The companies that are investing considerable effort into developing their biorefinery concepts integrated to their or their partners' pulp mills are of course making their own fact-based economical and technical feasibility studies to justify their investments. The machine and process suppliers are currently collaborating with their customers, not only in research and development but also in deployment side by building novel piloting plant for the ultimate test of their concepts. The industry is currently shaping their approaches towards sustainable biorefinery concepts of the future. The task is challenging and as with all new business developments endeavors demand a value chain approach.

3. Biorefinery Concepts

When discuss further the biorefineries, it is good to define the biorefinery concepts and then understand their fit into the current pulp mills. As a definition, biorefinery is an industrial manufacturing process in which biomass is converted into power, fuels, variety of chemicals and biomaterials using different biomass conversion processes and equipments.

As stated earlier, the biorefinery concepts are definitely not a new area of development. Research related to this field already has a long history in the industry. Also, when looking at some of the existing concepts in a deployment timeline, the biorefinery business development activities can be mapped as shown in Fig. 2.

The scope of this chapter is to look at the pulp mill related biorefinery platforms. In the past, the product from the process have included pulp

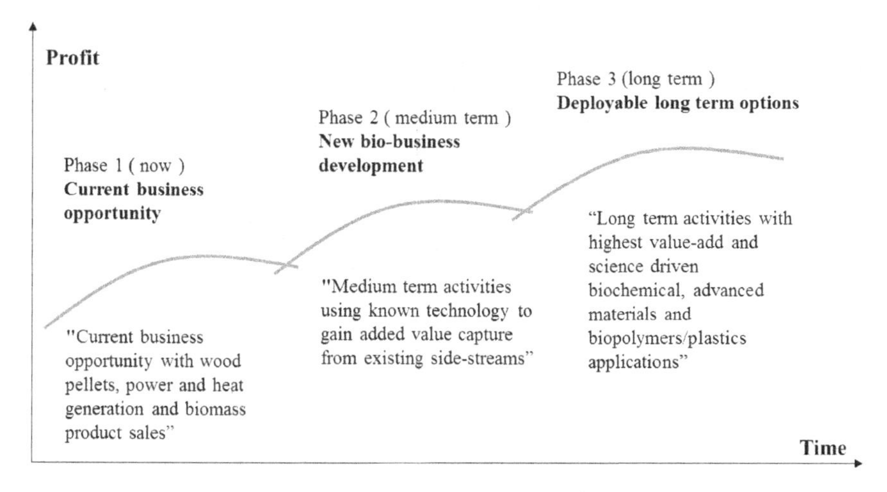

Fig. 2. Map of the deployment time line for biorefinery concepts.

and paper products, tall oil and turpentine in addition to electric power and "waste" heat. The future concepts are adding to the list, lignin and carbohydrate based chemicals and materials. The integrated concept of converting pulp and paper mills into biorefineries was for a few years seen as a great opportunity to ease the financial difficulties on the industry. This momentum has however slowed down and competing concepts of building stand-alone biomass power plants across the U.S. have emerged.[30–33] Unfortunately the newer approach will potentially create a price competition for renewable materials and therefore the resource sustainability may be an additional challenge.[34–36]

Integrated concepts have been also looked at from the benefit of using pre-existing infrastructure together with the existing supply chain and trained employees. In location selection it is important to understand the importance of for example process steam to the adjacent mill. If the paper mill disappears, this has to be compensated by another revenue source or targeted use of the steam.

The big challenge in an integrated concept is the assumption that there is enough usable feedstock material for the conversion to added value products, steam or electric power. Unfortunately, especially in the more modern mills the process and materials efficiencies and waste utilization technologies are so advanced that the business model may not be sustainable and additional investments may be difficult to justify.

In order for biorefinery concept to compete with the feedstock price hike that is potentially created by biomass power plants, a more diversified portfolio of low cost alternative feedstocks including tree trimmings, waste wood, mill sludge, cornstover, switchgrass and recycled paper should be looked at to increase the cost competitiveness of the biorefineries. However, the technological and logistic flexibility of utilization of multiple feedstocks in industrial process technologies is not a straightforward task and therefore it should be one of the areas in continuous research and development.

3.1. *The conversion pathways*

The two main pathways for the lignocellulosic feedstock conversion in biorefineries are biochemical and thermochemical processes. When looking at the biochemical side the processes are acid hydrolysis and fermentation or enzymatic hydrolysis and fermentation. In the thermochemical side solid biomass gasification, black liquer gasification and fast pyrolysis are the main processes to look at.

Lignocellulosic biomass constitutes four components, i.e., cellulose, lignin, hemicellulose and other extractives. In general as lignocellulosic material has crystalline form with hydrogen bonding[37] and is therefore more difficult to convert into ethanol than starches, sugars, starches and

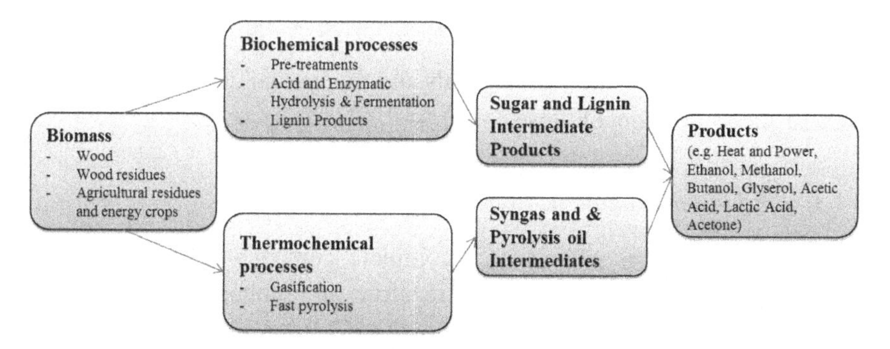

Fig. 3. Two pathways towards biorefinery concepts.

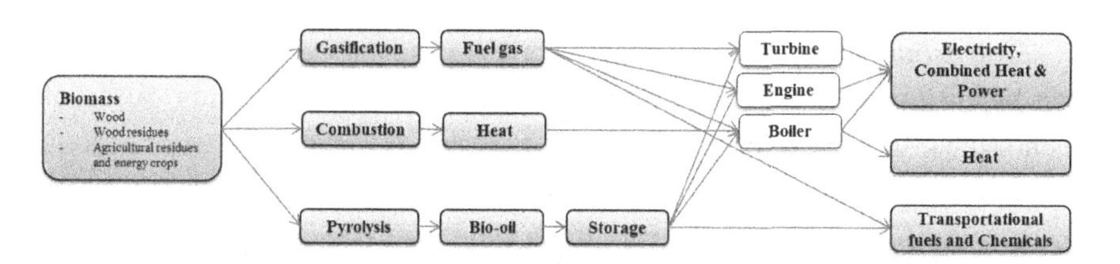

Fig. 4. Thermal conversion processes and product pathways.

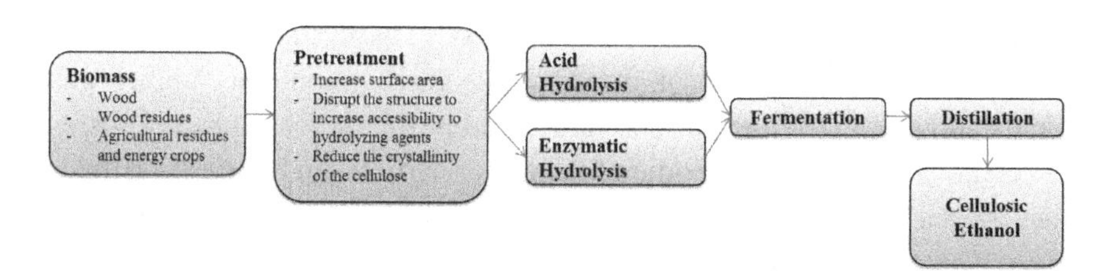

Fig. 5. Hydrolysis pathway to bioethanol.

oils. Lignocellulose itself is the fibrous material that forms the plant cell walls and consists of three major components 1. Cellulose, with high molecular weight polymers of glucose (repeated units of two anhydrous glucose rings joined via a β-1,4 glycosidic linkage) that are held together as bundles of fibers for material strength.[37] It typically accounts for 40 wt% of the lignocellulose. 2. Hemicellulose, which in general consists of short polymers made of different sugars gluing the cellulose bundles together. It typically accounts for about 25 wt% of the lignocellulose. 3. Lignin, consists of a tri-dimensional polymer of propylphenol.[38] It is embedded in and also bound to the hemicellulose. It creates structural rigidity to the lignocellulosic materials and the typical content is some 20 wt% of the total lignocellulose.

3.1.1. *Solid biomass gasification*

Solid biomass gasification is an incomplete combustion of biomass in temperatures typically above 800°C resulting in combustible gases composed of hydrogen (H_2), carbon monoxide (CO), and traces of methane (CH_4). The exact composition can be changed by altering the design of the gasification process. Additionally liquids and residue solids are formed in the process depending on the biomass used.[39]

The produced gases are typically called synthesis gases or syngas and can be further converted in to heat, power and chemicals using different techniques. In principle, gasifiers can use a large variety of carbonaceous feedstocks as long as a correct particle size and moisture content is used. In addition to moisture content, the particle size is a major factor in the solid biomass gasification. Due to the increasing particle surface area to volume ratio for smaller particles, the gas-to-particle heat transfer is strongly in correlation with the smaller particle sizes. After a certain size the heat transfer effect becomes dominative. These smaller particulates facilitate faster rates of heat transfer and gasification thus higher gas yields and energy efficiencies are attributed to the increased heat transfer in smaller size particles.[39] Some of the larger particles in biomass gasification process break into smaller sizes due to rapid expansion of gases during the pyrolysis phase. In general fuel quality fluctuation effects are a major influencer for the gasifier operations and the composition of the syngas. The gasification medium can vary and range from supercritical water to gaseous form. The medium has a great influence on the syngas composition and heating value. The real advantage of gasification is in that the burning of the syngas is more efficient than the direct combustion of the fuel. In principle it also gives more flexibility to the process. Based on the gas-solid contacting mode, gasifiers are classified into three principal types: fixed or moving bed; fluidized bed and entrained flow.

3.1.2. *Gasification-based biorefineries integrated with pulp mills*

In case of plugging a gasification biofuels system into an existing pulp mill, the basic technology is well-proven and include components such as black liquor gasifier system and steam generator and cooling system for produced syngas and a system for carbon dioxide and hydrogen sulfide removal from the raw syngas. Also conversion plant where syngas is transformed into liquid fuel and a distillation plant for purification of the fuel for desired product specifications is one of the obvious key components.

Black liquor is composed of spent cooking liquor and a mixture of dissolved lignin, hemicelluloses and other extractives from the wood after the digestion of wood chips in the Kraft pulping process.[41] It is an excellent gasification feedstock for syngas production and is traditionally burned by mills in order to produce steam for the pulp mill processes and also to recover the spent pulping chemicals. Black liquor is a mixture of organics, inorganics and water. The water amount is expressed as mass ratio of dried black liquor to unit of black liquor before drying. This ratio is called the black liquor dry solids and if it reaches levels below 20 % this means that the net heating value of black liquor is negative. That is the higher the dry solids the hotter the adiabatic combustion temperature and thus the throughput of the gasifier.

3.1.3. *Fast pyrolysis*

Fast pyrolysis is one of the three main thermal bioprocessing routes to provide a useful and valuable biofuel. As a process, pyrolysis is thermal decomposition occurring in the absence of oxygen. It also is the first step in any combustion and gasification process followed by partial or total oxidation of the material. Main process parameters tune the conversion towards different end product creation: charcoal production with low temperature and long vapor residence time, liquid fuels with moderate temperature and short vapor residence time, and gas, with high temperature and longer residence time.[45]

As fast pyrolysis process takes place in a timeframe of a few seconds or less, not only chemical reaction kinetics but also heat and mass transfer processes and phase transition phenomena play important roles additionally. In the process the biomass is decomposed and generates vapors, aerosols and charcoal. During the cooling and condensation process, a dark brown liquid is formed with a heating value of half of the conventional fuel oil. Critical parameters for successful fast pyrolysis based biomass liquid fuel production include, high reaction and heat transfer rates that typically can be accomplished only by using well ground biomass, well controlled pyrolysis temperature of around 500°C with vapor phase temperature of about 400–450°C, short vapor residence time (less than 2 seconds) and rapid cooling of the pyrolysis vapor to bio-oil product. Typical yield are oil 60–70%, Char 12–15%, Gas 13–25%.[46] In order to achieve controllable process parameters, the particle of the biomass to be pyrolyzed should be very small (about 100–200 microns) and the biomass should be pre-dried to about 10% level to avoid major reaction water creation.[45] The pyrolysis oil is dark low viscosity oil with up to 15–20% of water contents.[47]

There are several different approaches in fast pyrolysis processing and most potential for high volume pyrolysis oil production are bubbling fluidized bed, circulating fluidized bed and Auger reactor.

The main benefits of pyrolysis oil as a biorefinery product is that the equipment is operated in atmospheric pressures and moderate temperatures and the biomass yield are high exceeding 70 wt%. Also, a major benefit is the storage and transport as liquid. Disadvantages include high oxygen, water concentration, phase separation and polymerization of the pyrolysis oil of which high acidity and corrosive nature can be challenging for storage solutions.

3.1.4. *Acid hydrolysis and fermentation*

The basic hydrolysis reaction for cellulose conversion to sugars can be expressed as following:

$$(C_6H_{10}0_5) + nH_20 \rightarrow nC_6H_{12}0_6$$

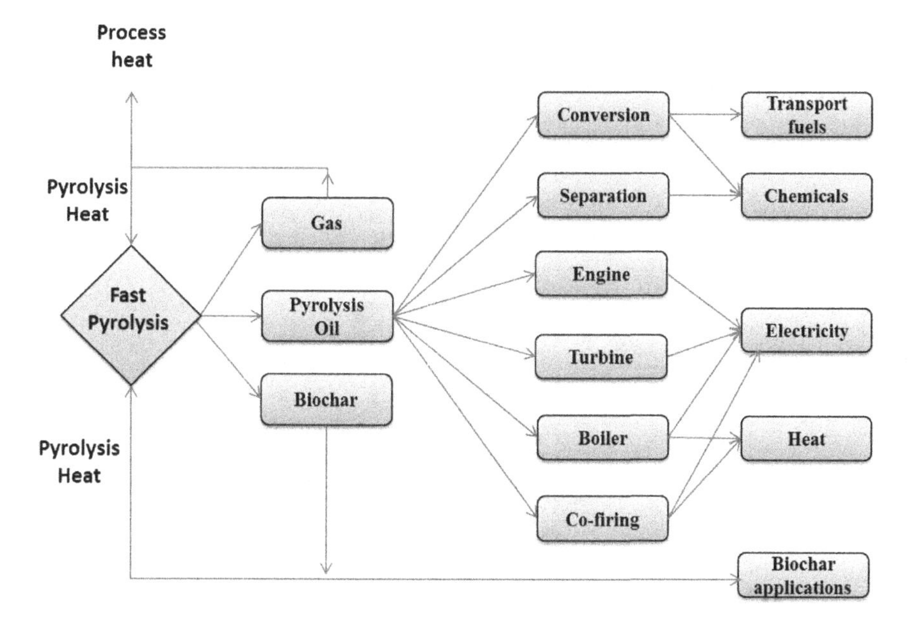

Fig. 6. Fast pyrolysis processes and routes to different products.

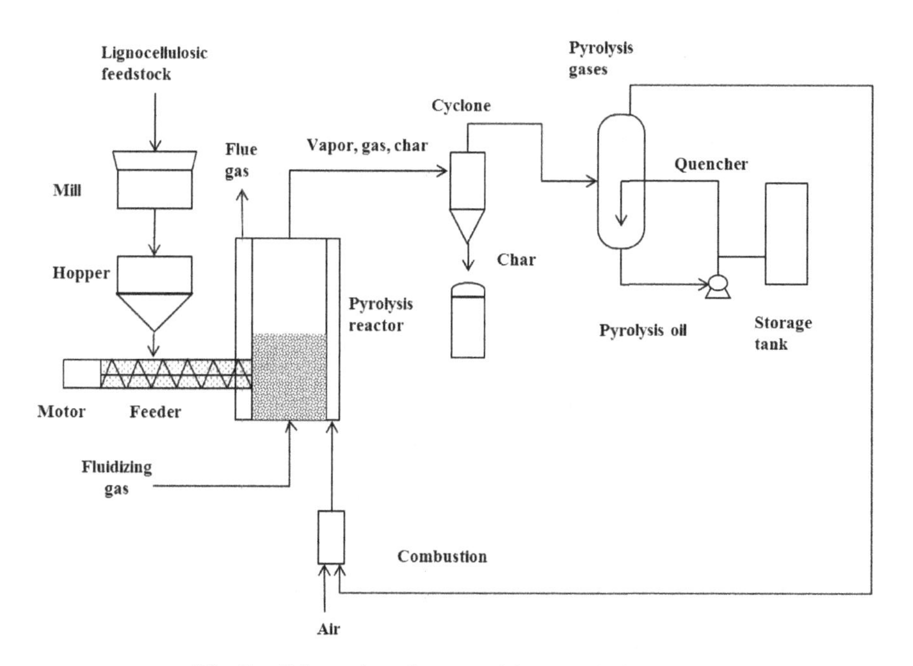

Fig. 7. Schematics of a general fast pyrolysis system.

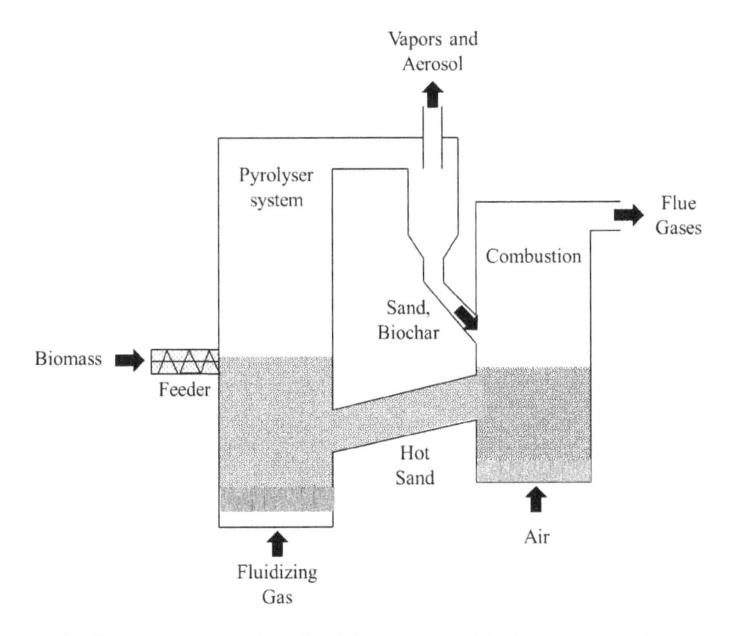

Fig. 8. Schematics of a circulating fluidized bed pyrolysis system.

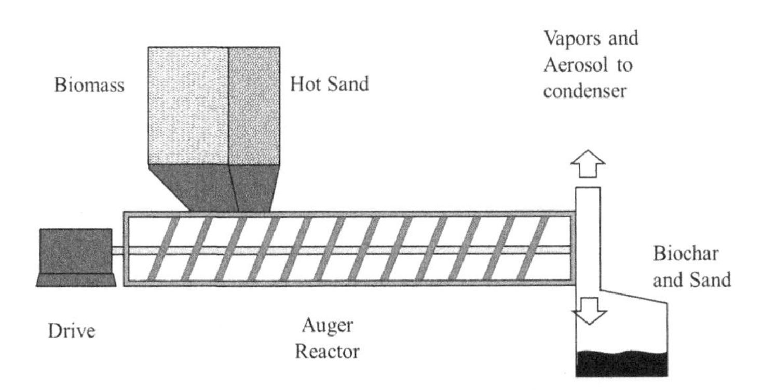

Fig. 9. Schematics of an Auger reactor.

The hydrolysis reaction can be catalyzed by either acid hydrolysis or fermentation. Acid hydrolysis has a long history; for example, dilute acid hydrolysis of biomass is, by far, the oldest technology for converting biomass to ethanol, dating back to end of 1800's in Germany.[48] The most common step is converting biomass through hydrolysis to glucose in addition to other monosaccharide by using acids treatments. This is typically done by using sulphuric acid or hydrochloric acid either in high (low temperature) or dilute (high temperature) concentrations. Still the economics in the biomass hydrolysis process is the main obstacle for large scale commercial success. In the acid hydrolysis process, certain monosaccharide derived products can further complicate the process as they can act as inhibitors to the next fermentation stage.

The complete hydrolysis of cellulose produces glucose, while the hemicellulose produces several pentoses and hexoses. Typically two stage hydrolysis is used as hemicellulose hydrolyses in a much easier manner compared to cellulose. This way one can limit the contamination of glucose, one of the desirable end products. When highly toxic hydrolyzates are formed, a detoxification phase is also needed prior to fermentation. Some of the desirable products after glucose fermentation include alcohols like ethanol, butanol, carboxylic acids like acetic acid, lactic acid and other products such as acetone and amino acids.

3.1.5. *Enzymatic hydrolysis and fermentation*

The biochemical conversion route involves breaking down of biomass into its component glucose and polysaccharides by using enzymatic hydrolysis. As the lignocellulosic material is very resistant to enzymatic digestion, commonly proper dilute acid or alkali or thermochemical pre-treatments are used to enhance the process. Due to the recalcitrance of lignocellulose structure, pre-treatments are generally harsh and combinations of physical, chemical, biochemical or thermal treatments are used.[49] It is well known that pre-treatment also represents a-significant cost element in lignocellulosic bioethanol conversion process. The main goal is to create fermentable sugars for bioconversion through fermentation. However, the process also produces monosacharaides usable for further processing towards other value added chemicals.[50]

The mere number of chemical building blocks reachable through fermentation is vast. Fermentation has been in use since the beginning of the industrial era and currently around 8 million tons of fermentation products are being produced annually.[51]

Major challenges in this route are the cost of the enzymes and the low conversion rates caused by the heterogeneous nature of the raw cellulosic material. That is enzymatic digestion and the rate of these processes is not just dependent on the enzymes but also on the chemical and morphological features of the raw material.[48] The enzymatic hydrolysis of wood and other cellulosic materials has been a major focus in research during the recent years due to the extremely high potential in possibly finding an economical solution to renewable energy source.

4. Lignin and Its Opportunities in Biorefineries

As reviewed earlier the field is divided into the gasification and other thermal conversion process related researches, developments and production of biofuels directly from wood or biomass through biochemical processes. Value prior to pulping is an approach in which carbohydrate components that do not enter the paper product are recovered before the pulping process is started. Thus the pulp mill simultaneously produces fibers and sugars by for example extracting hemicelluloses from wood chips using oxalic acid solutions or diethyl oxalate.

We have looked into the utilization of cellulose and hemicellulose, but the potential of lignin as a large scale biorefinery value added product is of interest to many companies. Lignin is discussed in the following as its role in the general energy balance and energy portfolio of the Kraft process is so important. In general lignin is the second most abundant natural amorphous polymer and the only biomass constituent based on randomly branched and cross-linked aromatic units.[49] There are many current and also emerging applications for lignin including precursors for carbon fiber, component of and catalyst for conducting polymers, components in polymer blends and alloys, replacement for phenol-formaldehyde and urea-formaldehyde foams, adhesive resins and sealants. Energy related applications include binders for foundry, charcoal briquettes, carbons and electrodes. Also, directly biofuels related applications include using methanol soluble nitrated lignin for ignition acceleration purposes in methanol fueled diesel engines. Main benefits for these and other applications are the renewable nature, low-cost, higher heat resistance and water resistance.

Major challenges in the continuous research and development of its use are the intrinsic color that limits its use mainly into industrial application and lignin's inability to be dispersed and associate well into polymers or plastics. This feature results in weakness or brittleness rather than reinforcing or strengthening effect. This issue is likely associated with lignin's polyelectrolyte character and therefore for certain applications the synthesis of lignin becomes an important part of its usability as polymer additive.[51]

4.1. *Lignin sources in biorefinery concepts*

There is quite a variety of ways to incorporate the lignin extraction and processing to biorefinery concepts. The feasibility of these approaches of course depends on the business model, feedstock availability and the product portfolio of the biorefinery.

4.1.1. *Lignin from Kraft pulping process*

The Kraft pulping process is highly potential source of lignin for a biorefinery. Major benefits include the fact that pulp and paper industry has a highly efficient infrastructure for growing, harvesting, transporting and mechanically and chemically processing the lignocellulosic feedstock material. However, the industry already has a major use for the lignin and therefore will not be able to sustain a sufficient supply of this material from its processes. Kraft pulping operations are highly integrated and depend highly on the lignin as an energy source for its chemical recovery process. If this energy source would be taken away other energy sources would be needed including natural gas or coal. Unless the value of the extracted lignin is high enough or if a better more energy efficient use of for example black liquor through gasification was implemented, the economics of the mill operations would be unrealistic.

The general approach to the extraction of lignin from Kraft process is through lowering the pH using for example carbon dioxide from the boiler's flue stack or simply using mineral acid. Major portion of the Kraft lignin is precipitated after lowering the pH and can be further processed by filtering and washing. This approach produces lignin that can be almost free of sugars and with relatively low sulfur content of 1–2%.[50–51]

4.1.2. *Lignin from sulfite pulping process*

Looking at another alternative, sulfite pulping, one must conclude that in the United States it is also not a sustainable source for lignin as only about 2% of the total pulp production is based on sulfite process. The total potential processing capacity is about 1.1×10^9 lb/yr.[52] From technological point of view, however, the sulfite pulping process is very potential as it generally does not recover pulping chemicals, which makes the lignin-rich black liquor stream available for conversion to chemicals. There is a considerable difference in the lignin produced from the sulfite pulping compared to that produced from the Kraft process.

The isolated lignin from the sulfite process contains considerable sulfur in the form of sulfonate groups. After stripping and recovery of the sulfur, extraction of the sulfite lignin, also called lignosulfonate, is typically made from waste pulping liquor concentrate.[52] There are a few approaches in the recovery techniques but the economically feasible is the precipitation of calcium lignosulfonate with excess lime it recovers up to 95% of the liquor's lignin.[50,53]

The produced lignosulfonate material has several disadvantages including the fact that they are quite impure materials and contaminated by the cations used during pulp production and recovery. They are also normally mixtures and the total content of lignin is only 70% with rest being carbohydrates, ash, and other inorganic materials.[52] Other disadvantages include sulfur content, solubility to water throughout almost the entire pH range.

4.1.3. *Other lignin production technologies*

There are several other lignin production technologies including organic solvents, pyrolysis lignin, steam explosion, dilute acid process, and a few other techniques.[54] One of the attractive future techniques is organic solvents based processing in which separation of lignocellulosic material is done through treatment with organic solvents. There are several approaches including a variety of solvents and their combinations including acids and alkali to enhance the process speed.

In general these processes are considered more environmentally friendly. This type of lignin is typically like unsulfonated kraft lignin and the lower molecular weight and lower sulfur content make it attractive for use in low-molecular weight phenols and aromatics.[54–55]

5. Future of Biorefining in Pulp Mills

The building of pulp mill integrated biorefineries is mainly applicable to non-integrated chemical market pulp mills, as for the integrated chemical and mechanical pulp mills have less excess of biomass.

As the chemical pulp mills are becoming more and more energy efficient, potentially there will be excess biomass available for alternative usage. Additionally the value of the internal biomass components such as black liquor, bark, forest residues like tree trumps can be boosted with biorefining technologies. This creates opportunities to fill the nationally increasing energy gap due to economic growth. The pulp and paper industry has the benefit of using the existing logistics channels so that excess biomass can be collected and fed into the process as long as the value of the refined products is high enough.

In the energy focused biorefinery the excess available lignin or hemicellulose is used for the production of additional electricity together with possibly biofuels. If gasification is used, methanol or dimethyleteher (DME) can be produced as biofuels. Another alternative route is the extraction of lignin from the black-liquor. In some mills in which there is a clear bottle neck in the capacity of the recovery boiler, this solution offers a benefit of increased production and produced lignin for mill use or as a value-added product.

There are two foreseen biorefinery products in the transportation fuels market: ethanol fuel and Fischer-Tropsch diesel. In a recent study[56] it was stated that the biomass gasification route will be commercialized later in the U.S. compared with Nordic countries like Finland and that enzymatic hydrolysis and fermentation are to be commercialized faster in the North America. According to the study, Sweden will be spearheading the black liquor gasification and related gas cleaning. The study also points out the differences in the feedstock focus. In North America and Brazil broader look at energy crops is of more interest compared with Nordic countries in which black liquor is seen as much more significant source of biomass. Clearly the greatest barriers for feedstock utilization in both North

America and Nordic countries are logistics and transportation related. The sustainability or environmental awareness and debate have recently increased in the North America and can be considered a new challenge in the area of biorefining.

The economics landscape of pulp mill biorefinery concepts should be looked at more a combination of energy, fuels and value-added renewable chemicals and materials. For example black liquor can be used as source for hemicelluloses and lignin. One can use extracted hemicellulose xylan and upgrade it to fibre additives for improving paper bulk, as thickeners, food additives, emulsifiers, all the way to adsorbants and adhesives. Lignin can as well be used as value-added products and chemicals ranging from food industry flavors to resins and sealants.

Of course one should not forget the opportunities for energy saving in the mill when black liquor gasification will in the future be more used for increasing the share of internal biomass of the total fuel consumption.

Some new topic areas to look at from energy perspective are rejects from recycled paper processing, which contain large quantities of combustible materials such as cellulose fibers and plastics. Plastics that contain only carbon, hydrogen, and oxygen are present no environmental risk if they are burned in small quantities in modern combustors. Plastics containing halogenated compounds such as polyvinylchloride or teflon do need pollution control devices when combusted so that it increases the cost of this alternative. Low-quality recycled paper also presents a potential valuable source of energy. It has a relatively high heating value and its processing and separation from waste streams is fairly established. It is also environmentally sound with low sulfur content, and low nitrogen oxides emissions.

The time of major excitement in pulp mill biorefinery concept development may have diminished during the past few years. A new development is the rising role of US natural gas as energy and feedstock source for chemicals and other industries. The latest development of new sources of shale gas has offset declines in production from conventional gas reservoirs, and has led to major increases in reserves of US natural gas. It will have an increasing importance in the general development of US manufacturing competitiveness. This development may potentially change biorefinery related investment decisions. Many of the pulp mill biorefinery concepts are mill specific and relevant to the existing equipment and machinery, feedstock and supply chain and the existing energy portfolio. For this reason there will continue to be interest in finding a more optimized business model for sustainable energy, renewable chemicals and materials production. In order to capture these opportunities,the industry needs to move faster and develop a less risk adverse innovation culture and a more productive innovation eco-system.

Acknowledgments

I would like to thank the Institute of Paper Science and Technology (IPST) at Georgia Institute of Technology for its support as the writing of this chapter started while working at IPST. I would also like to thank those who provided relevant material that was used in this article, and Dr. Markku Karlsson and Dr. Art Ragauskas for valuable comments and insight to the vast topic of biorefining and its future in pulp and paper industry and its value chain. The finalization of the writing of this chapter took place at Alabama Center for Paper and Bioresource Engineering at Auburn University.

References

1. IEA World Energy Outlook 2006, IEA PUBLICATIONS, 9, rue de la Fédération, 75739 PARIS CEDEX 15, PRINTED IN FRANCE BY STEDI, (612006231P1) ISBN 92-64-10989-7-2006.
2. Transesterification of triacetin with methanol on solid acid and base catalysts, D. Lopez, J. Goodwinjr, D. Bruce and E. Lotero, Applied Catalysis A: General (2005). Volume: 295, Issue: 2, Pages: 97–105.
3. Second-Generation Biofuels Economics and Policies, Miguel A. Carriquiry, Xiaodong Du, Govinda R Timilsina, Policy Research Working Paper 5406, The World Bank Development Research Group Environment and Energy Team, August 2010.
4. Second-generation biofuels and local bioenergy systems, Blanca Antizar-Ladislao, Juan L. Turrion-Gomez, Biofuels, *Bioprod. Bioref.*, **2**, 455–469 (2008).
5. A. Wingren, M. Galbe and G. Zacchi, Techno-economic evaluation of producing ethanol from softwood: Comparison of SSF and SHF and identification of bottlenecks, *Biotechnol Prog.*, **19**, 1109–1117 (2003).
6. EU Strategy for Biofuels, Offi cial Journal of the European Union. http://ec.europa.eu/agriculture/biomass/biofuel/com2006_34_en.pdf/ (March28, 2008).

7. L. R. Lynd, M. S. Laser, D. Bransby, B. E. Dale, B. Davison, R. Hamilton, M. E. Himmel, M. Keller, J. D. McMillan, J. J. Sheehan and C. E. Wyman, How biotech can transform biofuels, *Nature Biotechnol.*, **26**, 169–172 (2008).
8. C. N. Hamelinck, G. Van Hooijdonk and A. P. C. Faaij, Ethanol from lignocellulosic biomass: Techno-economic performance in short-, middle- and long-term, *Biomass Bioenerg.*, **28**, 384–410 (2005).
9. Reducing Greenhouse Gas Emissions From U.S. Transportation, David L. Greene, Andreas Schafer, May 2003.
10. Reducing Transport Greenhouse Gas Emissions: Trends & data 2010 — © oecd/itf 2010.
11. US EIA. International Energy Outlook, Washington, DC, USA: United States Energy Information Administration. (2009, May 27). 2009.
12. Congressional Budget Office (2008). Effects of Gasoline Prices on Driving Behavior and Vehicle Markets. Washington, D.C. http://www.cbo.gov/ftpdocs/88xx/doc8893/01–14-GasolinePrices.pdf.
13. Cambridge Systematics, Inc. (2009). Moving Cooler: An Analysis of Transportation Strategies for Reducing Greenhouse Gas Emissions. Urban Land Institute, Washington, D.C.
14. T. M. L. Wigley, R. Richels and J.A. Edmunds, "Economic and Environmental Choices in the Stabilization of Atmospheric CO2 Concentrations," *Nature*, **379**, 240–243 (1996).
15. National Energy Technology Laboratory (NETL), 2008. Reducing CO2 Emissions by Improving the Efficiency of the Existing Coal-fired Power Plant Fleet, DOE/NETL-2008/1329. U.S. Department of Energy, National Energy Technology Laboratory, Pittsburgh, PA. July 23, 2008.
16. Greenhouse Gas Emissions from Coal Gasification Power Generation Systems, John A. Ruether, Massood Ramezan, and Peter C. Balash. *J. Infrastruct. Syst.*, **10**, 111 (2004).
17. Climate change and the potential of coal gasification, Schon, Samuel C., and Arthur A. Small III. Geotimes 51.9 (Sept 2006): 20(4).
18. Climate change and the potential of coal gasification, N. P. Myhrvold and K. Caldeira, *Environ. Res. Lett.*, **7**, (2012).
19. Greenhouse gas emissions from building and operating electric power plants in the Upper Colorado River Basin, Pacca S. and Horvath A, *Environ. Sci. Technol.*, **36**, 3194–200 (2002).
20. Advanced Post-Combustion CO2 Capture, Howard Herzog, Jerry Meldon, Alan Hatton, Prepared for the Clean Air Task Force under a grant from the Doris Duke Foundation, April 2009.
21. Intergovernmental Panel on Climate Change (IPCC) Special Report, Carbon Dioxide Capture and Storage, New York: Cambridge University Press (2005).
22. K. Sanden, T. Ursin1, A-H. Haal and, H.A. Haugen, "CO2 capture from gas power plants — Just CatchTM Potential cost reductions," Proceedings of the 8th International Conference on Greenhouse Gas Control Technologies, Trondheim, Norway, O. Bolland, H. F. Svendsen, P. Zweigel and J. Gale. (eds.), Elsevier (2006).

23. US Crude Oil and Total Petroleum Imports Top 15 Countries, 2011 ftp://ftp.eia.doe.gov/pub/oil_gas/petroleum/data_publications/company_level_imports/current/import.html
24. Diversification of oil import sources and energy security: A key strategy or an elusive objective?, Vlado Vivoda, Energy Policy, Volume 37, Issue 11, November 2009, Pages 4615–4623.
25. BP, 2007. BP Statistical Review of World Energy June 2007, http://www.bp.com/conversionfactors.jsp.
26. Hakes, Jay, A Declaration of Energy Independence: How Freedom from Foreign Oil Can Improve National Security, Our Economy, and the Environment. Hoboken, NJ: John Wiley & Sons, Inc., 2008.
27. LaCasse, Chantale and Andre Plourde, On the Renewal of Concern for the Security of Oil Supply, *Energy Journal*, **16 (2)**, 1–23 (1995).
28. The Energy Independence and Security Act of 2007 (P.L. 110–140, H.R. 6)
29. Regulatory announcement: http://www.epa.gov/otaq/renewablefuels/420f10007.pdf
30. M. Berggren, E. Ljunggren, F. Johnsson, Biomass co-firing potentials for electricity generation in Poland-matching supply and co-firing opportunities. *Biomass and Bioenergy* September 2008;32:865e79.
31. R. V. D. Broek, A. Faaij, A. V. Wijk. Biomass combustion for power generation, *Biomass and Bioenergy*, **11**, 271e81 (1996).
32. A. Franco, N. Giannini. Perspectives for the use of biomass as fuel in combined cycle power plants, *International Journal of Thermal Sciences*, **44**, 163e77 (2005).
33. T. Proll, C. Aichernig, R. Rauch and H. Hofbauer. Fluidized bed steam gasification of solid biomass e Performance characteristics of an 8 MWth combined heat and ower plant, *International Journal of Chemical Reactor Engineering*, **5**, 54 (2007).
34. “Biomass Technology Review”, Biomass Power Association, October 21, 2010, 51 pages.
35. http://www.eia.gov/oiaf/analysispaper/biomass/
36. http://feedstockreview.ornl.gov/pdf/billion_ton_vision.pdf
37. D. Klemm, B. Philipp, T. Heinze, U. Heinze, W. Wagenknecht. Comprehensive cellulose chemistry. Chichester: Wiley VCH; 1998.
38. Lignocellulose conversion: An introduction to chemistry, process and economics, Jean-Paul Lange, Shell Global Solutions, The Netherlands, DOI: 10.1002/bbb.7; *Biofuels, Bioprod. Bioref.*, **1**, 39–48 (2007).
39. Thermochemical Biomass Gasification: A Review of the Current Status of the Technology, Ajay Kumar *et. al.*, Energies 2009, 2, 556–581; doi:10.3390/en20300556.
40. T. M Grace, “Chemical Recovery Process Chemistry,” Chemical Recovery in the Alkaline Pulping Processes 3ed, pp. 57–78, Atlanta: TAPPI press, 1992.
41. Fischer–Tropsch fuels refinery design, Arno de Klerk, *Energy Environ. Sci.*, **4**, 1177–1205 (2011).
42. Diesel Production from Fischer–Tropsch: The Past, the Present, and New Concepts, Dieter Leckel, *Energy Fuels*, **23 (5)**, 2342–2358 (2009).

43. D. Eric. Larson, Huiyan Yang, Dimethyl ether (DME) from coal as a household cooking fuel in China, Energy for Sustainable Development l Volume VIII No. 3l September 2004.
44. China's growing methanol economy and its implications for energy and the environment, Chi-Jen Yang, B. Robert Jackson, *Energy Policy,* **41**, 878–884 (2012).
45. Characterization of Lignocellulosic Materials, Thomas Q. Hu, PhD (Editor), ISBN: 978-1-4051-5880-0, April 2008, Wiley-Blackwell.
46. Review of fast pyrolysis of biomass and product upgrading, A.V. Bridgwater, *Biomass and Bioenergy*, Volume 38, March 2012, Pages 68–94.
47. Techno-Economic Analysis of Biomass Fast Pyrolysis to Transportation Fuels, M. Mark Wright, Justinus A. Satrio, and Robert C. Brown, Daren E. Daugaard, David D. Hsu, Technical Report, NREL/TP-6A20–46586, November 2010.
48. E.C. Sherrard, F.W. Kressman, "Review of Processes in the United States Prior to World War II." Industrial and Engineering Chemistry, Vol 37, No. 1, 1945, pp 5–8.
49. BSCL USE PLAN: Solving Biomass Recalcitrance, M. Himmel, T. Vinzant, S. Bower, and J. Jechura, Technical Report, NREL/TP-510–37902, August 2005.
50. Bio-based Chemicals, Value Added Products from Biorefineries IEA Bioenergy — Task42 *Biorefinery*, (2012).
51. D. Feldman,. in Chemical Modification and Usage of Lignin, Kluwer Academic Press, New York, Hu, T. H. ed., Kluwer AcademicPress, New York (2002), p. 81.
52. B. R Smith, R. W. Rice and P. J. Ince, Pulp Capacity in the United States, 2000, USDA Forest Service, General Technical ReportFPL-GTR-139; Pulp and Paper North American Factbook 2001, Paperloop Publications (2002).
53. "Lignin" — Kirk-Othmer Encyclopedia of Chemical Technology, Stuart E. Lebo Jr., Jerry D. Gargulak, Timothy J. McNally, Published Online: 20 DEC 2001 doi: 10.1002/0471238961.12090714120914.a01.pub2.
54. Recent Industrial Applications of Lignin: A Sustainable Alternative to Nonrenewable Materials, Jairo H. Lora and Wolfgang G. Glasser, Journal of Polymers and the Environment, Volume 10, Numbers 1–2 (2002), 39–48.
55. T. J. McDonough, J. Tappi. 1993, 76(8), 186; M. C. Schroeter, *J. Tappi*, **74(10)**, 197 (1991).
56. Forest biorefineries — A business opportunity for the Finnish forest cluster, Journal of Cleaner Production, Volume 19, issue 16 (November, 2011), p. 1884–1891.

CHAPTER 12

INTEGRATED POSSIBILITIES OF PRODUCING BIOFUELS IN CHEMICAL PULPING

RAIMO ALÉN

University of Jyväskylä
Department of Chemistry, Laboratory of Applied Chemistry
P.O. Box 35, FI-40014 Jyväskylä, Finland
raimo.j.alen@jyu.fi

The utilization of various organic fractions for biofuel production depends on several factors, such as the type of feedstock, energy and pulp prices, production scale, and available conversion technologies. In addition to combustion of various organic material residues, there are thermochemical processes available for producing biofuels from harvesting and debarking residues, tall oil, separated lignin, and some novel promising options are currently being developed. Moreover, integrated pulp mill biorefinery concepts offer opportunities to produce liquid biofuels (e.g., ethanol and butanol) based on fermentation of hemicelluloses. This chapter summarizes the most important existing as well as emerging technologies for incorporating biofuel production into an existing pulp mill.

1. Introduction

1.1. *Pulping processes*

In the simplest case, the biorefinery utilizes only one feedstock with a single process resulting in a single major product.[1] However, in a more complicated case, such as a modern pulp mill, the biorefinery utilizes one feedstock (wood or non-wood feedstock), but is capable of manufacturing several end products, including possibilities for producing various biofu-els. For this reason, it is important to clarify the formation of different feedstock-derived degradation products as pulping by-products. In addition, for outlining the overall production of pulping-based biofuels, it is essential to understand the efficient separation of by-products (e.g., the economic fractionation of cooking waste liquors) and, on the other hand, relevant techniques to convert these by-products into fuel.

Table 1 gives a broad classification of the commercial pulping processes combined with yield data. With respect to significant production of by-products only the chemical pulping is of importance. In general, the average yield value of chemical pulp is in the range 45–55%. Yields of the fiber products from dissolving pulping (i.e., acidic sulfite, multistage sulfite, and prehydrolysis kraft methods) are generally 35–45%, and these pulps are utilized in the manufacture of cellulose derivatives and related products. It should be pointed out that chemical pulping accounts for 70% of the total worldwide pulp production, and currently about 90% of chemical pulps (about 130 million tons) are produced by the dominant kraft (sulfate) process.[1]

Table 1. Commercial pulping methods.[1]

Method	Yield (% of wood)
Chemical pulping	35–60
Kraft, polysulfide kraft, prehydrolysis kraft	
Soda-anthraquinone (AQ)	
Acid sulfite, bisulfite, AQ alkali sulfite	
Multistage sulfite	
Semichemical pulping	65–85
Neutral sulfite semichemical (NSSC)	
Soda	
Chemimechanical pulping	80–90
Chemithermomechanical (CTMP)	
Chemigroundwood (CGWP)	
Mechanical pulping	91–98
Thermomechanical (TMP)	
Refiner mechanical (RMP)	
Stone groundwood (SGWP)	
Pressure groundwood (PGWP)	

Although the importance of sulfite pulping has clearly decreased during recent decades, pulping of wood and especially non-wood materials in the presence of organic solvents ("organosolv methods") may gradually increase in the future.[1] A great number of organosolv type methods based on the utilization of completely or partly organic solvents (mainly ethanol and methanol) have been reported, but so far most of these processes have been studied only on a laboratory scale. In general, the traditional delignification methods for non-wood feedstock are the soda and soda-AQ processes, although various sulfite and kraft processes are used, as well.

1.2. *Possibilities of pulping-based biofuel production*

The term "biofuels" covers a wide range of fuels including solid biomasses, liquid fuels, and different gases (biogases).[1,2] "First-generation biofuels" produced by well-known technologies are typically ethanol from fermentable sugars and biodiesel from vegetable oils (or from animal fats and recycled greases) after transesterification (i.e., the production of fatty acid methyl esters — FAMEs). In general, petroleum-based diesel fuels are blends of a very large number of different hydrocarbons and are principally rather similar to biodiesel, although the latter is mainly used as a diesel additive. However, the detailed composition of biodiesel is greatly dependent on the raw material source. For example, multicomponent biodiesel can be produced thermochemically from extractives in a kraft mill, but the integrated kraft mill-based production of pure single-component substances, such as ethanol and methanol, suitable for fuel purposes, is also possible by using alternative techniques.

The main chemical components (cellulose, hemicelluloses, lignin, and extractives) of wood and non-wood biomass behave differently during delignification.[1] For this reason, in addition to a thorough knowledge of the feedstock's structure, a good understanding of the chemical behavior of these constituents is not only useful for the delignification, but also for the development of pulping-based biorefinery concepts.

During kraft pulping, roughly half of the initial wood material is dissolved in the cooking liquor ("black liquor").[1,3] The organic solids in black liquor consist typically of degraded lignin and polysaccharides-derived low-molecular-mass degradation products (aliphatic carboxylic acids) (see Sec. 3.2. Lignin). Small amounts of extractives and polysaccharides, such as hemicellulose residues, are also present. In the kraft mill, after the separation of most of the extractives in the tall oil soap, the black liquor is concentrated in multiple-effect evaporators and then combusted in the recovery furnace for the recovery of cooking chemicals and the generation of energy.[4–6]

Kraft black liquor as such is the most important by-product of kraft pulping and the combustion of this heterogeneous biofuel is a significant source of energy.[1] In addition, the recovery boilers producing bioenergy have developed considerably in the past 70 years, culminating with units that are among the largest biofuel-fired boilers in the world. In general, black liquor has several combined features that make its combustion different from that of other fuels. For example, the high content of inorganic material (25–40%) and water (15–35%) decreases the heating value (12–15 MJ/kg dry solids) of the black liquor to a clearly lower level than that of other common industrial fuels. In this chapter, however, the versatile combustion chemistry and various phenomena taking place in the recovery furnace are not discussed in detail.

In principle, wood and harvesting residues, in addition to the material from debarking at a mill, can be processed in a large number of ways ("feedstock-based technology"), using various mechanical (e.g., pulverization and pelletizing), chemical (e.g., hydrolysis followed by fermentation), and thermochemical (including torrefaction, pyrolysis, gasification, liquefaction, and combustion) methods, to produce energy (i.e., in the form of biofuel, heat, or electricity) and chemicals.[1,7,8] Typically, in many countries roughly half of the initial dry wood material remains as harvesting residues (i.e., stumps, roots, tops, branches, needles or foliage) and debarking residues while the debarked wood yields only about 50% fiber (i.e., about 25% of the initial dry wood material) after delignification. In general, the combustion of bark represents typical integrated production of bioenergy since it normally takes place in the on-site bark-burning furnace.

In this chapter, some possibilities of producing bioenergy mainly from the potential by-products of chemical pulping ("kraft process-based technology" and "acid sulfite process-based technology") are briefly discussed; more detailed background data on particular topics and their current status can be obtained from several literature sources. In addition, special emphasis has been put on kraft pulping (i.e., the utilization of hemicelluloses, extractives, and lignin) and generally on the possible biorefinery unit processes to provide a general understanding of the versatile chemistry behind these conversions. It should be pointed out that in some cases the principles shown have been so far accomplished only on a laboratory and pilot scale.

2. Autohydrolysis of Wood Chips

2.1. *Basic considerations*

Various acid treatments of biomass have received considerable research attention over the years.[1] In the typical case, dilute H_2SO_4 has been initially added to remove hemicelluloses prior to more effective acid or enzymatic hydrolysis of cellulose, thus simultaneously enhancing the reactivity of cellulose in the residual solids.[9–12] An illustrative and analogous example of this kind of treatment is the commercial manufacture of furfural from varying xylan-containing agro-biomasses.

Among the pretreatment methods autohydrolysis is of special interest because, compared to prehydrolysis with dilute mineral acids, water is the only reagent, making it an environmentally friendly and inexpensive process.[1] The effluents (hydrolysates, pH 3–5) from conventional autohydrolysis operations contain a mixture of various carbohydrates (in addition to oligo- and polysaccharides, also some monosaccharides) together with a minor amount of other organics (a few inorganics are also present), including aliphatic carboxylic acids (acetic and formic acids) as well as furans — furfural and 5-(hydroxymethyl)-2-furaldehyde or HMF — and heterogeneous fractions of lignin and extractives-derived material.

In the chemical pulp industry, the most common and effective approach is to introduce an acidic pretreatment stage into the kraft pulping process (Fig. 1), mainly serving the purpose of removing hemicelluloses (the removal of cellulose is negligible) from wood chips.[1,13–15] These carbohydrates can then be further converted, for example, after a total hydrolysis to fermentable sugars, into value-added products such as ethanol, butanol, and other fermentation products.[16] This kind of approach is analogous to the conventional prehydrolysis kraft process in which

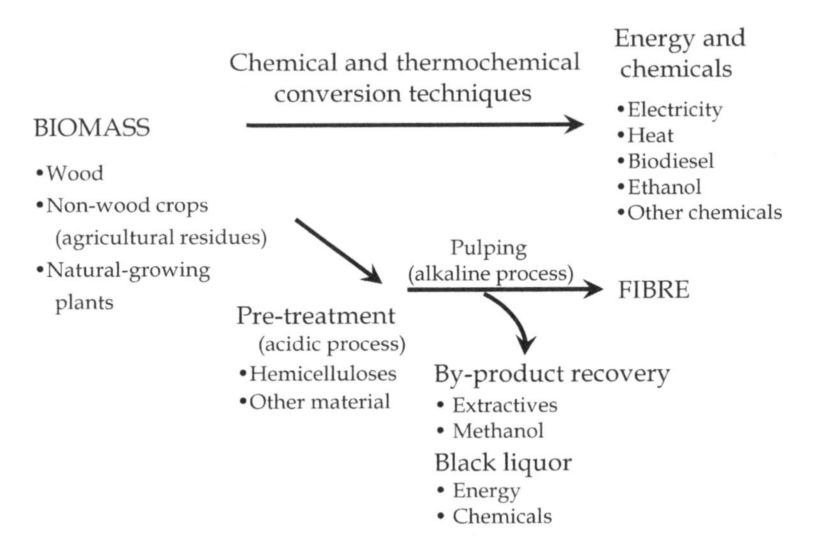

Fig. 1. Principles of typical biorefinery concepts in the forest industry.[1]

dissolving pulp is produced after an extensive removal of hemicelluloses during the first high-temperature acidic stage.

Prehydrolysis of wood chips has been investigated under a variety of conditions and from several points of view.[17–29] Table 2 shows one example of the chemical composition of softwood- and hardwood-derived hydrolysates (autohydrolysates). It has been claimed[30–33] that all the non-carbohydrate materials may be harmful, especially when considering the biochemical utilization of hydrolysate-based carbohydrates. For this reason, the removal of these inhibitors, at least partial if not complete, would be of benefit.

2.2. *Autohydrolysate-based products*

There are several drivers for the production of ethanol for transport, since the demand for fuel for the transport sector is rising in Asia, especially in China and India, as a result of an increasing number of vehicles.[16] A second driver for the growing interest in biofuels is generally that domestic production of transport fuel will reduce the dependence on foreign oil.

Ethanol has traditionally been used as fuel for vehicles in its pure form, but it is normally used as a petrol additive to increase the octane

Table 2. Examples of the chemical compositions of softwood- and hardwood-derived hydrolysates (% of the initial wood dry matter).[a]

Component	Birch	Pine
Carbohydrates	12.7	9.6
Monosaccharides	1.4	1.8
Oligo- and polysaccharides	11.3	7.8
Uronic acids	0.4	0.6
Furans	0.1	0.1
Volatile acids	1.6	0.5
Acetic acid	1.5	0.4
Formic acid	0.1	0.1
Lignin	2.1	1.1
Total	16.9	11.9

[a]The chips are treated at 150 °C for 90 min (liquor-to-wood ratio 5 L/kg).

number and to curb vehicle emissions.[1] However, the industrial production of ethanol is so far mainly based on inexpensive raw materials, such as corn (starch) and sugar cane (sucrose), rather than various autohydrolysates or sulfite spent liquors (see Sec. 5. By-Products of Acid Sulfite Pulping) from the pulping industry. Ethanol is conventionally purified by distillation, although new separation techniques have also been developed for removing it during or after fermentation. These methods are expected to consume less energy than conventional distillation and they may also offer new possibilities for autohydrolysate-based production.

In addition to ethanol, butanol (1-butanol) has been traditionally produced from carbohydrates.[16] Its biochemical production, "ABE (acetone-butanol-ethanol) fermentation", is among the first industrial-scale fermentation processes. At the beginning of the 20th century the ABE process was one of the most important fermentation processes after the ethanol fermentation by yeast. This process was almost completely abandoned in the 1960s, because it could not compete economically with butanol and acetone produced from oil. However, many recent manufacturing developments may also in this case restore the autohydrolysate-based production.[34]

Butanol has many favorable properties when used as fuel.[1] It can be used unmodified in internal combustion engines and blended in diesel compression and jet turbine engines. In addition, butanol has a higher energy content than ethanol (the carbon content in ethanol and butanol is 51.0 and 64.8 wt-%, respectively), low volatility, and high flash point and it is thus safer to handle than ethanol. Butanol is not corrosive like ethanol, and for this reason, it is compatible with existing logistics.

In principle, hexosans or hexose sugars (e.g., in softwood autohydrolysates) can be converted into HMF by dehydration in an analogous way as furfural from pentosans or pentose sugars.[16] HMF ethers, such as 5-(methoxymethyl)furfural, 5-(ethoxymethyl)furfural, 5-(*tert*-butoxymethyl)-furfural, have been considered suitable for biofuel purposes.[35] HMF can also be converted into 2,5-dimethylfuran, which is a liquid biofuel. However, since HMF is not steam-volatile (i.e., its removal from a reaction mixture is not simple) and the reaction readily goes further, the production of its degradation products, levulinic and formic acids, is possible on prolonged heating.[16] Although levulinic acid can be considered an important platform chemical and it can be further converted to a variety of other potential chemicals, it is also suitable for a potential petrol additive.

3. By-Products of Kraft Pulping

3.1. *Extractives*

Besides black liquor, the significant by-products of softwood kraft pulping are crude sulfate turpentine and tall oil soap.[1,36] The availability of these extractive-based by-products in a mill is strongly dependent on the wood species used for pulping, the method and time of storing logs and chips, and the growth conditions of the trees. However, even among extractive-rich pine species there is significant variation in the availability of these by-products.

In principle, crude turpentine is recovered from the digester relief condensates and purified in the distillation process.[1] However, batch digester recovery systems differ substantially from continuous digester recovery systems in the methods by which the relief vapors are captured. The average yield of crude turpentine from pine species is 5–10 kg/ton of pulp, being somewhat lower for spruce species.

The tall oil soap is removed from black liquor due to density differences during the black liquor evaporation process (the optimum dry solids content is 28–32%), and the organic acids (fatty and resin acids) in the tall oil soap are liberated by adding H_2SO_4 to yield crude tall oil (CTO).[1,7] The average yield of CTO is in the range of 30–50 kg/ton of pulp, corresponding to 50–70% of the initial amount in the raw material used for pulping. CTO is normally purified and fractionated by vacuum distillation (3–30 mbar, 170–290°C). The main fractions and their mass proportions are as follows[1]: light oil (10–15%), fatty acids (20–40%), rosin (resin acids, 25–35%), and pitch residue (20–30%). Of these fractions, only some parts of the light oil and pitch residue are generally used for combustion, other fractions and parts being utilized for chemical purposes.

Traditionally, the conversion of various vegetable oils (containing mainly triglycerides) into diesel-type fuels is known as alcoholysis (transesterification) and pyrolysis to produce, respectively, fatty acid esters (e.g., the reaction with acidic methanol yields FAMEs) and hydrocarbons.[37] However, commonly practiced transesterification has some drawbacks including the requirement of a large amount of methanol, which is usually derived from natural gas or other fossil fuel sources. The formation of a low value by-product in the form of glycerol is also a problem, at least until there are enough facilities for its refining. Additionally, the heating value of these ester mixtures is lower than that of conventional diesel fuels and there are also stability issues with this kind of biodiesel.[38] In contrast, reactions from pyrolysis (fast pyrolysis) of vegetable oils seem to offer a promising option for the production of renewable fuels.[39–43] These kinds of biofuels also have similar chemical components with conventional petroleum diesel fuels.[44]

Although it is possible to produce FAMEs based on the esterification of the rosin fraction in CTO, there has also been some research dealing with the utilization of CTO for producing biofuels by other methods.[45,46] For example, UPM-Kymmene Corporation has studied the industrial-scale production of biofuels from CTO by hydrogenation.[47] The product´s characteristics correspond to those of traditional oil-based fuels. An interesting approach is also the direct pyrolysis of the tall oil soap for producing biofuel.[48–50] In this case, the product quality is satisfactory and, compared to the normal refining process, no external H_2SO_4 is needed for acidulation, since during the pyrolysis the sodium bound to organic acids is liberated with the simultaneous formation of Na_2CO_3. It should be pointed out that saponified vegetable oils (soaps) have also been pyrolyzed (see Sec. 4.1. General aspects for pyrolysis) as such for producing biodiesel.[39,51–53]

3.2. *Lignin*

As already indicated in Sec. 1.2., possibilities of pulping-based biofuel production, lignin, as a significant component of black liquor (Table 3), is mainly burned for its high heat value as part of the kraft pulping recovery process. It is also known that lignin begins to separate in the form of a fine precipitate when the pH (about 13) of the black liquor is reduced to about 10 by acidification.[1] The precipitation of lignin has a long history and the general approach has been to use carbon dioxide from flue gases and from causticization with possible further addition of pure carbon dioxide and strong mineral acid for this purpose.[54–63] The partial removal of lignin is

Table 3. Typical composition of the dry matter of silver birch (*Betula pendula*) and Scots pine (*Pinus sylvestris*) kraft black liquors (% of the total dry matter).[1]

Component	Birch	Pine
Lignin	25	31
Aliphatic carboxylic acids	33	29
Volatile acids[a]	12	10
Hydroxy acids[b]	21	19
Other organics	9	7
Extractives	3	4
Carbohydrates[c]	5	2
Miscellaneous	1	1
Inorganics[d]	33	33

[a]Acetic and formic acids.

[b]Mainly glycolic, lactic, 2-hydroxybutanoic, 3,4-dideoxy-pentonic, 3-deoxy-pentonic, xyloisosaccharinic, and glucoisosaccharinic acids.

[c]Mainly hemicellulose-derived fragments.

[d]Including the sodium bound chemically to organics.

used in the cases where the overloaded recovery furnace is the process bottleneck. It also affords the production of solid biofuel with a high energy density and low ash content.[7] In addition, there have been various process concepts applied to non-wood soda pulping in which partial lignin precipitation and sodium recovery were accomplished with carbonation together with electrodialysis (e.g., $Na_2CO_3 \rightarrow NaOH + CO_2$ and $R\text{-}CO_2Na \rightarrow RCO_2H + NaOH$).[64–66]

In the typical process, the black liquor is first evaporated to a solids content of 25-30% and the tall soap skimmings are recovered.[59] Lignin is then precipitated by two-stage carbonation (first with impure CO_2 followed by pure CO_2 under a pressure of 1500 kPa), resulting in about 80% yield. To ensure the efficient filtration of the precipitated lignin, the carbonated liquor must be heated to about 90°C to coagulate the lignin. The hydrogen sulfide liberated is recovered by absorption in green liquor. The most well-known recent full-scale process of precipitating lignin by carbonation is known as the "LignoBoost process".[63] It includes an effective additional acidic washing stage needed for the utilization of the precipitated lignin. One of the latest process applications has been installed at the Thunder Bay kraft pulp mill, where a pilot-scale lignin extraction demonstration plant produces lignin under real-world conditions.[67] In practice, with respect to the mill's energy balance, only a partial recovery of lignin is possible with a separation process that does not much interfere with the recovery of cooking chemicals.

In principle, all alkali lignin-type materials (cf. lignosulfonates, in Sec. 5. By-Products of Acid Sulfite Pulping) can be treated under a number of different conditions, from thermochemical methods to chemical processes.[1,68–74] For this reason, a wide range of products can also be obtained, for example, solid and liquid fuels, carbon fibers, resins, and many other polymer modifications. In principle, lignin is thought to replace many chemicals that are currently derived from petroleum-based sources and is most likely to be used as a bio-based additive to polymers.[67] In addition to this chemical utilization, a potential alternative is that the lignin could be sold to external customers for use, for example, in a combined heat and power plant.[63] The lignin may be either in the form of powder, pellets, or mixed with other fuels. For example, in the case of the "LignoBoost process", the sodium content of the lignin is rather low and the corrosion of combustion equipment as well as the amount of ash produced during the incineration can be minimized. However, the kraft lignin still contains some chemically bound sulfur (about 2%) and for this reason, the combustion plant has to be equipped with a flue gas treatment system for handling the sulfur emissions.

4. Thermochemical Treatment of Black Liquor

4.1. *General aspects*

On heating biomass even to relatively low temperatures (>100°C), various chemical changes start to take place. The thermal or thermochemical conversion of lignocellulosic materials always results in three groups of substances[1]: gases, condensable liquids (tars), and solid char products. However, the relative proportions of these products depend on both the chosen method and the specific reaction conditions. The main thermochemical conversion processes are pyrolysis, gasification, liquefaction, and combustion, whereas torrefaction represents a typical pretreatment process.

Torrefaction of woody biomass is a process for drying biomass to remove volatile compounds and to decompose a part of its hemicelluloses.[1,7,8] It can be considered a mild pyrolysis performed at 200–300°C for several minutes in the absence of air. This pretreatment enhances the energy density of wood by increasing its carbon content and net caloric value and is therefore expected to become more important in the future, especially as a treatment preceding gasification. The torrefied material is more brittle than the feedstock wood material and it has intermediate characteristics between coal and the feedstock.

The two actual thermal conversion methods, pyrolysis and gasification, differ in that pyrolysis refers to thermal degradation in the complete or near complete absence of an oxidizing agent (air or oxygen), whereas in gasification, lignocellulosic materials are converted by heating in the presence of controlled amounts of oxidizing agents to primarily provide a simple gaseous phase (i.e., synthesis gas or "syngas").[74] In addition, pyrolysis processes are generally carried out at lower temperatures (about 500°C) than those of gasification (above 800°C). In spite of this clear distinction, pyrolysis is often used to describe any chemical changes brought about by the application of heat, even with air or other additives.[75]

In gasification, thermal conversion proceeds along various steps (e.g., drying, pyrolysis, and volatile combustion, as well as char combustion and gasification), which in practice occur sequentially and partly in parallel.[1,76] In general, pyrolysis is also the first step in gasification and combustion (>900°C) processes. The resulting syngas is mainly composed of nitrogen, steam, carbon monoxide, carbon dioxide, hydrogen, and methane, although its composition can be varied over an extreme range depending upon the basic design of the gasification process.

Syngas can be used either as fuel of for producing chemicals such as methanol and ammonia.[1] Methanol can be further processed, for example, to dimethyl ether (DME, a petrol additive) or formaldehyde. In addition, a wide range of aliphatic hydrocarbons (i.e., olefins and paraffins) together with oxygenated products can be produced from syngas by catalytic conversion according to the Fischer-Tropsch process. This process is based on an old technology developed in the early 1920s by Franz Fischer and Hans Tropsch who used alkalized iron catalysts to produce liquid hydrocarbons rich in oxygenated compounds mainly from coal-derived syngas (formerly known as the "Synthol process").

The gasification of lignocellulosic materials is currently reconsidered because fossil fuels can no longer fully satisfy our energy needs.[1] For this reason, it is readily understandable that biomass-based gasification technology is under extensive development. However, there are still some technical issues that should be clearly improved before this technology becomes fully competitive with the conventional use of fossil fuels.

In addition to pyrolysis and gasification, there are processes in which liquid fuels are obtained by liquefaction (i.e., the direct biomass to liquid (BTL) process).[1,76] This conversion can be accomplished at 300–350°C under a high pressure in the presence of carbon monoxide or hydrogen. In this case, it is often called "catalytic liquefaction", whereas a non-catalytic process is known as "direct liquefaction". In general, because of the need for more complicated and expensive reactors in liquefaction, there is less interest in liquefaction than in pyrolysis.

4.2. *Gasification*

In black liquor gasification (BLG) the black liquor is gasified in a separate reactor.[7] Two different process modifications have been studied[77]: at

600–850°C, where the inorganic compounds (mainly Na_2CO_3) are below their melting points, and at 800–1200°C, which produces a molten mixture of inorganics (smelt). In both cases, syngas is separated from ash and the syngas and inorganics are cooled to recover heat. The inorganics are then dissolved into water and introduced to the recovery of cooking chemicals. In principle, the recovery furnace could be replaced by the BLG process, although, due to many unsolved problems, this has so far not been realized on a full scale.

Various BLG technologies including, for example, Manufacturing and Technology Conversion International (MTCI), Direct Alkali Regeneration System (DARS), and the Chemrec processes, have been studied.[7,76–78] Although the recovery furnace and BLC processes have a similar overall efficiency, the latter process allows, for example, a syngas-based production of DME or methane.[79] However, in the BLC technologies there are still challenges dealing mainly with the feeding of black liquor against a high pressure and temperature, high requirements for material quality for reactor walls and connections (i.e., the influence of corrosion), high demand for syngas cooling, and the purification of syngas from ash.[7,76] In addition, the recovery of cooking chemicals is not straightforward and only a portion of the sulfur is converted to sodium sulfide.

4.3. *Liquefaction*

Because of several factors, including the low thermal efficiency in energy recovery, the limited capacity for electricity generation and the high capital cost of the recovery furnace, the possibility of enhancing black liquor conversion to high-value forms of energy by liquefaction has been also studied.[76] The basic idea has been to convert the organic matter in black liquor (Table 3) by a technique similar to that used in the liquefaction of wood.

In the typical process (i.e., liquid-phase thermal treatment), black liquor is treated under a reducing atmosphere (CO or H_2) at a high temperature (300–350°C) and pressure (about 20 MPa) with or without the addition of additives (e.g., NaOH).[80,81] During such a treatment an oil-like organic product formed mainly from lignin with a total yield 40–60% of black liquor organics separates out from an aqueous phase containing the inorganic material and the residual organics, mainly low-molecular-mass aliphatic carboxylic acids. The organic phase can be used after refining as biofuel for various purposes and the cooking chemicals can be recovered from the aqueous phase. So far, this kind of thermochemical conversion of black liquor has been carried out only in small-scale reactors and is at present economically not very attractive.

5. By-Products of Acid Sulfite Pulping

The majority of the organic material in acid sulfite spent liquors originates from lignin (lignosulfonates) and hemicelluloses (Table 4).[1,82] Old calcium sulfite mills had serious environmental problems caused by the evaporation condensates (i.e., due to a high biochemical oxygen demand, BOD) and sulfur emission from the destruction of the organic material for heat recovery.[83] The conventional cooking system containing calcium sulfite ($CaSO_3$) and calcium bisulfite ($Ca(HSO_3)_2$) from limestone ($CaCO_3$) was mainly used due to its low cost and, because of the absence of stringent environmental quality regulations, there was no need to recover calcium.[6] However, in modern acid sulfite (or bisulfite) cooking, sodium and magnesium are mainly used and the inorganic chemicals are efficiently recovered and regenerated. In addition, the organic solids are

Table 4. Typical composition of silver birch (*Betula pendula*) and Norway spruce (*Picea abies*) acid sulfite spent liquors (kg/ton pulp).[1]

Component	Birch	Spruce
Lignosulfonates	435	510
Carbohydrates	380	270
Monosaccharides	305	215
Oligo- and polysaccharides	75	55
Aliphatic carboxylic acids	130	70
Acetic acid	75	30
Aldonic acids	55	40
Extractives	40	40
Others	55	30

burned to generate energy. In this case, the heat value of the lignosulfonates is rather high, roughly half of that of oil, whereas the heat value of the carbohydrate fraction is much lower, roughly one-third of that of oil.[82]

A variety of useful products can be produced from acid sulfite spent liquors. Fermentation methods have traditionally played a dominant role in the industrial processing of the carbohydrate fraction of such liquors.[1] These liquors have mainly been used for producing ethanol and single-cell protein. Although this was once an effective way of reducing the pollution load from the mill, this kind of by-product utilization is currently not economically attractive. For this reason, since the dissolved organic solids represent a considerable fuel value, it is understandable that there are rather few industrial applications for the production of chemicals or other products from the soluble components from acid sulfite pulping.

With respect to the production of biofuels from the main components of sulfite spent liquors, the possible industrial utilization of the carbohydrates can be limited only to various fermentation processes. The obvious products would be ethanol and butanol, which can be separated afterwards by distillation. However, fermentation inhibitors such as sulfur dioxide, furans, and lignosulfonates must be at least partly removed from the liquor prior to the fermentation. In contrast, the high-molecular-mass lignosulfonates can be easily separated in a more or less pure form from the low-molecular-mass fragments by ultrafiltration and used in a powder form after spray-drying for a number of applications, especially because of their adhesion and dispersion properties.[1]

6. Conclusions

In general, a prerequisite for finding a realizable process concept is that all the constituents of the feedstock material are taken into account when planning target-oriented economic ways of producing potential biofuels. With this respect, especially a kraft mill offers attractive possibilities to produce alternative biofuels in addition to the traditional market of pulp and extractives (turpentine and tall oil). However, the utilization of various organic fractions for fuel purposes is complex and depends on several factors including the type of feedstock material, the energy and pulp prices, the scale of the production, and the conversion technologies needed. In addition to a straightforward combustion of various organic material residues and black liquor, there are thermochemical processes available for producing biofuels from harvesting and debarking residues, tall oil, and separated lignin, and some new promising options are under development. In addition, in the integrated pulp mill biorefinery concepts there are possibilities for producing liquid biofuels (e.g., ethanol and butanol) based on the fermentation of hemicelluloses, although such approaches do not seem currently very competitive.

References

1. R. Alén, Principles of biorefining, in Biorefining of Forest Resources, R. Alén (Ed.), *Paper Engineers' Association/Paperi ja Puu Oy*, Helsinki, Finland, 2011, pp. 55–114.
2. A. C. Hansen, D. C. Kyritsis and C. L. Lee, Characteristics of biofuels and renewable fuel standards, in Biomass to Biofuels, A. A. Vertès, N. Qureshi, H. P. Blaschenk, and H.Yukawa, (Eds.), John Wiley & Sons, Ltd., Chichester, United Kingdom, 2010, pp. 3–26.
3. R. Alén, J. Käkölä and H. Pakkanen, Monitoring of kraft pulping by a fast analysis of aliphatic carboxylic acids, *Appita J.*, **61(3)**, 216–219 (2008).
4. T. M. Grace, B. Leopold, E. W. Malcolm and M. J. Kocurek, (Eds.), Pulp and Paper Manufacture, Vol. 5, Alkaline Pulping, 3rd edn., *The Joint Textbook Committee of the Paper Industry, TAPPI&CPPA*, USA and Canada, 1989.
5. R. Alén, Analysis of degradation products: A new approach to characterizing the combustion properties of kraft black liquors, *J. Pulp Pap. Sci.*, **23(2)**, J62–J66 (1997).
6. R. Alén, Basic chemistry of wood delignification, in Forest Products Chemistry, P. Stenius (Ed.), *Fapet Oy*, Helsinki, Finland, 2000, pp. 58–104.
7. M. Hamaguchi, M. Cardoso and E. Vakkilainen, Alternative technologies for biofuels production in kraft pulp mills — Potential and prospects, *Energies*, **5**, 2288–2309 (2012).
8. B. Acharys, I. Sule and A. Dutta, A review on advances of torrefaction technologies for biomass processing, *Biomass Conv. Bioref.*, **2**, 349–369 (2012).
9. Y. Sun and J. Cheng, Hydrolysis of lignocellulosic materials for ethanol production: A review, *Biores. Technol.*, **83**, 1–11 (2002).
10. N. Mosier, C. Wyman, B. Dale, R. Elander, Y. Y. Lee, M. Holtzapple and M. Ladish, Features of promising technologies for pretreatment of lignocellulosic biomass, *Biores. Technol.*, **96**, 673–683 (2005).
11. P. Kumar, D. M. Barrett, M. J. Delwiche and P. Stroeve, Methods for pretreatment of lignocellulosic biomass for efficient hydrolysis and biofuel production, *Ind. Eng. Chem. Res.*, **48**, 3713–3729 (2009).

12. F. M. Gírio, C. Fonseca, F. Garvalheiro, L. C. Duarte, S. Marques and R. Bogel-Łukasik Hemicelluloses for fuel ethanol: A review, *Biores. Technol.*, **101**, 4775–4800 (2010).
13. H. Sixta, A. Potthast and A. W. Krotschek, Chemical pulping process, in Handbook of Pulp, H. Sixta, (Ed.), Wiley-VCH Verlag GmbH, *Weinheim*, Germany, 2006, pp. 109–509.
14. H. Sixta and G. Schild, A new generation kraft process, Lenzing. Berichte **87**, 26–37 (2009).
15. J. Kautto, K. Henricson, H. Sixta, M. Trogen and R. Alén, Effects of integrating a bioethanol production process to a kraft pulp mill, *Nordic Pulp Pap. Res. J.*, **25**, 233–242 (2010).
16. L. Viikari and R. Alén, Biochemical and chemical conversion of forest biomass, in Biorefining of Forest Resources, R. Alén (Ed.), *Paper Engineers´ Association/Paperi ja Puu Oy*, Helsinki, Finland, 2011, pp. 225–261.
17. K. D. Sears, A. Beélik, R. L. Casebier, R. J. Engen, J. K. Hamilton, and H. L. Hergert, Southern pine prehydrolyzates: Characterization of polysaccharides and lignin fragments, *J. Polymer Sci.*: Part C, **36**, 425–443 (1971).
18. G. Garrote, H. Domínguez and J. C. Mild autohydrolysis: an environmentally friendly technology for xylooligosaccharide production from wood, *Chem. Technol. Biotechnol.*, **74**, 1101–1109 (1999).
19. L. Ramos, E. M. Kristenson and U. A. T. Brinkman, Current use of pressurized liquid extraction and subcritical water extraction in environmental analysis, *J. Chromatogr. A.*, **975**, 3–29 (2002).
20. S.-H. Yoon and A. van Heiningen, Kraft pulping and papermaking properties of hot-water pre-extracted loblolly pine in an integrated forest products biorefinery, *Tappi J.*, **7(7)**, 22–27 (2008).
21. F. Carvalheiro, L. C. Duarte and F. M. Gírio, Hemicellulose biorefineries: A review on biomass pretreatments, *J. Sci. Industr. Res.*, **67**, 849–864 (2008).
22. W. J. Frederick, Jr., S. J. Lien, C. E. Courchene, N. A. DeMartini, A. J. Ragauskas and K. Iisa, Co-production of ethanol and cellulose fiber from southern pine: A technical and economical assessment, *Biomass & Bioenergy*, **32**, 1293–1302 (2008).
23. J. J. Paredes, R. Jara, S. M. Shaler and A. van Heiningen, Influence of hot water extraction on the physical and mechanical behavior of OSB, *Forest Prod. J.*, **58(12)**, 56–62 (2008).
24. M. S. Tunc and A. R. P. van Heiningen, Autohydrolysis of mixed southern hardwoods: Effect of P-factor, *Nord. Pulp. Pap. Res. J.*, **24(1)**, 46–51 (2009).
25. T. E. Amidon and S. Liu, Water-based woody biorefinery, *Biotechn. Adv.*, **27(5)**, 542–550 (2009).
26. C. C. Teo, S. N. Tan, J. W. H. Yong, C. S. Hew and E. S. Ong, Pressurized hotwater extraction (PHWE), *J. Chromatogr. A.*, **1217**, 2484–2494 (2010).
27. H. Li, A. Saeed, M. S. Jahan, Y. Ni and A. van Heiningen, Hemicellulose removal from hardwood chips in the pre-hydrolysis step of the kraft-based dissolving pulp production process, *J. Wood Chem. Technol.*, **30**, 48–60 (2010).

28. M.S. Tunc and A. R. P. van Heiningen, Characterization and molecular weight distribution of carbohydrates isolated from the autohydrolysis extract of mixed southern hardwoods, *Carbohydr. Polym.*, **83(1)**, 8–13 (2011).
29. J. Lehto and R. Alén, Purification of hardwood-derived autohydrolysates, *BioResources*, **7(2)**, 1813–1823 (2012).
30. L. Olsson and B. Hahn-Hägerdal, Fermentation of lignocellulosic hydrolysates for ethanol production, *Enz. Microb. Technol.*, **18**, 312–331 (1996).
31. J. P. Delgenes, R. Moletta and J. M. Navarro, Effects of lignocellulose degradation products on ethanol fermentations of glucose and xylose by *Saccharomyces cerevisiae*, *Zymomonas mobilis*, *Pichia stipitis* and *Candida shehatae*, *Enz. Microb. Technol.*, **19**, 220–225 (1996).
32. E. Palmqvist and B. Hahn-Hägerdal, Fermentation of lignocellulosic hydrolysates. I: Inhibition and detoxification, *Biores. Techn.*, **74(1)**, 17–24 (2000).
33. T. J. Schwartz and M. Lawoko, Removal of acid-soluble lignin from biomass extracts using Amberlite XAD-4 resin, *BioResources*, **5**, 2337–2347 (2010).
34. J. Cousin Saint Remi, T. Rémy, V. Van Hunskerken, S. van der Perre, T. Duerinck, M. Maes, D. De Vos, E. Gobechiya, C. E. A. Kirschhock, G. V. Baron and J. F. M. Denayer, Biobutanol separation with the metal-organic framework ZIF-8, *ChemSusChem.*, **4**, 1074–1077 (2011).
35. G. -J. Gruter and E. de Jong, Furanics: novel fuel options from carbohydrates, *Biofuels Technol.*, (Issue 1) 11–17 (2009).
36. B. Holmbom, Extraction and utilization of non-structural wood and bark components, in Biorefining of Forest Resources, R. Alén (Ed.), *Paper Engineers´ Association/ Paperi ja Puu Oy*, Helsinki, Finland, 2011, pp. 176–224.
37. K. A. Doll, Comparing biofuels obtained from pyrolysis, of soybean oil or soapstock, with traditional soybean biodiesel: Density, kinematic viscosity, and surface tensions, *Energ. Fuel*, **22**, 2061–2066 (2008).
38. X. Junming, J. Jianchun, L. Yanju and C. Jie, Liquid hydrocarbon fuels obtained by the pyrolysis of soybean oils, *Biores. Technol.*, **100**, 4867–4870 (2009).
39. C. Chang and S. Wang, China´s motor fuels from tung oil, *Ind. Eng. Chem.*, **39**, 1543–1548 (1947).
40. K. Kitamura, Studies of the pyrolysis of triglycerides, Bull. *Chem. Soc. Jpn.*, **44**, 1606–1609 (1971).
41. P. Nichols and R. Holman, Pyrolysis of saturated triglycerides, *Lipids*, **7**, 773–779 (1972).
42. J. W. Alencar, P. B. Alves and A. A. Craveiro, Pyrolysis of tropical vegetable oils, *J. Agric. Food Chem.*, **31**, 1268–1270 (1983).
43. K. D. Maher and D. C. Bressler, Pyrolysis of triglyceride materials for the production of renewable fuels and chemicals, *Biores. Technol.*, **98**, 2351–2368 (2007).
44. A. Demirbas, Biodiesel — A Realistic Fuel Alternative for Diesel Engines, *Springer*, London, United Kingdom, 2008.

45. S. Y. Lee, M. A. Hubbe and S. Saka, Prospects for biodiesel as a byproduct of wood pulping — A review, *BioResoures*, **1**, 150–171 (2006).
46. A. Demirbas, Production of biodiesel from tall oil, *Energy Sour.* Part A, **30**, 1896–1902 (2008).
47. S. Mannonen, UPM — producing fuels of the future from wood-based raw materials, *in Proc. The 4th Nordic Wood Biorefinery Conf.* (NWBC 2012), Helsinki, Finland, 23–25 October, 2012, pp. 121–124.
48. V. Arpiainen, Production of Light Fuel Oil from Tall Oil Soap Liquids by Fast Pyrolysis Techniques, Licentiate´s Thesis, University of Jyväskylä, Laboratory of Applied Chemistry, Jyväskylä, Finland, 2001, 51 p. (in Finnish).
49. H. Lappi, Production of Hydrocarbon-rich Biofuels from Extractives-derived Materials, Doctoral Thesis, University of Jyväskylä, Laboratory of Applied Chemistry, Jyväskylä, Finland, 2012, p. 111
50. H. Lappi and R. Alén, Pyrolysis of tall oil-derived fatty and resin acid mixtures, ISRN Renewable Energy, 409157, 8.p
51. I. C. P. Fortes and P. J. Baugh, Study of calcium soap pyrolysates derived from Macauba fruit (Acrocomia sclerocarba M.). Derivatization and analysis by GC/MS and CI-MS, *J. Anal. Appl. Pyrolysis*, **29**, 153–167 (1994).
52. A. Demirbas, Diesel fuel from vegetable oil via transesterification and soap pyrolysis, *Energy Sources* Part A, **24**, 835–841 (2002).
53. A. Demirbas, Biodiesel fuels from vegetable oils via catalytic and non-catalytic supercritical alcohol transesterifications and other methods. *A survey, Energy Convers. Manage.*, **44**, 2093–2109 (2003).
54. D. R. Dhingra, M. S. Bhatnagar and P. C. Nigam, Lignin recovery from soda black liquor, *Indian Pulp Paper*, **7**, 311–315 (1952).
55. K. Gray, H. L. Crosby and J. C. Steinberg, Recovery of chemicals in wood pulp preparation, U.S. Patent 24, 293, June 13, 1956.
56. J. W. T. Merewether, The precipitation of lignin from eucalyptus kraft black liquors, *Tappi*, **45(2)** 159–163 (1962).
57. V. M. Nikitin, A. V. Obolenskaya, V. M. Skachov and A. D. Ivanenko, Precipitation of alkali lignin with carbon dioxide under pressure, *Bumazhn. Prom.*, **38(11)** 14–15 (1963).
58. S. Basu, Studies on carbonation of black liquor for lignin precipitation and its subsequent separation, *IPPTA*, **8**, 207–214 (1971).
59. R. Alén, P. Patja and E. Sjöström, Carbon dioxide precipitation of lignin from pine kraft black liquor, *Tappi*, **62(11)**, 108–110 (1979).
60. R. Alén, E. Sjöström and P. Vaskikari, Carbon dioxide precipitation of lignin from alkaline pulping liquors, *Cellulose Chem. Technol.*, **19**, 537–541 (1985).
61. V. Uloth and J. Wearing, Kraft lignin recovery: Acid precipitation versus ultrafiltration. Part I: Laboratory test results, *Pulp Pap. Can.*, **90(9)**, 67–71 (1989).
62. V. Uloth and J. Wearing, Kraft lignin recovery: Acid precipitation versus ultrafiltration. Part II: Technology and economics, *Pulp Pap. Can.*, **90(10)**, 34–37 (1989).

63. H. Wallmo, Lignin Extraction from Black Liquor — Precipitation, Filtration and Washing, Doctoral Thesis, Chalmers University of Technology, Forest Products and Chemical Engineering, *Gothenburg*, Sweden, 2008, p. 73
64. K. Radhamohan and S. Basu, Electrodialysis in the regeneration of paper mill spent liquor, *Desalination*, **33**, 185–200 (1980).
65. M. E. L. N. Arulanantham and R. Shanthini, Recovery of sodium hydroxide from Embilipitiya black liquor by electrodialysis, *IPPTA*, **9(3)**, 1–8 (1997).
66. M. E. L. N. Arulanantham and R. Shanthini, Recovery of sodium hydroxide from Embilipitiya black liquor by electrodialysis, *IPPTA*, **9(3)**, 1–8 (1997).
67. C. McDonald, A proving ground for lignin, *Pulp Pap. Can.*, **113(3)**, 14–16 (2012).
68. K. V. Sarkanen and C. H. Ludwig, (Eds.), Lignins — Occurrence, Formation, Structure and Reactions, John Wiley & Sons, New York, NY, USA, 1971, 916 p.
69. T. Enkvist, Phenolics and other organic chemicals from kraft black liquors by disproportionation and cracking reactions, *J. Appl. Polym. Sci.: Appl. Polym. Symp.*, **28**, 285–295 (1975).
70. D. W. Coheen, Chemicals from lignin, in Organic Chemicals from Biomass, I.S. Goldstein, (Ed.), CRC Press, *Boca Raton*, FL, USA, 1981, pp. 143–161.
71. W. G. Glasser and S. Sarkanen, Lignin: Properties and Materials, American Chemical Society, Washington, DC, USA, 1989, 545 p.
72. E. K. Pye, Industrial lignin production and applications, in Biorefineries — Industrial Processes and Products, Volume 2, B. Kamm, P.R. Gruber and M. Kamm, (Eds.), Wiley-VCH, *Weinheim*, Germany, 2006, pp. 165–200.
73. J. Bozell, J. Holladay, D. Johnson and J. White, Top value added chemicals from biomass — Volume II: Results of screening for potential candidates from biorefinery lignin, U.S. Department of Energy, *Oak Ridge*, TN, USA, 2007, p. 79
74. R. Alén, Conversion of cellulose-containing materials into useful products, in Cellulose Sources and Exploitation — Industrial Utilization, Biotechnology, and Physico-Chemical Properties, J.F. Kennedy, G.O. Phillips and P.A. Williams, (Eds.), Ellis Horwood, *Chichester*, England, 1990, pp. 453–464.
75. E. J. Soltes and T. J. Elder, Pyrolysis, in Organic Chemicals from Biomass, 2nd printing, I.S. Goldstein, (Ed.), CRC Press, *Boca Raton*, FL, USA, 1981, pp. 63–99.
76. J. Konttinen, M. Reinikainen, A. Oasmaa and Y. Solantausta, Thermochemical conversion of forest biomass, in Biorefining of Forest Resources, R. Alén (Ed.), *Paper Engineers´Association/Paperi ja Puu Oy*, Helsinki, Finland, 2011, pp. 262–304.
77. M. Naqvi, J. Yan and E. Dahlquist, Black liquor gasification integrated in pulp and paper mills: A critical review, *Biores. Technol.*, **101**, 8001–8015 (2010).
78. J. Hrbek, Biomass gasification opportunities in forest industry, in Proc. IEA Bioenergy Task 33 Workshop, *Piteå*, Sweden, 19 October, 2011.
79. M. Naqvi, J. Yan and M. Fröling, Bio-refinery system of DME and CH4 production from black liquor gasification in pulp mills, *Biores. Technol.*, **101**, 937–944 (2010).
80. R. Alén, P. McKeough, A. Oasmaa and A. Johansson, Thermochemical conversion of black liquor in the liquid phase, *J. Wood Chem. Technol.*, **9**, 265–276 (1989).

81. R. Alén and A. Oasmaa, Thermochemical conversion of hydroxy carboxylic acids in the liquid phase, *Holzforschung*, **43**, 155–158 (1989).
82. E. Sjöström, Wood Chemistry — Fundamentals and Applications, 2nd edition, Academic Press, Inc., San Diego, CA, USA, 1993, pp. 237–240.
83. E. Sjöström, Alternatives for balanced production of fibers, chemicals, and energy from wood, *J. Appl. Polym. Sci.: Appl. Polym. Symp.* **37**, 577–592 (1983).

INDEX

Printed in the United States
By Bookmasters